Industrial Ecology and the Sustainable Development Goals (SDGs)

Syed Abdul Rehman Khan
Xuzhou University of Technology, China

Muhammad Tanveer
Imam Mohammad Ibn Saud Islamic University, Saudi Arabia

IGI Global
Scientific Publishing
Publishing Tomorrow's Research Today

Vice President of Editorial Melissa Wagner
Director of Acquisitions Mikaela Felty
Director of Book Development Jocelynn Hessler
Production Manager Mike Brehm
Cover Design Phillip Shickler

Published in the United States of America by
IGI Global Scientific Publishing
701 East Chocolate Avenue
Hershey, PA, 17033, USA
Tel: 717-533-8845
Fax: 717-533-7115
Website: https://www.igi-global.com E-mail: cust@igi-global.com

Library of Congress Cataloging-in-Publication Data

Names: Khan, Syed Abdul Rehman, 1987- editor. | Tanveer, Muhammad, editor.
Title: Industrial ecology and the sustainable development goals (SDGS) / edited by: Syed Abdul Rehman Khan, Muhammad Tanveer.
Description: Hershey, PA : IGI Global Scientific Publishing, [2025] | Includes bibliographical references and index. | Summary: "This compelling book aims to bridge the gap between academics and practitioners, facilitating the exchange of their valuable knowledge. It delves into existing sustainable practices and, importantly, explores innovative approaches to promoting sustainability across different organizational levels, inspiring potential for change"-- Provided by publisher.
Identifiers: LCCN 2024060483 (print) | LCCN 2024060484 (ebook) | ISBN 9798337301396 (hardcover) | ISBN 9798337301402 (paperback) | ISBN 9798337301419 (ebook)
Subjects: LCSH: Industrial ecology. | Sustainable Development Goals (Project)
Classification: LCC TS161 .I5438 2025 (print) | LCC TS161 (ebook) | DDC 658.7--dc23/eng/20250328
LC record available at https://lccn.loc.gov/2024060483
LC ebook record available at https://lccn.loc.gov/2024060484
British Cataloguing in Publication Data
A Cataloguing in Publication record for this book is available from the British Library.

Table of Contents

Detailed Table of Contents

Lost baggage is one big concern for air travelers and sustainable airport operations. Airlines lose billions of dollars in penalties due to lost baggage, affecting both the productivity and sustainability of operations. We report on a blockchain-based baggage tracking system that significantly contributes to achieving several of the UN SDGs by promoting efficiency, transparency, & sustainability in airport operations. The system employs RFID and blockchain technologies to deliver a secure baggage tracking system. We discuss the baggage tracking problem as well as the solution conceptualization, design, implementation, and validation of an airport baggage tracking system based on blockchain technology. We also provide a link to the design and implementation code for others to build on our work. The system will help travelers track their baggage while traveling by using an application installed on their handheld devices like mobile phones. The airlines using this application can know baggage status and locations as well as identify the right owner of the baggage.

Operations Research and Management Sciences (OR/MS) methodologies have shown significant promise as well as success in solving and unraveling a variety of

and regulatory framework, helping to build a clean, secure global energy system. This extensive study examines green hydrogen capacity and future enhancements that could change global energy strategies. Thus, further research and collaboration are needed to overcome these obstacles and realize the sustainable future of green hydrogen.

The chapter consists of three parts: Family Firms, Sustainable Human Resource Management, and Pathways to Family Firms. First, we discuss the importance of family firms and the theoretical perspective of family science theories. Secondly, the author explores the emerging concept of sustainable human resource management (SHRM) and introduces pathways to sustainable futures. What is sustainability, and how was it integrated into HRM? It draws on the wider literature on Strategic Human Resource Management (SHRM), Sustainable Resource Management (SRM), and organizational theory to offer insights into the relationship between sustainability and HRM. The key issue addressed is the balancing act between short-term economic efficiency and long-term sustainability in managing human resources. Last but not least, we discuss the pathways to family firms and how family firm sustainable practices contribute to Sustainable Development Goals (SDGs) 2030.

The Sundarbans, a UNESCO World Heritage Site, is crucial for biodiversity and the livelihoods of millions. To promote environmental sustainability in this unique mangrove ecosystem, a focus on clean air and water is paramount. Addressing pollution from industrial discharges and agricultural runoff is essential to protect the region's fragile ecology. Community awareness programs can educate locals about the importance of maintaining air and water quality. Implementing eco-friendly agricultural practices, such as organic farming, can reduce chemical pollutants. Collaboration with NGOs and governmental bodies can facilitate sustainable development initiatives and stricter regulations on waste disposal. Restoration projects for mangroves also play a vital role in improving water quality and sequestering carbon. By prioritizing clean air and water, we can ensure the health of the Sundarbans for future generations while supporting the diverse wildlife that thrives in this remarkable habitat.

The revolutionary role that robo-advisors play in Shariah-compliant Islamic finance is examined in this essay. The first section lays out the fundamental ideas of Shariah-compliant investing, which serve as a framework for moral financial judgment. Examined is the emergence of robo-advisors, which show how these AI-powered tools are improving accessibility and efficiency while enabling adherence to Shariah laws. Key characteristics of robo-advisors that adhere to Shariah are examined, along with particular difficulties including cultural sensitivity and regulatory compliance. The research integrates case studies that demonstrate effective implementations, providing valuable perspectives on their real-world uses. The paper's conclusion, which offers a look ahead, highlights how ethical AI and robo-advisors have the potential to transform Islamic finance by promoting sustainability, inclusion, and creativity in Shariah-compliant investments.

This chapter explores how Artificial Intelligence (AI) can be harmonized with Sustainable Human Resource Management (HRM) to foster eco-friendly, ethical, and efficient business practices. When AI technologies revolutionize conventional HR functions, they provide opportunities for reduced resource use, improved workplace sustainability and concordance with the UN Sustainable Development Goals (SDGs). This chapter provides an in-depth analysis of AI’s role in recruitment, training, resource optimization, and employee well-being while emphasizing the potential for AI to drive both environmental and social sustainability within organizations.

Preface

The Sustainable Development Goals (SDGs) of the United Nations present an ambitious blueprint for a just, equitable, and environmentally secure future. Achieving these global targets necessitates integrated and interdisciplinary approaches that transcend traditional economic, social, and environmental boundaries. *Industrial Ecology and the Sustainable Development Goals (SDGs)* addresses this imperative by exploring diverse intersections between sustainability strategies, technological innovations, policy-making, and industrial transformation.

This volume brings together a rich array of scholarly contributions that reflect the evolving discourse on sustainable development. From blockchain applications in aviation logistics to the role of ethical artificial intelligence in finance, the chapters delve into both emergent technologies and systemic frameworks for driving global change. The compilation emphasizes practical insights, case-based research, and conceptual frameworks that connect industry practices with the wider sustainability agenda.

The themes span green human resource management, hydrogen energy technologies, circular economy solutions, and localized environmental actions—each serving as a building block in the pathway toward the SDGs. The contributors offer global perspectives, rooted in regional contexts from the Sundarbans to smart Islamic finance systems, ensuring relevance to academics, professionals, policy-makers, and students alike.

As industrial ecology continues to evolve as a vital discipline for sustainable systems design, this volume contributes critical knowledge, integrative methodologies, and thought leadership necessary for fostering resilient, inclusive, and low-carbon futures.

CHAPTER OVERVIEW

Chapter 1 introduces a blockchain-based baggage tracking system designed to enhance transparency and efficiency in airport operations. Through RFID integration, the system empowers travelers and airlines with real-time information, contributing to more sustainable and secure aviation practices aligned with the SDGs.

Chapter 2 critically examines the use of Operations Research and Management Sciences (OR/MS) in tackling complex socio-technical challenges. It advocates for integrating AI and systems thinking into OR/MS to improve decision-making outcomes that are sustainable, ethical, and equitable.

Chapter 3 explores Green Human Resource Management (Green HRM) as a bridge between workforce management and environmental sustainability. The chapter discusses practices such as green recruitment and training, emphasizing the alignment of organizational strategies with eco-conscious values.

Chapter 4 provides a comprehensive analysis of green hydrogen as a clean energy frontier. It compares electrolyzer technologies, highlights current global initiatives, and maps out the technical, economic, and regulatory landscape essential for large-scale hydrogen adoption.

Chapter 5 focuses on family firms and their unique challenges in integrating sustainability into human resource practices. By exploring theoretical and strategic perspectives, it presents pathways through which family-owned enterprises can support SDG 2030 goals while balancing economic and environmental objectives.

Chapter 6 highlights the environmental significance of the Sundarbans and outlines actionable strategies to preserve clean air and water in the region. The chapter emphasizes community engagement, sustainable farming, and regulatory collaboration to protect this fragile ecosystem.

Chapter 7 delves into the transformative role of ethical AI in Islamic finance. It examines Shariah-compliant robo-advisors and their potential to democratize access to ethical investment options while maintaining adherence to religious and cultural principles.

Chapter 8 explores the synergy between AI technologies and sustainable HRM practices. It illustrates how AI can revolutionize recruitment, resource allocation, and employee welfare, offering scalable solutions to align HR operations with sustainability imperatives.

Chapter 9 analyzes the motivators and inhibitors of green purchasing behaviors across global contexts. It identifies socioeconomic, cultural, and informational factors that shape consumer choices, and proposes strategic interventions such as eco-labeling and AI-driven transparency tools.

Chapter 10 investigates the water-energy nexus within the framework of a circular economy. It presents a nuanced discussion on the disparities between developed and developing economies and emphasizes the need for global cooperation to build a resilient, circular infrastructure that sustains both water and energy resources.

Chapter 1
Blockchain-Based Baggage Tracking System for Sustainable Airport Operations: Concept, Design, and Implementation

Asrar U. Haque
https://orcid.org/0000-0002-6651-0906
King Faisal University, Saudi Arabia

Abdul-Rahim Ahmad
https://orcid.org/0000-0001-7923-0105
King Faisal University, Saudi Arabia

Yousuf Ahmad
University of British Columbia, Canada

Hawraa A. Alsleh
King Faisal University, Saudi Arabia

Zahraa T. Bokhamis
King Faisal University, Saudi Arabia

Fatimah T. Bokhamis
King Faisal University, Saudi Arabia

ABSTRACT

Lost baggage is one big concern for air travelers and sustainable airport operations. Airlines lose billions of dollars in penalties due to lost baggage, affecting both the productivity and sustainability of operations. We report on a blockchain-based baggage tracking system that significantly contributes to achieving several of the UN SDGs by promoting efficiency, transparency, & sustainability in airport operations. The system employs RFID and blockchain technologies to deliver a secure baggage tracking system. We discuss the baggage tracking problem as well as the solution conceptualization, design, implementation, and validation of an airport baggage

DOI: 10.4018/979-8-3373-0139-6.ch001

tracking system based on blockchain technology. We also provide a link to the design and implementation code for others to build on our work. The system will help travelers track their baggage while traveling by using an application installed on their handheld devices like mobile phones. The airlines using this application can know baggage status and locations as well as identify the right owner of the baggage.

1 INTRODUCTION

Losing baggage is one of the most common concerns that travelers might face. Usually, this problem happens because of improper dealing with baggage, such as bags getting exchanged between travelers, or getting wrong ticket numbers on bags. Since multiple parties like airlines, airports, and airport security handle baggage at airports, the problem of effectively and efficiently handling baggage has become a complex inter-party undertaking. Complex tactical and operational issues related to baggage handling results in millions of lost pieces of baggage and billions of dollars in penalties and lost revenues. For instance, approximately 25 million bags are mishandled each year globally, which includes bags that are lost, delayed, damaged, or stolen. The mishandling rate has shown improvement over the years, with the latest SITA Baggage IT Insights report indicating a reduction in the global mishandling rate to 6.9 bags per 1,000 passengers in 2023, down from 7.6 in 202, due to which the airports lost more than US$ 2 billion (Alsyouf et al. 2018; Dike et al. 2024; United Nations).

Blockchain could enable a decentralized and transparent network for items of luggage (Puthal et al. 2018; Kshetri, 2019). It would help improve the bottom line of airlines and airports by reducing baggage losses. Also, the airport will pay less to an airline company for lost luggage. One advantage of blockchain is that it does not need a third party to complete its transaction. Not only that, blockchain is like a distributed ledger which helps reduce paperwork and processing time because all data will be in one system. In this chapter, we report on the concept, design, and implementation of an efficient and efficacious blockchain based solution for the lost baggage problem. The blockchain-based airport baggage handling system significantly contributes to achieving several of the United Nations Sustainable Development Goals (SDGs) by promoting efficiency, transparency, and sustainability in airport operations. More specifically, it aligns with SDG-8, SDG-9, SDG-11, SDG-12, SDG-13, and SDG-16.

1.1 Background

One of the most difficult challenges that airlines and train stations face is when delivering bags to their owners. That is due to several factors such as bags getting switched, errors in ticketing, mistakes in loading and offloading (Kato, 2015). As a result, many bags get lost. Imagine that a traveler has packed his bags with all the things that are needed during a trip. Upon arrival at the destination, the traveller tries to find his bag on the belt which might be a crowded area with many people searching for their bag. The situation worsens if the traveler cannot find the bag. If the bag is lost, the traveller may feel it like a nightmare.

In the past, transferring baggage from airplanes to the arrival terminal was made manually, resulting in many bags getting lost. In addition to lost baggage, the manual process of transferring was not only labor-intensive but also time-consuming. Conveyor belts were used to increase efficiency of baggage handling. However, it was still difficult for tourists to find their bags and lose of baggage continued to occur (Kato, 2015). In this chapter, we discuss concept, design, and implementation of an efficient blockchain based solution for the lost baggage problem. To help identify bags RFID (Radio Frequency Identification) tags are used in the proposed solution. RFID tags allow reading data without physical contact (Zhang et al. 2008).

Blockchain is made up of blocks containing the transaction that occurred within the network (Want, 2006). A "blockchain" as the name implies, is in its simplest terms a "chain" of previously validated "blocks" of transactions that constitutes an immutable digital ledger and the distributed, resilient foundation for transfers of value. Each block is divided into two parts, the header and the body. The body section contains the transaction information while the header contains the ID of the previous block (Dorri et al., 2017). The blocks in the chain-linked to each other and the first block in the chain is called "Genesis Block".

Blockchain is transnational because it is across national borders anyone who has an internet connection can access it (Foroglou & Tsilidou, 2016). All data written in the blockchain will be protected from change and from unauthorized access. That is done by using a cryptographic hash function. If the input encrypts using hash function, it is difficult to retrieve to the original one (Pierro, 2017).

Each user in the blockchain is identified by an address that is cryptographically generated. The private key is only known by the user while the public key is known to everyone in the network to apply transactions (Dorri et al., 2017). Moreover, blockchain uses digital signatures to make sure the data is not corrupted. The typical digital signature algorithm used in blockchains is the elliptic curve digital signature algorithm (ECDSA) (Zheng et al. 2017). In this process, each user must sign the transaction with their private key which is authenticated by using the user's public key. This ensures that data exchange in the blockchain is secure between different peers.

Blockchain network is a peer-to-peer network. It is generated by the collection of groups of devices or computers known as "Node". These blockchain nodes are responsible for approving and verifying the transaction via a process called mining which is a record-keeping service done semitonically through using the computer. Miners broadcast repeatedly to update the blockchain which makes blockchain consistent. So, for a new block to be accepted as a part of blockchain, miners need to work on it to generate a proof of work (Azaria, 2016).

1.2 Motivation

Many travelers lose their bags while traveling by sending them to the wrong destination. A poor baggage handling system can lead to significant downsides, including increased rates of lost, damaged, or delayed luggage, which frustrate passengers and erode trust in the airline. This results in customer dissatisfaction and negative travel experiences, potentially leading to complaints and a loss of clientele. Operational inefficiencies may cause bottlenecks, disrupting overall airport operations and increasing wait times. Additionally, airlines face higher costs due to compensations for mishandled luggage and the need for manual interventions. Such issues can damage the airline's reputation, reduce customer loyalty, and create security risks, while also contributing to staff burnout due to increased pressure. Overall, these factors collectively harm customer experience, operational effectiveness, and airline profitability. Nevertheless, a Blockchain-based baggage handling system may ameliorate many of these pain points, as discussed later.

The International Air Transport Association (IATA) has established a range of rules, regulations, and policies to enhance baggage handling systems in the aviation industry. Key elements include IATA Resolution 600, which sets standards for the acceptance and delivery of passenger baggage, and comprehensive guidelines that promote best practices in tagging and tracking. The Baggage Improvement Program (BIP) aims to reduce mishandled baggage through collaboration and the adoption of technologies like RFID. Additionally, IATA's Digital Transformation Initiatives encourage the integration of digital solutions for improved operational efficiency. Conferences such as the Passenger Service Conference focus on enhancing customer satisfaction in baggage handling, while standards for baggage tags ensure consistency in tracking. Collectively, these measures aim to improve the efficiency, accuracy, and overall passenger experience in baggage handling.

The European Union (EU) promotes improvements in baggage handling systems through various laws and policies that enhance efficiency, safety, and passenger rights. Key regulations, such as Regulation (EU) No 261/2004, ensure compensation for delays and mishandling, while Regulation (EC) No 1107/2006 mandates effective assistance for disabled passengers. Safety is addressed through rigorous screening requirements in Regulation (EC) No 300/2008. The Single European Sky initiative aims to optimize airport operations, and the Digital Single Market encourages the adoption of advanced technologies for better tracking. Additionally, the EU's Green Deal promotes sustainable practices, and funding for research projects supports innovations in baggage handling.

North American aviation rules and policies aimed at improving baggage handling systems involve regulations from the Federal Aviation Administration (FAA) and the Transportation Security Administration (TSA), alongside industry standards from organizations like Airlines for America (A4A) and IATA. The FAA establishes safety and operational standards, while the TSA enforces security protocols that impact baggage screening. Best practices for efficient tagging and tracking are encouraged, and A4A promotes initiatives to enhance communication among stakeholders. Airlines implement customer service commitments, including compensation for mishandled baggage, to foster better handling practices.

We are interested in this idea to support the 2030 vision for the Kingdom of Saudi Arabia as the number of tourists will be increase. The vision 2030 calls for making Saudi Arabia a transportation and logistical hub connecting the three continents of Asia, Africa, and Europe. As we know Saudi Arabia has the two holy mosques, so the number of travelers will increase during the Hajj and Umrah season, so our system facilitates and increases the number of travelers for our country. Saudi Arabia's regulations for improving airline baggage handling are primarily governed by the General Authority of Civil Aviation (GACA) and focus on safety, efficiency, and customer service. GACA establishes comprehensive regulations that align with international standards set by ICAO and IATA. Enhanced security measures for baggage screening are mandated, and airlines must adopt standardized procedures for tagging and tracking to minimize mishandling. Customer service policies encourage improvements by offering compensation for lost or damaged baggage.

Understandably, there is a push for technological advancements, such as automated systems, blockchains, and RFID tracking, along with requirements for staff training in effective baggage management. A good baggage handling system provides numerous benefits, including increased efficiency and reduced wait times for passengers, which enhances overall customer satisfaction. By minimizing the instances of lost or damaged luggage, it improves the travel experience and fosters trust in the airline's services. Additionally, efficient systems lead to cost savings by reducing operational expenses related to mishandling. Enhanced safety and security measures ensure the integrity of baggage handling, while better communication through real-time updates promotes transparency for both airlines and passengers. Overall, a robust baggage handling system contributes to smoother airport operations and encourages customer loyalty.

1.3 Problem Statement

As shown in Figure 1 below, the traditional baggage handling systems raise the distress and exasperation among travelers regarding losing baggage at airports. In addition, mishandling baggage by the airline will cause significant losses for the airline companies and the airport. When bags are entered into the baggage handling system, there could be various potential problems such as bags loaded onto the wrong carrier, bags are given wrong destination code, bags getting switched between travelers due to similarity of the bags, or routing labels getting damaged during the trip. Furthermore, bags may get lost when they pass through multiple parties handling it such as airline staff, airport staff, security staff, ground staff, etc. Using RFID and blockchain techniques will solve all the problems that are mentioned. These technologies will facilitate travelers, airline companies, airport, security and ground staff to track baggage (Kosba, 2016).

In order to underscore the importance of the problem, here we highlight some of pain points in existing/traditional baggage tracking systems. Addressing these pain points is essential for improving the efficiency and reliability of baggage tracking systems. These pinpoints include:

- **Lack of Real-time Tracking:** Inaccurate or delayed updates often lead to confusion, frustration, dissatisfaction, trust erosion, and financial losses for customers.
- **Lack of Integration:** Difficulty in integrating with different airlines and airport systems can hinder effective tracking.
- **Poor User Experience:** Complicated interfaces or lack of mobile compatibility can hinder passengers from accessing tracking information.
- **Loss, Misplacement, and Delays:** Frequent loss or misplacement of baggage often erode customers' trust in the tracking system.
- **Weak Data Privacy:** Concerns about personal data security and privacy can deter users, operators, passengers, etc. from fully engaging with the system.
- **Communication Gaps:** Lack of timely communication from airlines regarding baggage status can increase passenger anxiety and erode confidence.
- **Inefficiency in Manual Processes:** Reliance on manual tracking can lead to errors and inefficiencies in the system.
- **High Cost of Implementation:** High costs associated with implementing, maintaining, and upgrading advanced tracking technology can be a barrier for some airlines.
- **Limited Coverage:** Inadequate tracking capabilities for smaller airports or less popular routes can lead to inconsistencies.

- **Inadequate Customer Service:** Poor customer support for baggage issues can exacerbate passenger frustration.

In the subsequent subsection, we provide an overview of how a blockchain-based baggage handling system would improve tracking by providing real-time updates via a secure app, facilitating data sharing between airlines and airports, and offering a user-friendly interface. We also highlight how it would reduce loss through a tamper-proof record, ensure data privacy, and enable automated notifications. In addition, we describe how smart contracts would minimize errors, increase efficiency, engender long-term savings. In a nutshell, we describe how blockchain improves reliability, efficiency, and customer service in baggage handling.

Figure 1. Most frequent baggage handling problems encountered at airports.

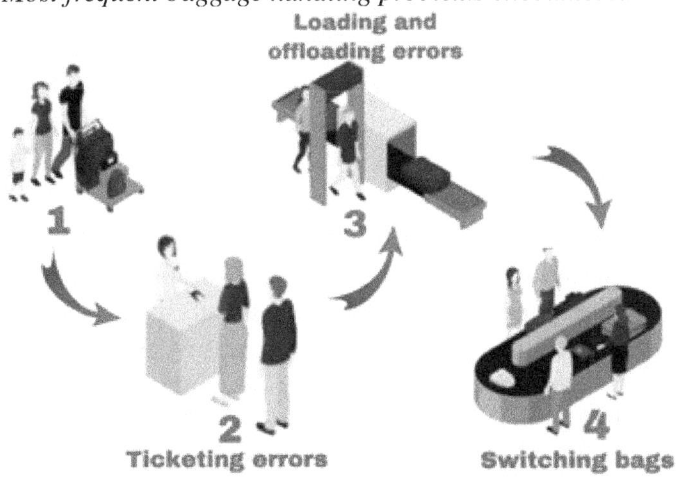

1.4 Proposed Solution

Using blockchain will keep all information about baggage such as baggage status and airport information for many airports in one system, so lost baggage will be easily identified. RFID tag which will be attached to the traveler's bags that will help to track baggage by passing bags through RFID reader which will be in belt converter in the airport and there will be a sorter at end of the belt that will match with information of RFID reader to sort bags and direct the bags to correct airplane.

These technologies will reduce check-in time, increase customer satisfaction, reduce cost, and enhance operational efficiency.

The main aim of our system is to get rid of the financial, emotional, and temporal costs of losing baggage while traveling through the airport. Our system will help people to track their bags. Furthermore, it will put limitations on losing bags as much as possible. Our focus will be on tracking bags in the airport. The proposed system is highly secure, as access to its data is restricted and tailored to individual privileges based on personal needs. This enhances both security and performance. A key feature of the system is its ability to connect multiple airports into a single integrated network, simplifying baggage tracking, especially in cases where a bag is lost or mistakenly sent to the wrong airport. Unlike barcodes or QR codes, the system does not require a scanner, reducing the likelihood of human error and eliminating the need to manually pass bags through scanning devices. This is particularly beneficial since barcodes can be damaged during baggage handling. RFID tags, on the other hand, can be embedded inside the bag, as data is read wirelessly. RFID readers are strategically placed throughout the airport, particularly in areas where baggage loss is more likely to occur. Additionally, the system eliminates the need for third-party involvement. Travelers can make reservations directly with airlines via blockchain, bypassing travel and tourism offices.

The system allows authorized parties—such as airport management, airlines, security personnel, and travelers—to track bags, with each party granted specific privileges based on their role. For example, travelers can only track their own bags, airlines can monitor bags registered with their company, and security personnel can access baggage information only for the specific area they are assigned to. The airport has the authority to define and assign these privileges. The system also assists governments in tracking travelers' bags. Furthermore, the system supports mobile connectivity, enabling travelers to easily track their bags, check their status, and receive notifications when their bags arrive at the destination. This streamlined approach improves efficiency and enhances the overall travel experience.

A blockchain-based airport baggage handling system significantly contributes to achieving several of the United Nations Sustainable Development Goals (SDGs) by promoting efficiency, transparency, and sustainability in airport operations. More specifically, it aligns with SDG-8, SDG-9, SDG-11, SDG-12, SDG-13, and SDG-16 (United Nations).

Regarding SDG-8 (Decent Work and Economic Growth), the system will help create new jobs. With airports adopting new technologies like blockchain, there may be an increase in demand for skilled workers in technology, logistics, and data management. The system will also improve economic efficiency by reducing the costs associated with lost baggage and improving operational efficiency can enhance economic growth for airlines and airports. Regarding SDG-9 (Industry,

Innovation, and Infrastructure), the system will enhance efficiency by streamlining baggage handling processes, blockchain technology reduces delays and improves overall airport operational efficiency. In addition, the system will advance and foster innovation and technological advancement in the aviation industry, encouraging the adoption of innovative solutions for logistics and operations.

Regarding SDG-11 (Sustainable Cities and Communities), the system will improve passenger experience as a more efficient baggage handling system enhances the travel experience, making airports more user-friendly and accessible. In addition, streamlining baggage processes can reduce congestion at airports, contributing to smoother operations and improved air quality in airport environments. Regarding SDG-12 (Responsible Consumption and Production), by reducing waste because improved tracking and management of baggage can reduce the incidence of lost luggage, minimizing waste associated with reprocessing and compensating for lost bags. In addition, efficient baggage handling leads to better resource management and optimization, reducing unnecessary fuel consumption and emissions from repeated handling of lost baggage. Regarding SDG-13 (Climate Action), the system will lower the carbon footprint by minimizing delays and optimizing baggage handling, blockchain can contribute to reduced emissions associated with aircraft waiting times and ground operations. In addition, the adoption of blockchain encourages airports to integrate more sustainable practices into their operations.

Regarding SDG-16 (Peace, Justice, and Strong Institutions), the system will promote transparency and accountability because Blockchain's tamper-proof nature ensures that all transactions related to baggage handling are transparent and accountable, enhancing trust among stakeholders. In addition, enhanced tracking and management of baggage increase security measures, reducing the risk of tampering or theft

1.5 Chapter Organization

The rest of the chapter is organized as follows. Section 2 provides some recent related work. Section 3 delineates the methodology we adopted and illustrates various components and ultimate integration into the final system. Section 4 outlines the scope of the system including functional and nonfunctional requirements as well as hardware and software requirements. Section 5 discusses various alternative solutions and blockchain technologies and provides justification for the adopted technologies. Section 6 provides details of the systems conceptualization and systems analysis using context diagrams, use case diagrams, UML class diagrams, entity relationship diagrams, activity diagrams, sequence diagrams, and data flow diagrams. Section 7 provides the system architecture and implementation framework. Section 8 includes discussions on the system outcomes. Section 9 concludes the chapter with some interesting future directions.

2 OVERVIEW OF THE RELATED WORK

In this section, we will discuss various technologies and applications which are relevant to our system. First, we discuss blockchain technology and various types of it. Second, we discuss object identification. Third, we discuss different kinds of platform for blockchain. Fourth, we discuss four applications used for tracking.

2.1 Blockchain Technology

Blockchain technology is a secure and resilient mechanism for conducting peer-to-peer electronic transfer of anything of value with a resulting immutable transaction record. It is a decentralized digital ledger system that records transactions across multiple computers in such a way that the registered transactions cannot be altered retroactively (Dorri et al. 2017). The key features of blockchain technology include:

1. **Decentralization**: Decentralization in Information Systems implies the distribution of data/information across multiple locations or nodes rather than relying on a single central server, database, or repository. Unlike traditional databases that are controlled by a central authority, a blockchain is maintained by a network of nodes (computers) that collectively validate and record transactions. Decentralization in information systems aims to create more resilient, efficient, and secure information management environments by facilitating redundancy, scalability, security, etc.
2. **Transparency**: Transparency in Information Systems implies the clarity and openness with which data and processes are managed, allowing all stakeholders understand how and why information is collected, stored, and used. It fosters trust among users by ensuring that actions and data are visible and verifiable. Since transactions on a blockchain are visible to all participants in the network, it helps to build trust among users.
3. **Immutability**: Immutability in Information Systems implies the property of data that ensures once it is created and stored, it cannot be altered or deleted. is crucial for maintaining data integrity and trustworthiness by providing audit trails, accountability tools, and backup/recovery options. Once a transaction is recorded on the blockchain, it cannot be changed or deleted without the consensus of the network. This makes blockchain highly secure against fraud and tampering, promoting reliability, security, accountability, and data governance.

4. **Consensus Mechanisms**: Consensus Mechanisms in blockchain networks are protocols that ensure all participants in the network agree on the validity of transactions and the state of the blockchain. They are essential for maintaining the integrity and security of the decentralized system by validating transactions, preventing double spending/fraud, and ensuring that all nodes have a consistent view of the blockchain. Blockchain networks use various consensus algorithms (like Proof of Work, Proof of Stake) to validate transactions and maintain the integrity of the ledger without the need for a central authority.
5. **Smart Contracts**: These are self-executing contracts with the terms of the agreement directly embedded/written into the code. They automatically enforce and execute the terms when conditions are met, enhancing efficiency, security, and transparency in agreements and transactions.

The aforementioned features of are provided in Blockchain technology through the following main components:

- **Blocks**: A Block in blockchain technology is a fundamental component that stores a collection/list of transactions and other relevant data like timestamp, reference to the previous block, etc. Each block is essentially a data structure that is linked to the previous block, forming a continuous chain.
- **Nodes**: These are the individual devices or computers that participate in the network, maintaining a copy of the entire blockchain. Nodes facilitate transaction validation, network security, and the overall functioning of the decentralized system.
- **Cryptography**: Cryptography involves techniques and mechanisms for securing communication and information by transforming it into a format that is unreadable to unauthorized users. It involves the use of mathematical algorithms and protocols to ensure data confidentiality, integrity, authentication, and non-repudiation. Blockchain uses cryptographic techniques to secure transactions and control the creation of new blocks.

Blockchain has many characteristics that make it optimal technique to solve many problems encountered in baggage handling. First, it is decentralized. Second, it provides pseudonymity that keeps the real-world identity of consumer's address hidden. Third, it offers transparency by which everyone on the network can access to the blockchain. Fourth, it promotes democracy in which every activity record is recorded as apart to the blockchain is agreed by consensus of all nodes. Fifth, it offers immutability so any small change in the transaction or record of blockchain can be detected by hash function and digital signature. Sixth, it offers audibility by which all peers in the blockchain hold a copy of all transactions. Seventh, it improves fault tolerance where any fault or data leakage can be identified through decentralized consensus. Eighth, it augments security as the blockchain uses many cryptographic techniques such as symmetric and asymmetric key cryptography to provide security.

2.2 Blockchain Types

Blockchains can be categorized into multiple types based on their usage and distinct attributes. Here we provide an overview of the main types of blockchain technologies and their differences.

2.2.1 Public (Permissionless) Blockchains

A public blockchain is a distributed and public ledger, which contains the records of all the transactions (Turkanović et al., 2018). It does not require a centralized authority to authenticate transactions (Ibid.). Therefore, it establishes trust in a decentralized network through consensus mechanism rather than relying on a third party such as banks. This would make the transactions on the network immutable once they are stored on the network. The word "public" means that anyone can join and access the blockchain network with read and write access permission (Dujak & Sajter, 2019).

2.2.2 Private (Permissioned) Blockchains

Private blockchain is like public blockchain in structure and mechanism. However, a private blockchain is a little bit different. The block with read and write access are controlled by an organization and access permission is limited in this type of blockchain (Dujak & Sajter, 2019). Most often, organizations may not want their users to have access to the details of every transaction (Ibid.). The owner of the private blockchain, which is usually an organization, may wish to give specific access to their users and control who can read or write to the private blockchain (Hölbl et al. 2018). To achieve this mechanism, the owner needs to know the identity of their users, so that the owner can determine what information should be accessed and what information should be restricted. The reason that organizations use private blockchains is to avoid anonymity. Sometimes organizations do not want to share their information with everyone, so a private blockchain is used to prevent the misuse of information.

2.2.3 Consortium (Federated) Blockchains

It is a semi-decentralized blockchain. A consortium blockchain is not granted to a single organization as a private blockchain instead of that it is granted to a group of approved organizations (Hölbl et al. 2018). Consortium blockchains is a permission network, it is like a private blockchains. Often, consortium blockchains are used in an enterprise, where a group of organizations collaborate by using blockchain technology to improve businesses (Ibid.). Nevertheless, this type of blockchain may give certain participants access (Dujak & Sajter, 2019). For instance, the root hash and its Application Program Interface (API) may be open to the public. Consequently, external entities can use the API to make a certain number of inquiries and obtain certain information related to blockchain status.

2.2.4 Hybrid Blockchains

Hybrid blockchains combine aspects of both public and private blockchains, allowing for private transactions while also providing public verification. Dragonchain is a good example of hybrid blockchains. Hybrid blockchains offer flexibility in data privacy and public accessibility. These are more suitable for businesses that require both secure private transactions and the ability to publish data publicly.

2.2.5 Sidechains

Sidechains are separate blockchains that are linked to a main blockchain, allowing assets to be transferred between them. Liquid Network is a good example of sidechains. Sidechains can operate independently but can interact with the main chain. These are useful for scalability and testing new features without affecting the main blockchain.

2.3 Object Identification

QR Codes and RFID (radio frequency identification) convert large amounts of data in a compact format. They have common features for facilitating operations by labor savings, reducing cost and increasing speed. However, there are some differences discussed in the following.

2.3.1 QR Code

The QR (Quick Rresponse) code is a two-dimensional barcode that enables it to contain more amount of data (Kieseberg et al. 2010). It consists of black square dots arranged in a white square (Sahu & Gonnade, 2013). It is a matrix symbol with high capacity and density. QR code reads by mobile phones/devices. There are many applications used for QR scanners. The scan line captures data with the sensor and stored it in memory. Then, the data will be analyzed using the software. It is used in several fields such as helping users to access URLs effectively and used to advertise electronic products.

The difference between QR code and a barcode holds information only in the horizontal direction, a QR code holds information in both horizontal and vertical directions so, QR code holds hundreds of times more information than a barcode. QR code has many advantages. It can be easily scanned from any angle, quick response and fast setup and ease of design. On the other side, it has many disadvantages like broken link or change of the link may lead to QR code failure, the scanning process may be delayed if the QR code is stained by something or if the label is damaged.

2.3.2 RFID Tag

Radio frequency identification (RFID) technology will enable us to uniquely identify an object from a distance because it is transferred data wirelessly, using radio waves. It will reduce efforts and eliminate human errors. The approaches for transferring power from the reader to the tag will use magnetic induction and electromagnetic (EM) which is related to RF antenna that might be near-field or the far-field (Want, 2006). Near-field will allow exchanging data in a short-range using high frequency. Far-field emissions capture EM waves propagating from a dipole antenna attached to the reader and a smaller dipole antenna in the tag receives this energy. There are three main types of RFID systems these types are low frequency (LF), high frequency (HF), ultra-high frequency (UHF) and microwave RFID. The types of RFID systems will depend on the communication distance involved in a specific application.

The RFID system consists of three main components transceiver with decoder, antenna and transponder RF tag. The transceiver (transmitter/receiver) and antenna are usually combined as an RFID reader (Want, 2006). **Figures 2** and **3** show how a fixated or portable RFID reader is connected to the network. The RFID reader uses radio frequency waves to transfer signals that activate the tag. When the tag is activated, it will send a wave to the antenna. Then it will be translated into data. The transponder is located in the RFID tag itself. There are many types of RFID, but we can divide them into two main categories active and passive according to types of transponders.

Figure 2. How the reader read RFID tag in near-field

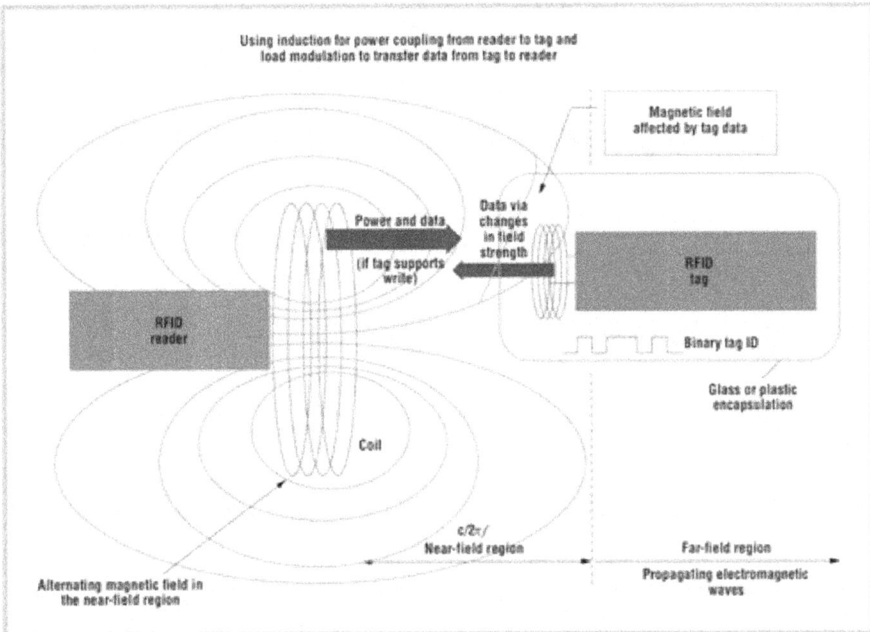

(Want, 2006)

Figure 3. How the reader reads RFID tag in Far-field

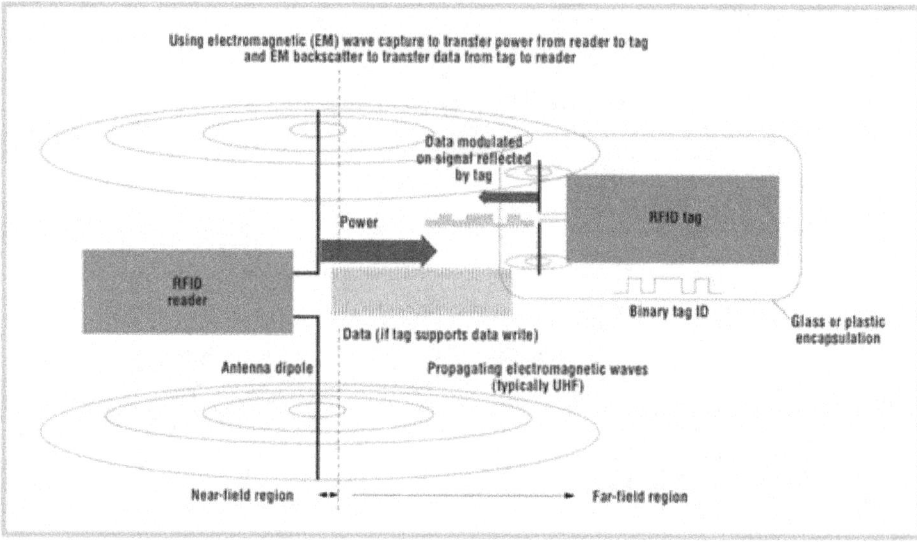

(Want, 2006)

2.3.2.1 The Active RFID Tags

The active RFID tag needs to power source by either connection to a powered infrastructure or using energy stored in an integrated battery, so its lifetime will depend on stored energy that will make it limited. In addition to its size and its cost will be very high (Want, 2006).

2.3.2.2 The Passive RFID Tags

The passive RFID tag does not need to power source. It will receive power from antenna. As such, its lifetime will be indefinite also its size will be small enough according to the active tag. Also, it is cheaper than an active RFID tag and its size is smaller than an active RFID tag. The most direct way to implement the passive RFID tag is through near-field communication (Want, 2006).

2.3.3 RFID for Effective Baggage Tracking in the Airline Industry

Many airlines have tested RFID technology, with Delta being the first to implement it successfully. RFID has proven to save costs by minimizing delays in operations, as it can quickly retrieve information and read multiple objects simultaneously. Implementing RFID in airports yields positive results for several reasons: it enhances traveler safety by making it easier to locate bags, reduces manual labor and human errors, and improves security measures. Unlike barcodes, which take significantly longer to read and often lead to traveler dissatisfaction, RFID offers a faster and more efficient solution. However, adopting RFID requires substantial infrastructure changes, such as installing RFID readers on conveyor systems. When Delta tested RFID, one of the biggest challenges was interference, which affected the ability to read tags or caused conflicts with other wireless systems. From a privacy perspective, the passive tags attached to bags are disposable, and travelers can disable them at the end of their journey using a kill switch (DeVries, 2008).

Organizations like IATA have explored how blockchain can be applied to various aspects of the aviation industry, including baggage handling. Blockchain technology enables secure and confident electronic value transfers between individuals or organizations without the need for third parties (Léopold, 2018). It provides a reliable and immutable method for tracking the status and location of bags, fostering trust among multiple parties. Additionally, IATA has leveraged blockchain for digital currency initiatives, incorporating more airlines, currencies, and smart contracts. Blockchain also supports digital finance by using smart contracts to reduce airlines' operational back-office costs. The Travel Grid architecture facilitates real-time digital interactions, lowers financial transaction costs, and enhances digital identity management and asset utilization (Léopold, 2018).

2.4 Review of Existing Tracking Applications and Technologies

Tracking applications have become an essential component of modern systems, offering efficient and secure solutions for monitoring and managing various processes. These applications leverage advanced technologies such as RFID, blockchain, and IoT to provide real-time tracking, reduce errors, and enhance user convenience. In the context of baggage tracking, such systems ensure seamless operations by connecting multiple stakeholders, including travelers, airlines, and airport authorities, while maintaining strict access controls and privacy measures. The integration of these technologies not only improves operational efficiency but also significantly enhances the traveler experience by enabling real-time updates and notifications. The following discussion explores the features and benefits of such tracking systems in greater detail.

2.4.1 Air Baggage Handling System Based on RFID

This is a traceable baggage system implemented at Beijing Capital International Airport (BCIA) for baggage tracking (Zhang et al. 2008). The system utilizes ultra-high frequency RFID (UHF) with anti-collision technology to prevent signal interference between RFID readers. The results from its implementation at the airport have been positive, improving user satisfaction and reducing issues related to lost or stolen baggage. The use of RFID for tracking at the airport offers several advantages, including contactless operation, multi-object recognition, non-line-of-sight reading, long-range detection, large memory storage, programmability, and durability. Additionally, it supports the use of universal loading devices (ULDs), which are essential for storing and transporting passenger baggage to its destination.

While the RFID tags are highly efficient, capable of reading up to 1,600 tags per second, the system has some limitations. Not all data for every airport is stored in a single system, and third-party involvement is often required to complete transactions. Furthermore, there are potential security and privacy concerns, as anyone with a compatible reader can access the information stored on the tags.

2.4.2 Trace Me Luggage Tracker

This mobile application integrates with the computer systems of airports and airlines globally to reduce the risk of lost baggage. It allows travelers to track their bags using a unique barcode placed on each bag. Travelers can monitor multiple bags, with each bag having its own distinct barcode. The barcode can be added in two ways: by scanning it or manually entering the tag code. Once scanned, the data encoded in the barcode is decoded and stored in memory. The software then analyzes this data. Travelers can view the status of all their tags in one place. The app shows whether a barcode is active, expired, or expiring soon, along with the number of days remaining before expiration. Since the application does not use blockchain, it relies on third-party systems for its operations.

2.4.3 BagTrack

BagTrack enables travelers to track their bags using iBeacon technology. To use the system, travelers only need the BagTrack app and an iBeacon device attached to their bag. iBeacons are a form of active RFID that operates using Bluetooth Low Energy (BLE) (Lin et al., 2015). Like active RFID, iBeacons do not rely on an external RFID reader for power. However, iBeacons are not highly accurate due to potential interference, which is common in any radio-based system. Additionally, Bluetooth tends to drain a phone's battery faster compared to passive RFID. While Wi-Fi-based technologies also consume battery power, most travelers keep their Wi-Fi enabled, making it a more practical option. iBeacons transmit data wirelessly over short distances, eliminating the need for a scanner. However, the system does not integrate all airport databases into one unified platform and requires third-party involvement to function effectively. Ref: https://github.com/robin24/bagtrack

2.4.4 Provenance

Provenance utilizes blockchain technology to track products (Provenance, 2025). It can be applied across industries that require reliable and unchangeable tracking information for their products. Provenance application is particularly useful for those involved in the production, distribution, and sale of goods. It enables users to trace and monitor the journey of each recorded product, identifying all parties involved in every stage of distribution—from producers to consumers and every step in between—using immutable ledger technology. Additionally, the system allows real-time updates to product information, shared with all participants via a decentralized and distributed ledger. This provides significant advantages, such as the ability to locate and, if necessary, halt any product in the supply chain.

2.4.5 ProxiAir

ProxiAir is a blockchain-based smart IoT cloud software designed to track travelers' baggage (Shull & Sękowska, 2020). It integrates IoT and RFID (Radio Frequency Identification) technologies. RFID tags are tracking devices used to monitor the progress, movement, and location of items they are attached to. These tags help track vehicles carrying luggage, waiting times, baggage transfer routes, and luggage locations. RFID is ideal for error-free identification of bags and suitcases, enabling efficient sorting and control checks as bags enter conveyor belts. The system uses the Ethereum Blockchain and requires antennas and RFID readers to be installed in zones where bags are frequently lost, allowing travelers to track their bags via a mobile app (Shull & Sękowska, 2020).

3 BLOCKCHAIN METHODOLOGY FOR AIRPORT BAGGAGE HANDLING

Each airport handles its information in one closed system which it operates in the intranets environment. In order to improve the performance of the airports, one of the best solutions is building one system for many airports but, that will raise problems in privacy and security. **Figure 4** illustrates the perfect solution which uses blockchain for storing data of the airport in private space. The airport will be storing its data information on the airport's server. For each airport there is a special gateway which is very secure to allow the airport to upload the data to the blockchain space. When all airport data are centralized in one system in blockchain, it will allow the travelers to track their bags via applications.

RFID is the abbreviation for radio frequency identification. It is used for identification of objects. **Figure 5** illustrates how the system uses electromagnetic waves to identify and track tags which will be attached to bags. The tags include chips on which information of the bag has been stored. Since the RFID has short wavelength, it needs access points (antennas) at every location where baggage may get lost in the airport. For instance, access points need to be installed on every conveyor belt at the airport. The tags will send information to the access point which transmits information to RFID reader. The reader then transmits the information to the airport's server.

Figure 4. Show how bags are tracked by RFID.

Figure 5. Airports and user application connected to the blockchain.

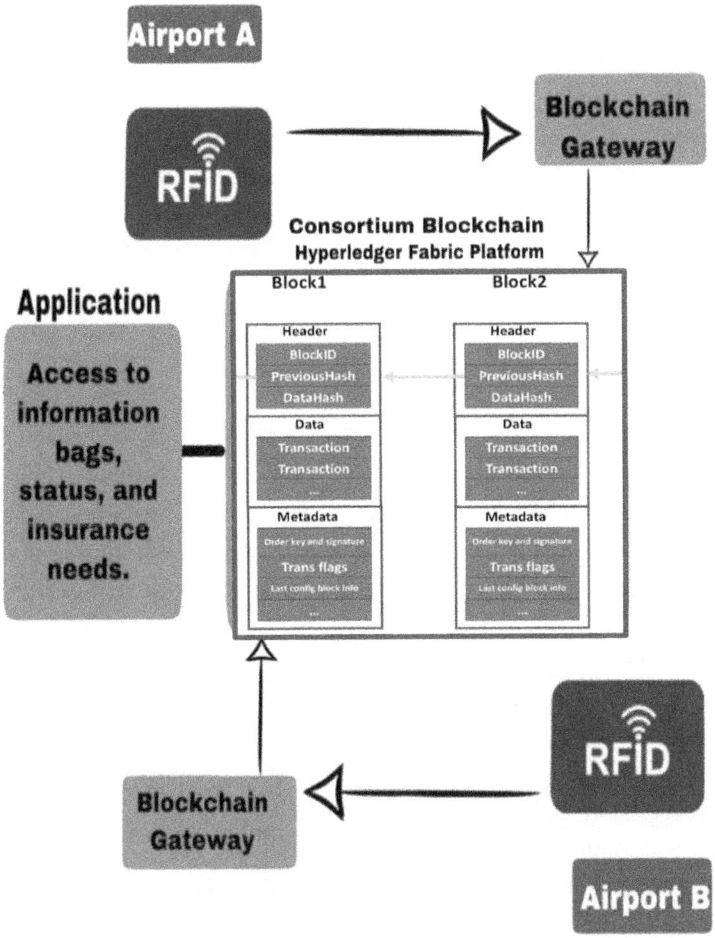

4 SYSTEM REQUIREMENTS

4.1 Software and Hardware Requirements

This section discusses the hardware and software used to implement the system. The main hardware component included are RC522 RFID Card Reader Module and Arduino Uno. The primary software used was the Oracle Blockchain Platform based on the Hyperledger Fabric platform for building the consortium blockchain.

In addition, we used Android studio for developing mobile phone applications. We used an online server to store external database.

4.2 Functional Requirements

Functional requirement is how the system should react to a particular input and how it should behave in a particular situation. The application has the following functional requirements:

4.2.1 The Traveler

1- The traveler shall be able to access to the blockchain.
2- The traveler shall be able to track baggage by using the RFID tag.
3- The traveler shall be able to see the history of the travels.
4- The traveler shall be able to see the status of the baggage like a bag not yet reached or bag has reached.
5- The traveler shall be able to receive the notification when his bag arrives.
6- The traveler shall be able to see the history of tracking bags information.

4.2.2 The Airport Admin

1- The airport admin shall be able to login to the system by entering the username and password.
2- The airport admin shall be able to view all airports.
3- The airport admin shall be able to view all airlines.
4- The airport admin shall be able to access the information in the blockchain.
5- The airport admin shall be able to edit employees and be able to update the information in the database.
6- The airport admin shall be able to add a new employee, who will be able to enter information in the database.
7- The airport admin shall be able to view his/her profile.
8- The airport admin shall be able to view travelers.

4.2.3 The Airport Security Staff

1- The security staff shall be able to login to the system by entering the username and password.
2- The security staff shall be able to view their profile.
3- The security staff shall be able to view all trips.
4- The security staff shall be able to view the airport.

5- The security staff shall be able to search by using RFID so the result will be the information for the owner's bag.
6- The security staff shall be able to search by using ticket number and travel ID so the result will be all the bags that traveler has for that specific ticket number.

4.2.4 The Airport Management Staff

1- The management staff shall be able to login to the system by entering the username and password.
2- The management staff shall be able to view all airports.
3- The management staff shall be able to view all airlines.
4- The management staff shall be able to view their profile.
5- The management staff shall be able to search by using RFID so the result will be the information for the owner's bag.
6- The management staff shall be able to search by using ticket number and travel ID so the result will be all the bags that traveler has for the given ticket number.
7- The management staff shall be able to search by using the reached state of the bag and travel ID, so the result will be all the bags that reached but not delivered to the traveler.

4.2.5 The Airline Admin

1- The airline admin shall be able to login to the system by entering the username and password.
2- The airline admin shall be able to edit employees by deleting the employee, so he will be able to update the information in the database.
3- The airline admin shall be able to add a new employee, so he will be able to enter information in the database.
4- The airline admin shall be able to view their profile.
5- The airline admin shall be able to view their travels.
6- The airline admin shall be able to view airports.

4.2.6 The Airline Staff

1- The airline staff shall be able to view all travels.
2- The airline staff shall be able to login to the system by entering the username and password.
3- The airline staff shall be able to view the company profile.
4- The airline staff shall be able to view the airport.

5- The airline staff shall be able to edit travel information, and he will be able to update the information in the database.
6- The airline staff shall be able to delete travel, and he will be able to enter information in the database.
7- The airline staff shall be able to edit traveler information, and he will be able to update the information in the database.
8- The airline staff shall be able to add a traveler, and he will be able to enter information in the database.
9- The airline staff shall be able to delete the traveler, and he will be able to update the information in the database.
10- The airline staff shall be able to edit bags for traveler, and he will be able to update the information in the database.
11- The airline staff shall be able to delete bags for traveler, and he will be able to update the information in the database.
12- The airline staff shall be able to add new bags for travelers, and he will be able to enter information in the database.
13- The airline staff shall be able to add a travel, and he will be able to enter information in the database.

4.3 Non-Functional Requirements

A non-functional requirement is the constraints on the service or function offered by the system. The application will have the following non-functional requirements:

4.3.1 Usability

The system should be easy to use and easy to learn by the user(s). Usability refers to the ease with which users can interact with a system, product, or service to achieve their goals effectively, efficiently, and satisfactorily. In the context of airport baggage handling, usability encompasses how well the systems and processes facilitate the tracking, management, and retrieval of passengers' luggage. Usability plays a critical role in airport baggage handling systems by enhancing operational efficiency, reducing errors, and improving the overall passenger experience. By focusing on usability the baggage handling system, airports can create a more efficient and pleasant travel experience, ultimately benefiting both passengers and airline operations.

4.3.2 System Availability

System availability in the context of an airport baggage handling system refers to the degree to which the system is operational and accessible to users when needed. It encompasses the uptime of hardware, software, and network components that facilitate the tracking, sorting, and delivery of baggage throughout the airport. The system should be available all the time as all the airports connect in one system. The system availability results in higher operational efficiency, greater passenger satisfaction, lower costs, improved real-time tracking, improved security, and improved safety.

4.3.3 Security

System security in the context of an airport baggage handling system refers to the measures and protocols implemented to protect the system from unauthorized access, data breaches, and other potential threats. This encompasses both physical security (protecting the infrastructure) and cyber security (protecting the data and software systems).

System security is critical for an airport baggage handling system as it protects both the physical assets (the luggage) and the sensitive data associated with it. Ensuring robust security measures is essential for maintaining operational integrity, passenger trust, and compliance with legal requirements, ultimately contributing to a safe and efficient travel experience. The following are some of the key security features in an airport baggage handling system:

1- The user shall be identified and authenticated in blockchain to access.
2- The user shall use a digital signature for each transaction.
3- The system shall not allow unauthorized access.

4.3.4 Confidentiality

Confidentiality refers to the obligation to protect and keep specific information secure from unauthorized access or disclosure. Confidentiality focuses on the handling of information once it has been shared, ensuring that it is not disclosed to unauthorized parties. Confidentiality in an airport baggage handling system refers to the protection of sensitive information related to passengers, luggage, and operational processes. This includes ensuring that only authorized personnel have access to data that could compromise passenger privacy or security. Confidentiality is important in protecting personal information, building trust, regulatory compliance, fraud prevention, operational integrity, and passenger experience. The following are some of the key confidentiality features in an airport baggage handling system:

1- The system shall keep data confidential between the parties concerned, not allowing the third party to access.
2- The system shall allow the user to access information based on the privilege of the user.

4.3.5 Privacy

Although privacy and confidentiality are related concepts, they have distinct meanings and implications, especially in contexts like data protection and security. Privacy refers to an individual's right to control their personal information and to decide how, when, and to whom that information is disclosed. Privacy is concerned with the individual's ability to manage their own information and maintain control over personal data.

Privacy in the context of an airport baggage handling system refers to the protection of passengers' personal information collected and processed during the baggage handling process. This includes data such as names, addresses, travel itineraries, and any other identifying information associated with luggage. Privacy in an airport baggage handling system is essential for protecting personal information, building trust, complying with legal standards, and enhancing overall security and passenger experience. By prioritizing privacy, airports can create a safer and more efficient environment for travelers.

Within the context of an airport baggage handling system, the system should ensure that transaction details are to be shared only among parties connected to each transaction with access based on permission.

4.3.6 Enhance Performance

Authorization in the context of an airport baggage handling system refers to the process of granting or denying access to specific resources, data, or functions within the system based on the identity and permissions of users. This ensures that only authorized personnel can perform certain actions, such as accessing sensitive information, managing baggage, or operating equipment. Authorization is essential in an airport baggage handling system as it safeguards security, maintains operational integrity, protects sensitive data, and ensures compliance with regulations. By effectively managing who can access what within the system, airports can enhance security, efficiency, and accountability, ultimately contributing to a safer and more efficient travel experience. The system shall be authorized to access only those who have permission, thus reducing verifications and that shall improve the performance

5 ALTERNATIVE BLOCKCHAIN TECHNOLOGIES

In this section, we discuss various alternatives regarding blockchain technologies and provide our rationale for the alternative we have chosen. The section contains three main subsections. The first subsection is about blockchain. The second subsection is about object identification. The third subsection is related to platform of the blockchain.

5.1 Blockchain Technology

As discussed, there are different types of blockchain based on their usage and distinct attributes. In our system, we use consortium (Federated) blockchains because there are multiple parties like traveler, airport management, and airlines etc. all whom can access to the blockchain. In addition, the consortium blockchain combines between public and private.

5.2 Platform Blockchain

There are many frameworks available for any organization deciding to implement blockchain such as Ethereum, Hyperledger Fabric, Quorum, etc. The decision will be based on the suitability of each organization. In this section, we will mention the most popular blockchain platform are used in most applications.

5.2.1 Ethereum

Ethereum is a framework used in a public blockchain (Hirai, 2017). It is flexible and considered efficient smart contract platforms (Ibid.). It is used for building and developing most of the decentralized applications. The important benefits offered by Ethereum are stability, safety and security (Ibid.). Ethereum provides greater transparency compared with other platform frameworks. It provides flexibility by allowing any developer to create applications using inbuilt programming language.

5.2.2 Hyperledger Fabric

Hyperledger Fabric is a Permissioned network that gives authority to register each user by providing identity access (Sousa et al. 2018). The transaction authority will define a range of areas within which each user can work. It is the best choice for the enterprise (Ibid.). It provides confidentiality by encrypting the transaction which can be modified by only an authorized person. As it is built on the modular approach, it requires fewer levels of verification and thus optimizes the entire software performance. This kind of framework on the blockchain allows enterprise to protect highly sensitive data by permitting access to only the parties concerned.

5.2.3 Quorum

Quorum supports both types of public and private blockchain transactions (Turkanović et al., 2019). It is a Permissioned network, and it is easily accessible. It is suitable for application demanding high speed and fast processing of private transactions (Ibid.). It is mostly used in financial services organizations.

In fact, the Ethereum was built as a B2C application, and it is used in public blockchain. Furthermore, it offers the greatest transparency thus it is not suitable for our system. Even though the Quorum supports both types of public and private transactions. It mostly based on Ethereum and it is more suitable used in financial services enterprise. On the other hand, in Hyperledger Fabric allows only users who have a specific legitimate interest to access the network and prevent others who haven't authorized access. It is most suitable for our system where the airports management are only responsible for giving security staff, passengers, and airlines specific access to the blockchain.

5.3 Object Identification

As discussed earlier, there are different types of object identification. In our system, we will use RFID especially a passive RFID tag because we will just implement it at a small distance in the near-field communications. That is because in the implementation we will make a prototype to simulate the environment in the airport. In addition, the passive RFID tag is smaller than active tags which will be easy to attach it to the bag. Moreover, there are a lot of features that make it better than the barcode. The RFID will not require human intervention which will reduce employment costs and eliminate human errors that reduce the possibility of losing a bag. Also, it can read hundreds of tags per second because no line-of-sight is required and the data will be read wirelessly. Moreover, in case of losing a bag, it will help to find it easily because it has extensive search capabilities. In addition, it has tags that can store large amounts of data and the possibility to update and modify.

6 SYSTEM CONCEPTUALIZATION

In this section, we show several different diagrams of the system. These diagrams include the use case diagram and context diagram. The section contains two main subsections. The first subsection is about use context diagram. The second subsection discusses the use case diagram.

6.1 Context Diagrams

A context diagram is a high-level visual representation of a system that illustrates its boundaries, external entities, and the relationships between them. It provides an overview of how the system interacts with outside elements, such as users, other systems, and external data sources. There are three key components of a context diagram: System Boundary, External Entities, and Data Flows. System Boundary represents the limits of the system, distinguishing what is within the system from what is outside. External Entities are the actors or systems that interact with the system being modeled. They can be users, other systems, or organizations. Data Flows are depicted using arrows or lines, showing the flow of information between the system and external entities as well as inputs to and outputs from the system.

Context diagrams are essential tools in system design and analysis, providing a clear, high-level representation of a system's interactions with external entities. They enhance communication, facilitate requirements gathering, and serve as a foundation for further analysis, ultimately contributing to a more effective and organized development process.

Figure 6 below shows how the data flows between the system and for external actors who are airlines, airports, and travelers. Consortium blockchain tracking bag system represents the entire system which includes all the transactions records, it allows external entities which are airlines, airports, and travelers to access specific information depending on the privileges of external entities. More context diagrams and details on context diagrams are provided in the online repository (Haq & Ahmad, 2025).

Figure 6. Context diagram shows how the data flow in the system.

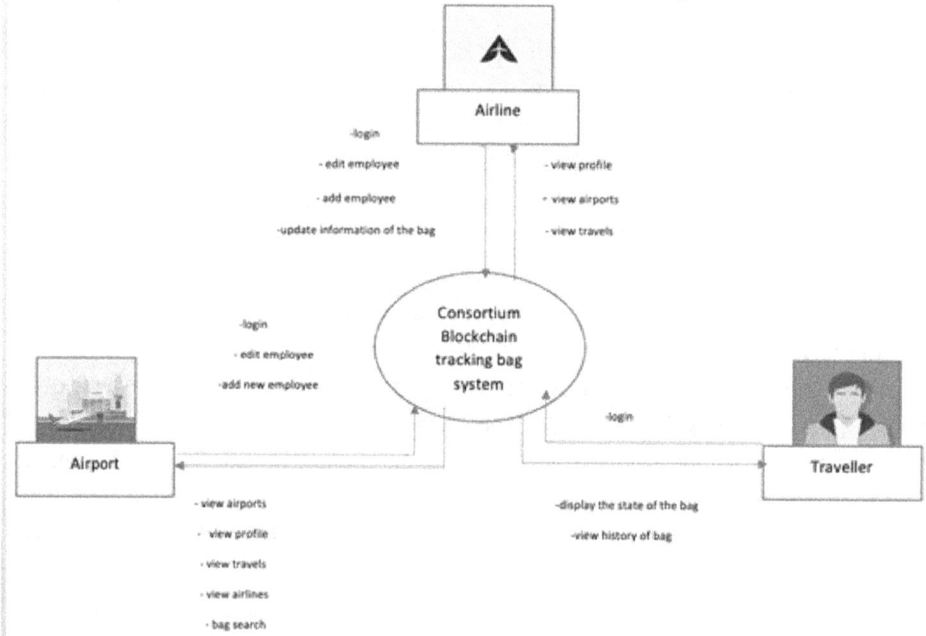

6.2 Use Case Diagrams

A Use Case diagram is a visual representation of the interactions between users (or "actors") and a system, illustrating the system's functionality from the user's perspective. It captures the various use cases, which are specific scenarios in which the system will be used and shows how those use cases relate to the actors involved. Use case diagram is the simplest representation of a user's interaction with the system. It shows the relation between different types of users with different cases they involved.

Use case diagrams are vital tools in system analysis and design, providing a user-centric view of system functionalities and interactions. They facilitate requirements gathering, enhance communication among stakeholders, and lay the groundwork for detailed design and testing, ultimately contributing to the development of effective and user-friendly systems. Use case diagrams are important to represent the functional requirements of a system. **Figure 7** shows how the use case diagram illustrates the relationship between the actors and the system. There are six external actors interact with a consortium blockchain system are traveler, Airline Admin, Airport Admin, Airline employee, Airport security member and Airport management member. Use case specification is a way to analysis and design the phases of the use case diagram as iterative way or manner. More use case diagrams and details on use case diagrams/ specifications are provided in the online repository (Haq & Ahmad, 2025).

Figure 7. Use case diagrams represent the functional requirements of a system.

6.3 UML Class Diagrams

A UML Class Diagram is a type of static structural diagram in Unified Modeling Language (UML) that depicts the classes, their attributes, methods, and the relationships between them within a system. Class diagrams are widely used in software engineering to model the structure of a system and provide a visual representation of its components. UML Class Diagrams are essential tools in software design and development, providing a structured and visual way to represent the classes and relationships within a system. They facilitate communication, support object-oriented principles, and serve as a guide for implementation and testing, ultimately contributing to the development of robust and maintainable software systems.

The UML class diagram demonstrates the different resources or classes of the system and relationship between them. **Figure 8** shows eight classes which are airport, airline, traveler, bag, travel, airport employee, airline employee, and blockchain. Each class has attributes, members, and methods. The main class in which all classes depend on is blockchain class that has four private attributes and one operation. More UML class diagrams and details on UML class diagrams are provided in the online repository (Haq & Ahmad, 2025).

Figure 8. UML class diagram

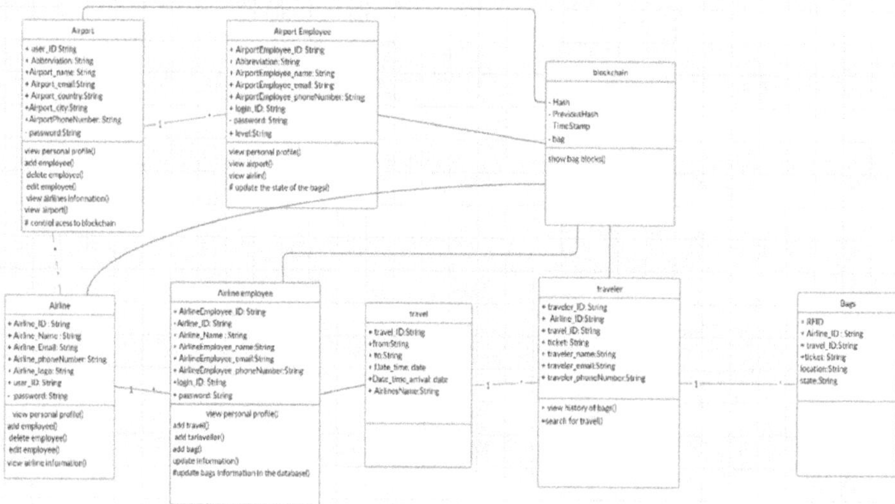

6.4 Entity Relationship Diagrams

An entity-relationship diagram (ERD) is a graphical representation of an information system that shows the relationships of entity sets stored in a database. Key Components of an ERD include entities, attributes, relationships, cardinality, primary keys, and foreign keys. Entities are represented as Rectangles, entities are objects or concepts that have data stored about them (e.g., Customer, Order, Product). Attributes are represented using Ovals are used to represent attributes, which are the data points or properties associated with an entity (e.g., Customer Name, Order Date). Relationships are represented using Diamonds to illustrate the relationships between entities, indicating how they interact with each other (e.g., a Customer places an Order). Cardinality indicates the nature of the relationship between entities, such as one-to-one, one-to-many, or many-to-many. Primary Keys are attributes that uniquely identify an entity (e.g., CustomerID, OrderID); whereas Foreign Keys are attributes that create a link between two entities, typically referencing the primary key of another entity.

Entity-Relationship Diagrams (ERDs) are crucial tools in database design and data modeling. They provide a clear and structured way to represent entities, their attributes, and relationships, facilitating communication among stakeholders and guiding the creation of efficient and organized databases. By identifying relationships and supporting normalization, ERDs ultimately contribute to the development of robust data systems. **Figure 9** shows the entity-relationship diagram for our system. It shows how the external agents which are traveler, airport, airline, airport employee, and airline employee interact with the system, and it shows how entities interact with each other. More ER diagrams and details on ER diagrams are provided in the online repository (Haq & Ahmad, 2025).

Figure 9. The Entity-Relationship (ER) diagram

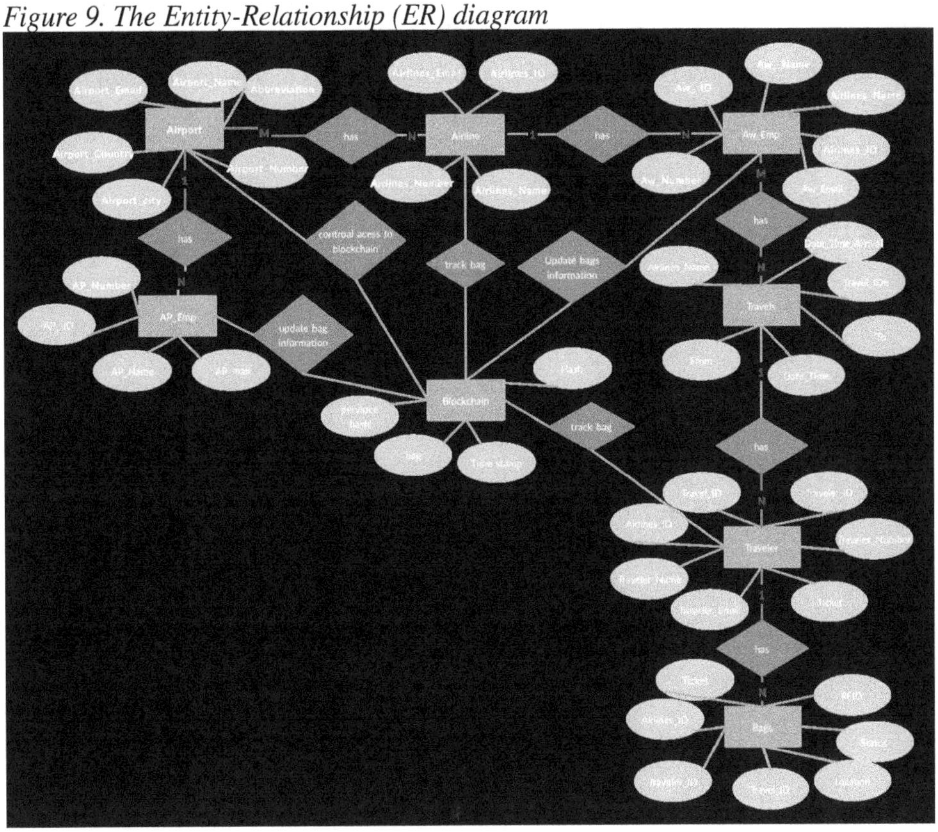

6.5 Activity Diagrams

An activity diagram is a behavioral diagram in the UML diagram. It is used to describe the dynamic aspects of the system. An activity diagram visually presents a series of actions or flow of control in a system like a flowchart or a data flow diagram. Key components of an Activity Diagram include activities, transitions, and decision nodes. Activities are represented as rounded rectangles and show the tasks or actions that need to be performed. Transitions are represented by arrows and indicate the flow from one activity to another, showing the sequence in which tasks are executed. Decision Nodes are shown as diamonds and represent points where decisions are made, leading to different paths based on certain conditions.

Activity diagrams are essential tools in process modeling and system design. They provide a clear, visual representation of workflows, facilitate communication among stakeholders, and support requirements gathering and process improvement. By clarifying roles and responsibilities, activity diagrams ultimately contribute to the development of efficient and effective systems. **Figure 10** shows the activity diagram for the traveler in our system. **Figure 11** shows the activity diagram for the airline include the airline employee and airline admin. **Figure 12** shows the activity diagram for the airport include the airport admin, airport management member and airport security member. More activity diagrams and details on activity diagrams are provided in the online repository (Haq & Ahmad, 2025).

Figure 10. Activity diagram for a traveler

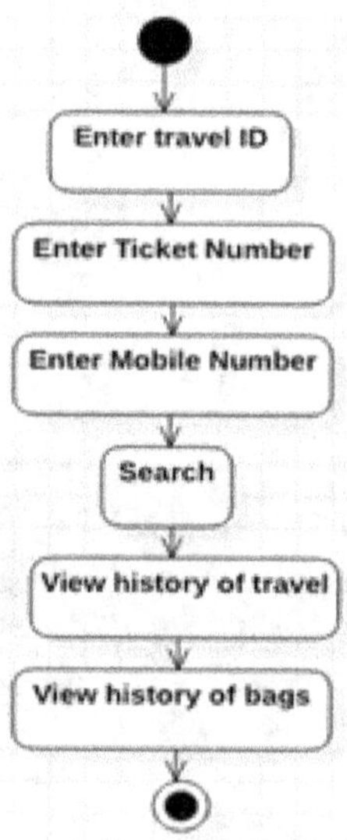

Figure 11. Activity diagram for airline

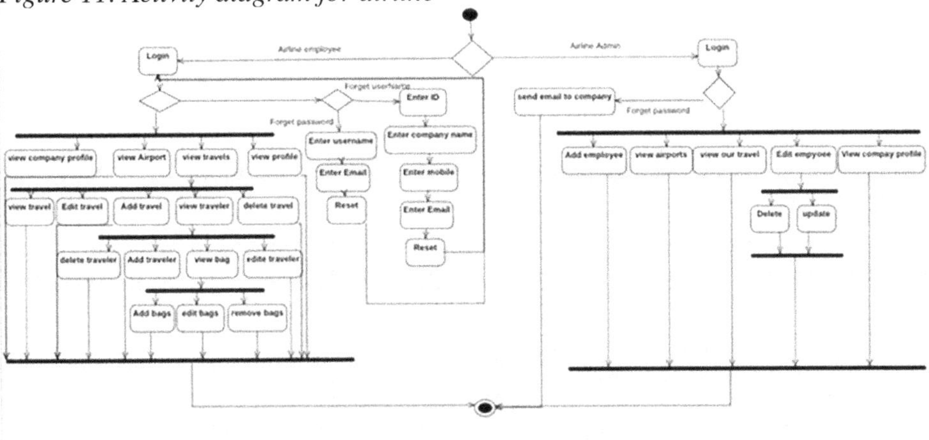

Figure 12. Activity diagram for airport

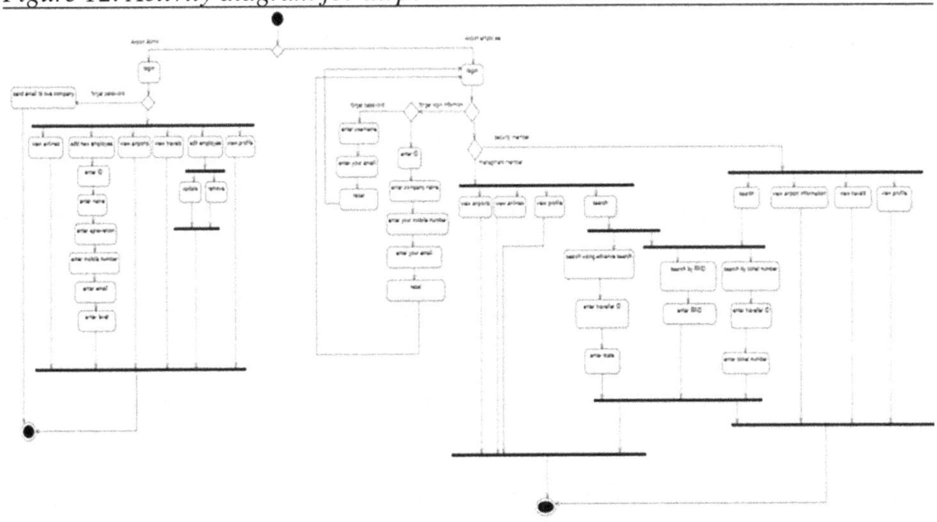

6.6 Sequence Diagram

Sequence Diagram is an interaction diagram in the UML diagram. A sequence diagram represents object interactions detail arranged in time sequence. The object as they interact over time, and they are organized according to object (horizontally) and time (vertically) also called lifelines. The key components of a Sequence Diagram are Lifelines, Activation Boxes, Messages, Combined Fragments, and Destruction Notation. Lifelines are vertical dashed lines that represent the existence of an object over time. Each lifeline corresponds to a participant in the interaction. Activation Boxes are rectangles on lifelines that indicate the period during which an object is active or controlling the flow of messages. Messages are arrows between lifelines that represent communication between objects. The arrows indicate the direction of messages. Combined Fragments are boxes that can encapsulate specific behaviors, such as alternative paths or loops within the interaction. Destruction Notation I an "X" at the end of a lifeline indicating that the object is destroyed at that point in the sequence.

Sequence diagrams are crucial tools in system design and analysis, providing a detailed representation of interactions among components in a system. They clarify use cases, facilitate communication, and support requirements gathering and testing, ultimately contributing to the development of robust and effective software systems. **Figure 13** shows the sequence diagram of the traveler. The sequence diagrams of the airport admin, airline admin, airline employee, airport security member, and airport management member are shared via an online repository (Haq & Ahmad, 2025).

Figure 13. Sequence diagram for traveler

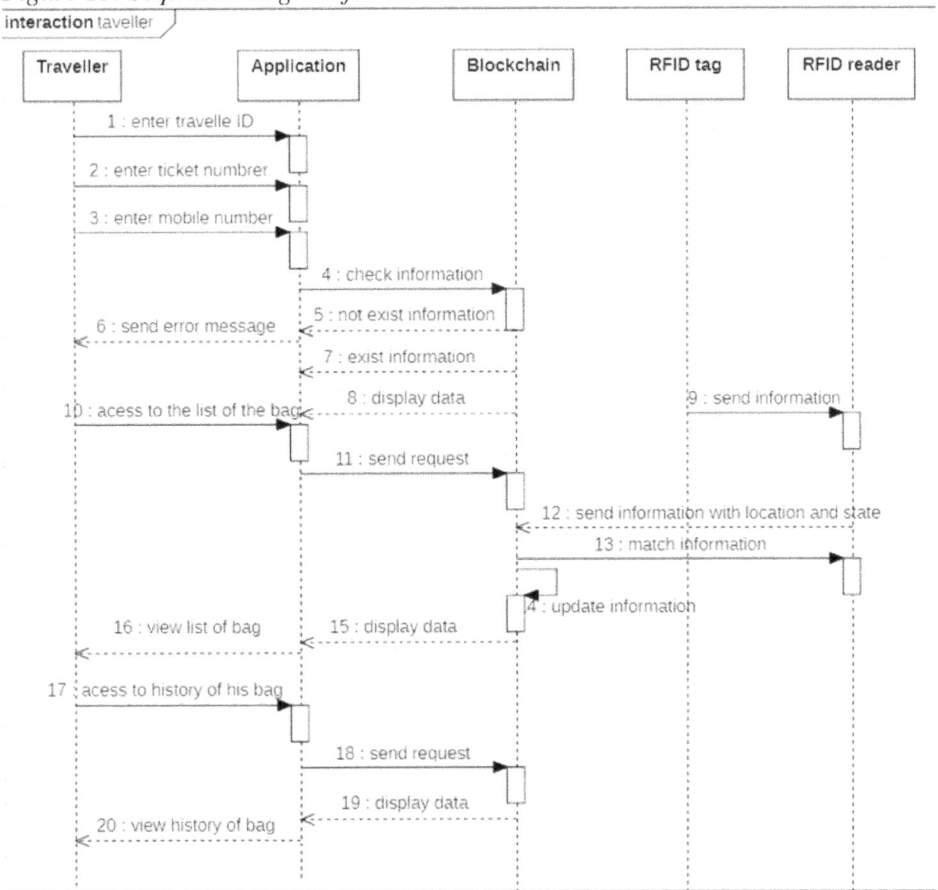

6.7 Data Flow Diagram

A data-flow diagram (DFD) is a way of visually representing the flow of data of a process system. It shows what kind of information will input and output from the system. It also shows how data moves from one part of the system to another, including inputs, outputs, storage points, and the processes that transform data. DFDs are commonly used in systems analysis and design to understand how data is processed and managed.

DFDs provide a structured and visual way to represent data movement within a system, facilitating communication, supporting requirements gathering, and aiding in the design and development of effective data management processes. By clarifying data flows and processes, DFDs contribute to the creation of robust and efficient systems. **Figure 14,Figure 15** and **Figure 16** show the proposed data flow diagrams for the system. contains six external agents that interact with the system. These agents are traveler, Airline Admin, Airport Admin, Airline employee, Airport security member and Airport management member. More DFDs and details on DFDs are provided in the online repository (Haq & Ahmad, 2025).

Figure 14. Data flow diagram for Level-0

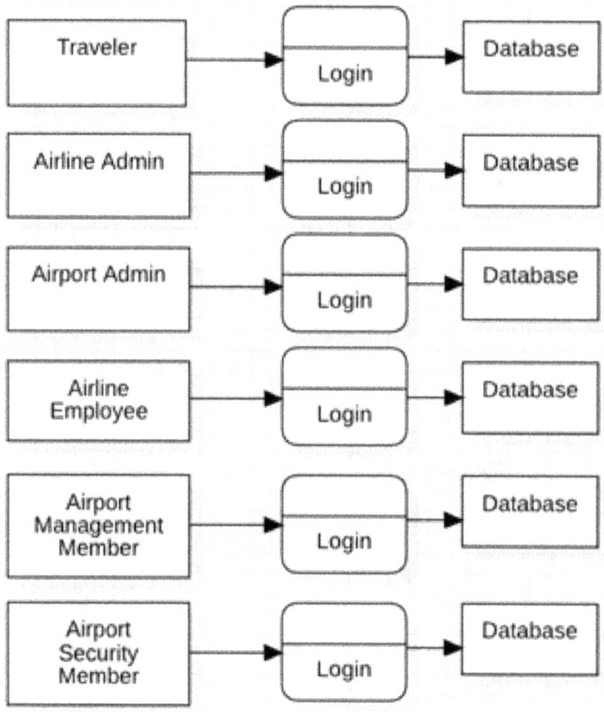

Figure 15. Data flow diagram Level-1

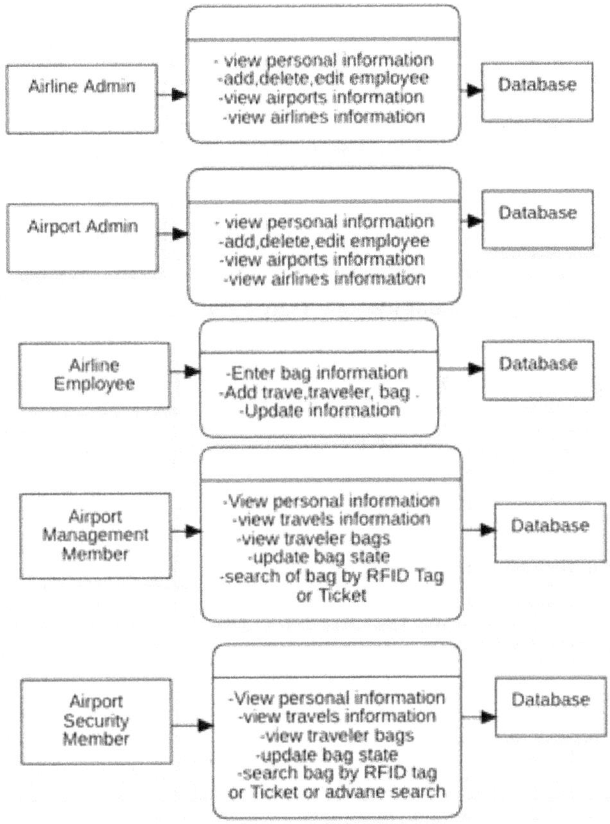

Figure 16. Data flow diagram for Level-2

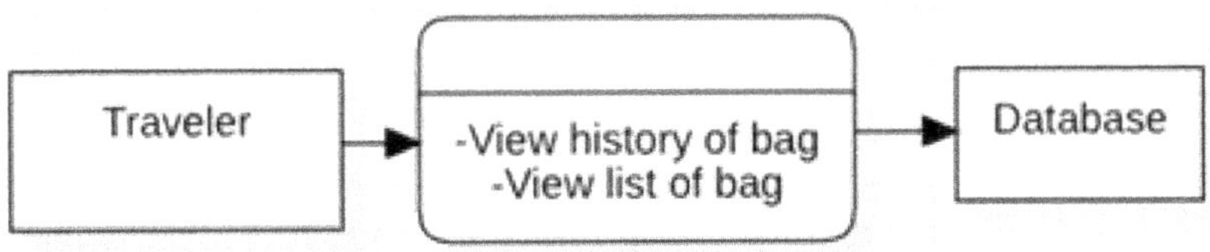

7 SYSTEM FRAMEWORK AND ARCHITECTURE

This section outlines the foundational design and components of the proposed system. A System Framework or Architecture refers to the foundational structure that defines how different components of a system interact, operate, and are organized. It encompasses the design principles, patterns, and guidelines that guide the development and integration of various elements within a system, ensuring that they work together effectively. A System Framework or Architecture is critical for understanding, building, and maintaining complex systems. By providing a structured approach to design and implementation, it facilitates communication, guides development, and ensures that all components work together effectively. A robust architecture ultimately contributes to the success and longevity of a system, enabling it to adapt to future needs and challenges.

Our subsequent discussions encompass the overall system architecture, the schema defining resource relationships, and a prototype of the application interface. Together, these elements provide a comprehensive understanding of the system structure, data flow, and user interaction design.

7.1 System Architecture

A system architecture is the conceptual model that defines the structure, behavior, and various views of a system. Figure 17 shows the architecture diagram for the system. Each bag has an RFID tag which is read by the RFID reader located at each airport. This data along with the name of the airport are transmitted to the server of the airport through the access points (antennas). The server of the airport updates the location and status of the bag in the blockchain. The traveler then can track their bag using the mobile app.

Figure 17. System architecture

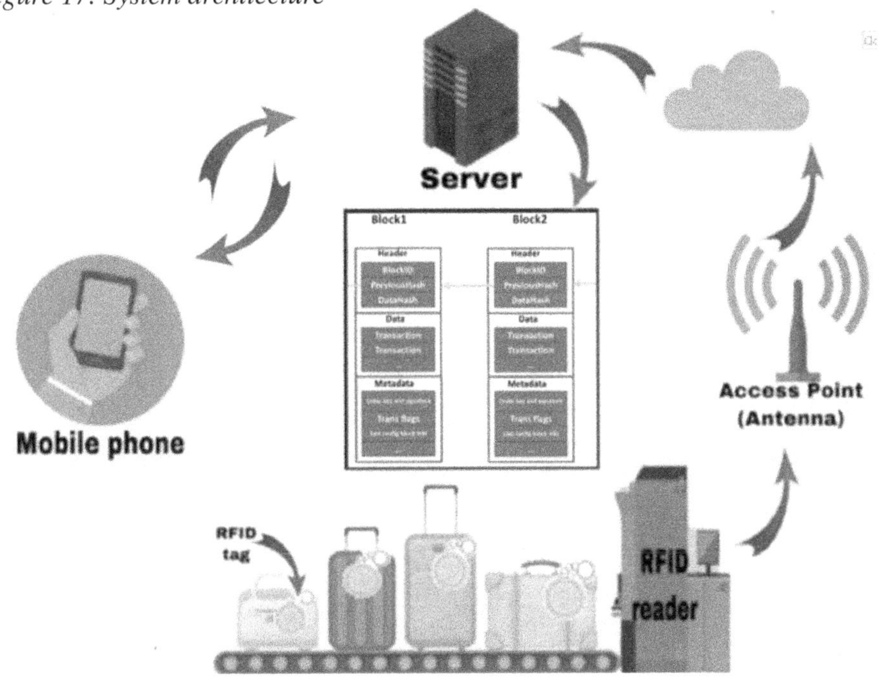

7.2 Resources Relationships Schema

A relational schema for a database is an outline of how data is organized. It includes the columns for each table and how tables are connected to each other. **Figure 18** shows the tables will be in the database and each table represents one actor. The airlines company table for the airline admin, the airport table for airport admin, the airline employee table for the employee in the airline, bags table for traveler's bags, travel table for the airline's travels, the traveler table for the travelers, the airport employee table has two different employees we can distinguish one from the other by the level column in the table. The two types of employees are security staff and management staff.

Figure 18. Relationships schema

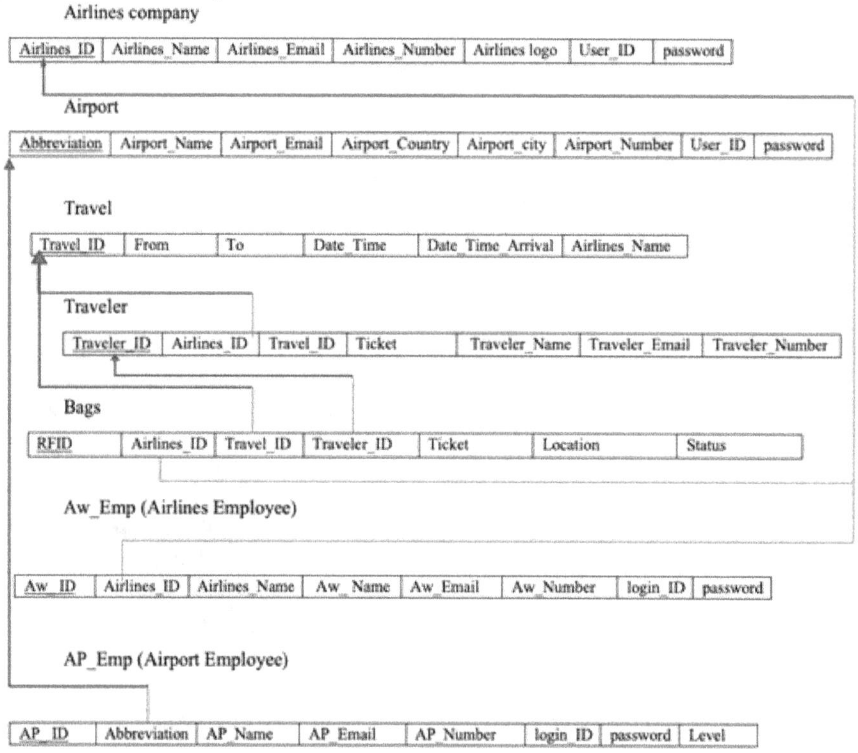

7.3 Application Interface Prototype

An Application Interface Prototype (often referred to as an Interface Prototype or UI Prototype) is a preliminary version of an application that focuses on the user interface (UI) and user experience (UX) design. It allows stakeholders to visualize and interact with the interface before the actual development begins. Prototypes can range from low-fidelity wireframes to high-fidelity interactive models that closely resemble the final product. The key components of an Application Interface Prototype include wireframes, interactive elements, visual design, and user flows. Wireframes are basic layouts that outline the structure of the user interface, showing the placement of elements like buttons, menus, and content areas. Interactive Elements are features that allow users to interact with the prototype, such as clickable buttons and navigational links, simulating the user experience. Visual Design in hi-fi prototypes may include colors, typography, images, and graphical elements to give a more realistic representation of the final product. User Flows are diagrams or paths showing how users will navigate through the application, highlighting the steps taken to complete specific tasks.

This section demonstrates the implementation of application interfaces for dashboard. The home interface, airport admin interfaces, airline admin interfaces, airline employee interfaces, traveler interfaces, management member interfaces, and security member interfaces are shown in the online repository (Haq & Ahmad, 2025). As an example, **Figure 19** shows the airport admin dashboard page. In this page, the admin can view airports, airlines, travels, employees, etc.

Figure 19. Airport admin dashboard

7.4 Tools Used for Implementation

This subsection discusses some of the major tools used in system implementation. The code as well as details of system implementation are available via online repository (Haq & Ahmad, 2025).

7.4.1 Hyperledger Fabric Platform

Hyperledger Fabric is an open source for distributed ledger technology (DLT) platforms to develop a blockchain network. Hyperledger Fabric offers the ability to create channels. Hyperledger Fabric also provides privacy and confidentiality of transactions. Also, it is the first distributed ledger platform that provides smart contracts authored in general-purpose programming languages such as Go and Node.js instead of domain-specific languages (DSL). Smart contracts ("chain code") run within a container environment like a Docker for isolation. Chain code operates as a trusted distributed application that gains its security and trust from the blockchain and the underlying consensus among the peers. Hyperledger Fabric blockchain network has a ledger consists of two distinct parts a blockchain which is the transaction logs and the world state. (Hyperledger Fabric, 2024).

7.4.2 Docker Container

Docker helps develop apps with multiple languages, frameworks, architectures. Docker facilitates workflow by giving developers the freedom to choose their tools and deployment environments. A container is a standard unit of software that packages code with its dependencies. So, the application runs quickly and reliably from one computing environment to another. There are many docker Partnerships such as Amazon Web Services (AWS), Microsoft Azure, and Snyk. We use AWS because we made the Virtual machine in AWS. Using Containers and VMs together provide a great deal of flexibility in deploying and managing app. We use a docker container to run the Blockchain Service. (Docker, 2024).

7.4.3 NodeJS / ExpressJS: API's Endpoint Service

Node.js is an open-source cross-platform that provides JavaScript runtime environment. It executes JavaScript code outside a web browser. It lets developers use JavaScript to write command line tools, and for server-side scripting. It can run scripts in the server side to produce dynamic web page content before the page is sent to the user's web browser. It unifies web application development around a single programming language, rather than different languages for server and client-side scripts. It is used to programmatically connect the application with a blockchain network. (Node.js, 2024)

7.5 Testing Outcome Functions of System

This section represents the testing and validation methodologies to chick whether these parts of the system work in an appropriate way. The test has been done by applying the test cases method to ensure there are no gaps in the system. In addition, to ensure the completeness and correctness of our system functions. A sample test case result is shown in tables 3. Detailed test cases are shown in the online repository (Haq & Ahmad, 2025).

7.5.1 Login/Logout for All Users

Table 1. Test case login

Test Case Name: login					
Test Case ID: TC# 1					
Step	Test Step	Input Data	Expected Result	Actual Result	Status (Pass/ Fail)
1	Password				
1.1	Enter right password	Fffff@88	Accepte the password and show message “login succssfull”		Pass
1.2	Enter wrong password	Fatimah88	Error message will appear that the password is wrong	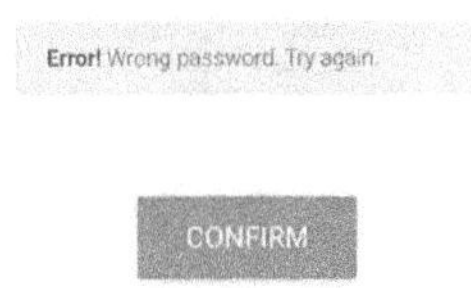	Fail
1.3	Leave password empty.		Error message will appear that “Required field missing-password is missing”	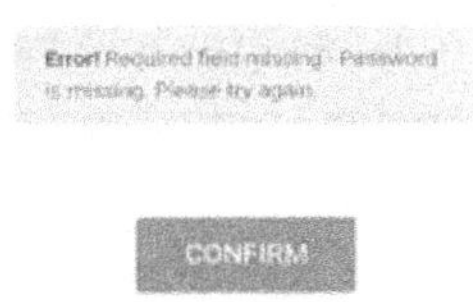	Fail

continued on following page

Table 1. Continued

Test Case Name: login					
2	Email				
2.1	Enter right Email	Fatimahtaher @hotmail.com	Accepte the password and show message "login succssfull"		Pass
1.2	Enter wrong Email	ss@.com	Error message will appear that "you are not registered user kindly register with us"	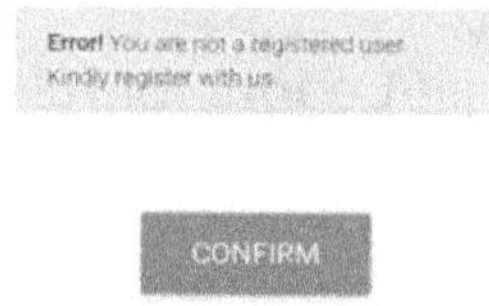	Fail
2.3	Leave Email empty.		Error message will appear that "Required field missing-email is missing"		Fail
3	Password and email				
3.1	Leave Email and Password empty.		Error message will appear that "Required field missing-email is missing, password is missing"	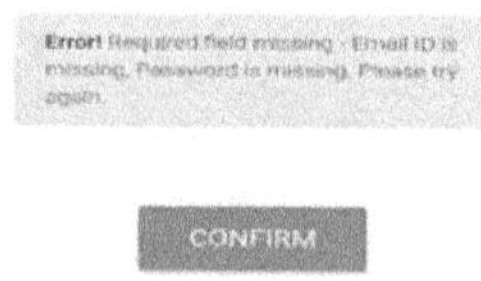	Fail

8 EXPECTED IMPACTS AND LIMITATIONS

Implementing a blockchain-based baggage handling system at airports can provide numerous benefits and have significant impacts on various aspects of airport operations and passenger experience. Here are some key benefits:

- **Enhanced Transparency:** Blockchain provides a tamper-proof record of each baggage journey, allowing all stakeholders (airlines, airports, ground handlers) to access real-time data about baggage status and location.
- **Improved Tracking:** Passengers can track their baggage in real-time, reducing anxiety about lost or delayed bags. This can be achieved through mobile apps that provide updates on baggage status.

- **Increased Security:** Each baggage transaction can be securely logged on the blockchain, minimizing the risk of tampering or unauthorized access. This enhances the security of the baggage handling process.
- **Streamlined Processes:** Blockchain can automate various processes through smart contracts, reducing manual interventions, paperwork, and potential errors associated with traditional baggage handling systems.
- **Reduced Lost Baggage:** With improved tracking and transparency, the incidence of lost or mishandled baggage can be significantly reduced, leading to higher passenger satisfaction.
- **Data Integrity:** The immutable nature of blockchain ensures that all records are accurate and verifiable, which helps in audits and compliance with regulatory requirements.
- **Cost Efficiency:** By reducing lost baggage and streamlining operations, airlines and airports can lower costs associated with compensation, tracking, and handling lost bags.
- **Collaboration Across Stakeholders:** A shared blockchain platform fosters collaboration among airlines, airports, and ground handling services, improving communication and coordination.

Nevertheless, a blockchain-based baggage handling system, while highly promising, offer some limitations. These limitations must be addressed carefully in order to realize the system benefits fully. Here we list some of these limitations:

- **Scalability Issues**: As the number of transactions increases, blockchain networks may face challenges related to scalability, leading to slower processing times.
- **Integration Complexity**: Implementing blockchain requires integrating with existing systems, which can be challenging, complex, and costly for airlines and airports.
- **Data Privacy**: While blockchain offers transparency, it can also raise privacy issues, as sensitive passenger information may be exposed on a public ledger.
- **Regulatory Challenges**: The legal and regulatory framework surrounding blockchain technology is still evolving, potentially complicating its adoption in the aviation industry.
- **User Adoption**: Effective implementation relies on all stakeholders (airlines, airports, and passengers) adopting the technology, which may face resistance or lack of understanding.
- **Cost of Implementation**: The initial costs associated with developing and deploying a blockchain system can be high, posing a barrier for some organizations.

- **High Energy Consumption**: Some blockchain protocols, particularly those using proof-of-work, can consume significant amounts of energy, raising sustainability concerns.

9 CONCLUSION

We report conceptualization, design, implementation, and validation of a blockchain-based baggage tracking system at airports. Blockchain technology has immense potential to significantly enhance baggage tracking in the airline industry by providing a secure, transparent, and efficient system for managing luggage information. This system can remove of ameliorate numerous pain points present in traditional baggage handling systems installed at airports.

A poor baggage handling system can lead to significant downsides, including increased rates of lost, damaged, or delayed luggage, which frustrate passengers and erode trust in the airline. Operational inefficiencies arising from poor baggage handling systems cause bottlenecks, disrupting overall airport operations, increasing wait times, increasing penalty/correctional costs, etc. Such issues can damage the airline's reputation, reduce customer loyalty, and create security risks, while also contributing to staff burnout due to increased pressure. Overall, these factors collectively harm customer experience, operational effectiveness, and airline profitability.

In contrast, a good baggage handling system provides numerous benefits like increased efficiency, reduced wait times, higher overall customer satisfaction, lower count of lost or damaged luggage, improved the travel experience, and greater trust in the airline's services, etc. Such efficient systems lead to cost savings by reducing operational expenses, enhancing safety and security measures, ensuring the integrity of baggage handling, improving communications, and promoting transparency. Overall, a robust baggage handling system contributes to smoother airport operations and encourages customer loyalty. These use of Blockchain technology could result in such improved baggage handling systems, delivering most of these highly sought-for benefits. Blockchain technology can furnish many benefits in baggage tracking systems. These benefits include immutable records, real-time updates, decentralized access, smart contracts, enhanced security, efficiency in claims processing, amenability to better data analytics, and even passenger empowerment. Indeed, a blockchain-based airport baggage handling system significantly contributes to achieving several of the United Nations Sustainable Development Goals (SDGs) by promoting efficiency, transparency, and sustainability in airport operations.

There are some interesting future enhancement opportunities, which can positively improve the functionality and applicability of our system. For instance, the system could contain multi-lingual options to cater the needs to wider global stakeholders.

This will help the user work/communicate in language the user is more comfortable. In addition, the system can be adapted for baggage tracking scenarios other than airports, like trains, buses, taxis, and container shipping companies.

ACKNOWLEDGEMENT

This work was supported by the Deanship of Scientific Research, Vice Presidency for Graduate Studies and Scientific Research, King Faisal University, Saudi Arabia (Grant No. KFU251161).

REFERENCES

Node.js. (2024). About Page. Available: https://nodejs.org/en/about/. (Accessed: 1-Oct- 20240.

Alsyouf, I., Kumar, U., Al-Ashi, L., & Al-Hammadi, M. (2018). Improving baggage flow in the baggage handling system at a UAE-based airline using lean Six Sigma tools. *Quality Engineering*, *30*(3), 432–452. DOI: 10.1080/08982112.2018.1437180

Azaria, A., Ekblaw, A., Vieira, T., & Lippman, A. (2016, August). Medrec: Using blockchain for medical data access and permission management. In 2016 2nd international conference on open and big data (OBD) (pp. 25-30). IEEE.

DeVries, P. D. (2008, January). The state of RFID for effective baggage tracking in the airline industry. *International Journal of Mobile Communications*, *6*(2), 151–164. DOI: 10.1504/IJMC.2008.016574

Dike, S. E., Davis, Z., Abrahams, A., Anjomshoae, A., & Ractham, P. (2024). Evaluation of passengers' expectations and satisfaction in the airline industry: An empirical performance analysis of online reviews. *Benchmarking*, *31*(2), 611–639. DOI: 10.1108/BIJ-09-2021-0563

Docker. (2024). Available: https://www.docker.com/company. (Accessed: 3-Oct-2024).

Dorri, M., Steger, M., Kanhere, S. S., & Jurdak, R. (2017, December). Steger, S. S. Kanhere and R. Jurdak, "BlockChain: A Distributed Solution to Automotive Security and Privacy,". *IEEE Communications Magazine*, *55*(12), 119–125. DOI: 10.1109/MCOM.2017.1700879

Dujak, D., & Sajter, D. (2019). Blockchain Applications in Supply Chain. In Kawa, A., & Maryniak, A. (Eds.), *SMART Supply Network. EcoProduction (Environmental Issues in Logistics and Manufacturing)*. Springer. DOI: 10.1007/978-3-319-91668-2_2

Foroglou, G., & Tsilidou, A. L. (2015, May). Further applications of the blockchain. In 12th student conference on managerial science and technology (pp. 1-8).

Haq, I., & Ahmad, A. R. (2025) "Shared Resources for Blockchain-based Baggage Tracking System for Sustainable Airport Operations: System Design, Implementation, and Testing", https://drive.google.com/file/d/1OpKgNGvLDa2wbpyHdidofSQrTOULBZ-k/view

Hirai, Y. (2017). Defining the Ethereum Virtual Machine for Interactive Theorem Provers. In Brenner, M. (Eds.), Lecture Notes in Computer Science: Vol. 10323. *Financial Cryptography and Data Security. FC 2017*. Springer. DOI: 10.1007/978-3-319-70278-0_33

Hölbl, M., Kompara, M., Kamišalić, A., & Nemec Zlatolas, L. (2018). A Systematic Review of the Use of Blockchain in Healthcare. *Symmetry*, *10*(10), 470. DOI: 10.3390/sym10100470

Hyperledger Fabric. (2024). Available: https://hyperledger-fabric.readthedocs.io/en/latest/whatis.html. (Accessed: 7-Oct-2024).

Kato, Z. (2015). SEA: Practical Application of Science. Academic Search Ultimate. Vol. 3 Issue 2, p229-234. 6p.

Kieseberg, P., Leithner, M., Mulazzani, M., Munroe, L., Schrittwieser, S., Sinha, M., & Weippl, E. *Proceedings of the 8th International Conference on Advances in Mobile Computing and Multimedia*, 430-435, 2010

Kosba, A. Miller, E. Shi, Z. Wen and C. Papamanthou, "Hawk: The Blockchain Model of Cryptography and Privacy-Preserving Smart Contracts," *2016 IEEE Symposium on Security and Privacy (SP)*, San Jose, CA, 2016, pp. 839-858. DOI: 10.1109/SP.2016.55

Kshetri, N. (2017). Can Blockchain Strengthen the Internet of Things? *IT Professional*, *19*(4), 68–72. DOI: 10.1109/MITP.2017.3051335

Léopold, E. (2018). Blockchain in Aviation White Paper. Retrieved April 26, 2024, from https://www.iata.org/en/publications/blockchain/

Lin, X. Y., Ho, T. W., Fang, C. C., Yen, Z. S., Yang, B. J., & Lai, F. (2015, August). A mobile indoor positioning system based on iBeacon technology. In 2015 37th Annual International Conference of the IEEE Engineering in Medicine and Biology Society (EMBC) (pp. 4970-4973). IEEE.

Pierro, M. D. (2017). What Is the Blockchain? *Computing in Science & Engineering*, *19*(5), 92–95. DOI: 10.1109/MCSE.2017.3421554

Provenance. (2025). Transparency for Your Products. Available online: https://www.provenance.org. (Accessed: 7-Feb-2025)

Puthal, D., Malik, N., Mohanty, S. P., Kougianos, E., & Yang, C. (2018, March). The Blockchain as a Decentralized Security Framework – Future Directions. *IEEE Consumer Electronics Magazine*, *7*(2), 18–21. DOI: 10.1109/MCE.2017.2776459

Sahu, S. K., & Gonnade, S. K. (2013). QR code and application in India. *Int J Recent Technol Eng*, *2*, 26–28.

Shull, C., & Sękowska, A. (n.d.). Proxiair. Retrieved April 27, 2020, from https://proxiair.com/

SITA. (2023). Baggage IT Insights 2023. Retrieved from https://www.sita.aero/resources/reports/baggage-it-insights-2023

Sousa, J., Bessani, A., & Vukolic, M. (2018, June). A byzantine fault-tolerant ordering service for the hyperledger fabric blockchain platform. In 2018 48th annual IEEE/IFIP international conference on dependable systems and networks (DSN) (pp. 51-58). IEEE.

TRACE ME Luggage Tracker Ltd. (2025). TRACE ME Smart ID. https://www.tmlt.co.uk. (Accessed: 7-Feb-2025).

Turkanović, M., Hölbl, M., Košič, K., Heričko, M., & Kamišalić, A. (2018). EduCTX: A Blockchain-Based Higher Education Credit Platform. *IEEE Access : Practical Innovations, Open Solutions*, *6*, 5112–5127. DOI: 10.1109/ACCESS.2018.2789929

United Nations. (n.d.). Sustainable Development Goals. United Nations. https://sdgs.un.org/goals

Want, R. (2006, January-March). An introduction to RFID technology. *IEEE Pervasive Computing*, *5*(1), 25–33. DOI: 10.1109/MPRV.2006.2

Zhang, T., Ouyang, Y., & He, Y. (2008). Traceable Air Baggage Handling System Based on RFID Tags in the Airport. *Journal of Theoretical and Applied Electronic Commerce Research*, *3*(1), 106–115. DOI: 10.3390/jtaer3010011

Zheng, Z., Xie, S., Dai, H., Chen, X., & Wang, H. "An Overview of Blockchain Technology: Architecture, Consensus, and Future Trends," 2017 IEEE International Congress on Big Data (BigData Congress), Honolulu, HI, 2017, pp. 557-564. DOI: 10.1109/BigDataCongress.2017.85

Chapter 2
Deploying OR/MS Tools for Decision Making in the Age of Artificial Intelligence and Sustainable Development Goals:
Social, Behavioral, and Technological Perspectives

Abdul-Rahim Ahmad
https://orcid.org/0000-0001-7923-0105
King Faisal University, Saudi Arabia

Yousuf Ahmad
University of British Columbia, Canada

Juwairiah Ahmad
Southern New Hampshire University, USA

Omar Ahmad
Payactiv, USA

Ibtihaj Ahmad
University of Nevada, USA

DOI: 10.4018/979-8-3373-0139-6.ch002

ABSTRACT

Operations Research and Management Sciences (OR/MS) methodologies have shown significant promise as well as success in solving and unraveling a variety of operational and tactical problems. However, this success is not matched in solving complex political, social, socio-technical, and strategic decision problems. Such an apparent lack of efficacy has been attributed to a variety of factors. We take behavioral and technological perspectives of opportunities and impediments in the adoption of OR/MS tools in a wider realm of problem-solving and decision-making. We also look at various limitations, and sources of limitations, of OR/MS approaches. We discussed some ideas on how to make OR/MS tools more effective through a combination of OR/MS with systems thinking, soft decision-making, and Artificial Intelligence (AI). We discuss how using AI in conjunction with OR/MS can help make decision-making more efficacious, leading to fair, equitable, robust, and sustainable development choices.

1. INTRODUCTION

The disciplines of Operations Research and Management Sciences (OR/MS) have been defined in various ways by various researchers and organizations. One concise definition is: "[OR/MS are the disciplines] of applying advanced analytical methods to help make better decisions" (INFORMS, 2025). For about a century, OR/MS has been growing in complexity and usefulness. Indeed, OR/MS brought with it new technologies and paradigms to aid effective decision-making (Taylor, 2018). More importantly, various decision-making paradigms and technologies have enhanced the effectiveness of interdisciplinary problem-solving teams and added value and efficacy of OR/MS tools (INFORMS, 2025). The proliferation of information technology further facilitated the use of OR/MS tools by more people, rendering an increased capability to add value through OR/MS tools. This impact of information technology on OR/MS discipline is aptly expressed by the manner in which The Institute for Operations Research and the Management Sciences (INFORMS) defines OR/MS: "[OR/MS are] the professional disciplines that deal with the application of information technology for informed decision-making" (INFORMS, 2025).

Factually, OR/MS disciplines are deemed responsible for substantial improvement in efficiency, efficacy, and sustainability of industry, government, and R&D (INFORMS, 2025). Nevertheless, the traditional style of preaching and teaching OR/MS made many people perceive OR/MS as a panacea in real-world problem-solving. Ironically, some OR/MS practitioners may even advocate it as the rational and scientific problem-solving regime. Conceivably, when people discover through hard experience that the OR/MS discipline has its constraints in dealing with the majority of omnipresent interdisciplinary and composite behavioral, socio-technical, and other ill-structured problems, a negative emotion prevails. Nevertheless, the OR/MS tools in tandem with other soft decision-making and problem-solving approaches, assisted by autonomous and intelligent agents, could deliver a lot more than what some people perceive.

In this chapter, our objective is to explore ways of enhancing the efficacy of OR/MS techniques within a broader context. We will also examine differences and complementarities between a pure OR/MS and a Systems approach to problem-solving. It is not our intention to cover these techniques in depth because of space and scope limitations. However, we have tried to identify some potent impediments in adoption of OR/MS tools from social, behavioral, and technological perspectives. Furthermore, our discussions are generalized in the sense that we would not be wading into any lists or details of either OR/MS tools and techniques or artificial intelligence (AI) domain, most of which can be found in any OR/MS or AI texts.

The rest of the chapter is organized as follows. Section 2 offers a brief historical perspective of the OR/MS discipline. Section 3 discusses the value provided by OR/MS tools in various arenas. Section 4 delineates some core aspects of OR/MS tools and techniques. Section 5 outlines some of the problems with traditional OR/MS approaches. Section 6 provides some recommendations for interdisciplinary and synergistic approaches needed in advancing applications of OR/MS as well as its adoption in more effective problem solving and decision-making. Section 7 elaborates on some of the salient benefits of using AI and Soft Computing in conjunction with OR/MS. Section 8 concludes the chapter.

2. OR/MS: THE HISTORICAL PERSPECTIVE

We believe that a brief glimpse of the history and success of OR/MS as a discipline would enable putting some relevant arguments in a better perspective (Gass & Assad, 2005).

2.1. Origins of OR/MS in the Second World War

The OR/MS has its early development during the WW-II, when British military authorities faced pressing problems that were not amenable to any known scientific methods (Pidd 2001; Thomas, 2015). These problems arose from new technologies and applications to organizing operations. In order to maximize utilization of military resources, especially radar installations, and to analyze the available data, the military formed teams of scientists from a wide variety of backgrounds, named the Operational Research Group. The eventual success of the OR group in improving resource utilization and effective decision-making is credited for the creation of a similar group in the US military. In fact, it can be argued that the victory of allies in the Second World War is attributable to efficient and effective deployment of a myriad of scarce resources made possible through the application of OR.

This historical context of OR/MS highlights its foundational role in addressing complex problems through scientific analysis and interdisciplinary collaboration. The discipline's origins in military operations during the Second World War underscore its importance in resource optimization and decision-making, which have since been applied across various sectors beyond the military.

2.2. OR/MS in the Post-War Era

In the post-war scenario, British and US scientists realized that this approach of creating diverse teams for building and optimizing models could be used to great effect and benefit in business and the public sector (Pidd 2004). The practice of employing OR/MS in civilian domain soon spread throughout the developed world. By the end of the 1950s, OR/MS had established a set of problems and methods that it generally dealt with. Lists of such tools and applications can be found in any OR/MS text (Taha, 2022; Taylor, 2018). The spectacular rate of increase in computing power coupled with the development of more advanced modeling and optimization tools meant that more complex models and problems have been progressively tackled with OR/MS techniques.

2.3. OR/MS in 1980s

The modeling/optimization approach remained a major constituent of OR/MS theory and practice throughout the 1960s and 70s. However, starting in the early 1980s, many OR/MS researchers and practitioners realized and acknowledged that the traditional OR/MS approach was not adequate in all situations (Ackoff, 2001). Several researchers and practitioners expressed concerns about the ever-increasing mathematical bias and narrow focus on optimization in the OR/MS. These people suggested that the systems thinking, the soft systems methodology, and other related techniques were more appropriate than mathematical programming for many complex, ill defined, and difficult-to-model issues that OR/MS practitioners were confronting (Ackoff 2001, Ahmad 2005). They extended the narrow quantitative optimization focus to incorporate thoughts from such other disciplines as philosophy, sociology, and organizational behavior. Indeed, this multidisciplinary approach envisioned by proponents was in line with the earliest applications of OR during wartime when the OR Group consisted of experts from a wide array of disciplines and work domains.

2.4. OR/MS in 1990s

The 1990s saw significant changes in computer and information technology that greatly impacted the OR/MS field. During this time, there was a noteworthy increase in user-friendly spreadsheets and affordable OR/MS software, making advanced analytical tools more accessible to a wider audience. The concurrent development of simpler communication and computational tools, as well as the widespread sharing of OR/MS knowledge via the internet, more people were able to engage with OR/MS applications.

However, this increase in participation often came with its own downsides, including the fact that many such individuals lacked formal training or education in OR/MS principles and struggled to effectively apply these tools without proper guidance, oversight, or intellectual support (Alford et al., 2025; Daellenbach & McNickle, 2012). Some practitioners, often termed as pseudo-practitioners, with only basic understanding of a small number of OR/MS tools, has been creating unwanted examples and cases of less efficacious problem-solving and decision-making via OR/MS (Daellenbach & McNickle, 2012). Additionally, such practitioners struggled to integrate OR/MS tools with other disciplines like AI and soft computing, limiting their ability to tackle complex real-world problems that required more comprehensive methods. Consequently, the OR/MS discipline came under increased pressure to not only prove its value and relevance but also ensuring adequate knowledge and skills are imparted on practitioners.

2.5. OR/MS in 2000s

In the 2000s, OR/MS discipline experienced significant advancements and transformations. These advancement and transformations mainly resulted from rapid technological progress and evolving business needs. The integration of more sophisticated computing power and software capabilities allowed practitioners to tackle more complex problems with greater speed, efficiency, efficacy, and accuracy. There was a surge in the development and deployment of advanced analytical tools like optimization algorithms, simulation techniques, and data mining tools. A notable trend in 2000s was the rise of big data analytics. This involved leveraging on the ability to collect and store vast amounts of data to obtain actionable insights, resulting in improved forecasting, planning, and management as well as improvements in efficiency and efficacy. Additionally, the growing importance of real-time data led to the development of dynamic modeling techniques, which allowed for more responsive and agile decision-making processes.

The 2000s also marked a welcomed emphasis on interdisciplinary collaboration and interdisciplinary problem-solving methods in OR/MS. Practitioners started working more intently with experts from fields such as computer science, engineering, business, and economics, fostering a more holistic understanding of complex socio-technical systems. The rise of open-source software and online platforms also supported and promoted knowledge sharing and collective problem-solving. Moreover, the globalization and the prevalence of e-commerce engendered new challenges and opportunities for using OR/MS. Organizations needed to augment operations across diverse geographical locations, make tactical advanced logistics, and strategize supply chains. The focus on sustainability and corporate social responsibility also emerged as significant themes, prompting OR/MS to address environmental and social factors in decision-making processes.

2.6. OR/MS in 2010s

In the 2010s, OR/MS experienced significant advancements driven by the explosion of big data and the integration of artificial intelligence (AI) and machine learning (ML). Organizations harnessed vast datasets to enhance decision-making through advanced analytics, enabling more accurate forecasting and insights (Feroz & Stines, 2024). The ability to analyze data in real-time became increasingly important, allowing OR/MS practitioners to develop adaptive models that responded to changing conditions, particularly in sectors like supply chain management and healthcare (Helo & Hao, 2025).

The decade also emphasized sustainability and social responsibility, with practitioners incorporating environmental and social factors into their decision-making processes. Collaboration between OR/MS professionals and experts from various fields fostered innovative solutions to complex problems, while the rise of cloud computing and Software as a Service (SaaS) made sophisticated OR/MS tools more accessible to organizations of all sizes. As a result, the 2010s solidified OR/MS's role as a critical discipline in addressing contemporary challenges and driving efficiency across diverse industries.

2.7. OR/MS in 2020s and Beyond

In the 2020s and beyond, OR/MS is expected to evolve significantly, driven by rapid advancements in AI, automation, and big data analytics (Feroz & Stines, 2024). Such advancement in technologies would enhance decision-making processes, enabling real-time analytics predictive modeling, etc. across times, industries, organizations, and more. In addition, there has been a heightened focus on resilience and risk management, particularly in response to such global challenges as the COVID-19 pandemic, prompting practitioners to develop even more advances models for addressing uncertainties and disruptions (Chen, 2020).

Moreover, sustainability and ethical considerations should remain central to OR/MS, guiding organizations and industries toward responsible practices aligned with environmental and social responsibility/equity. Interdisciplinary, interorganizational, and inter-government collaboration would be essential as complex and composite global challenges require insights from diverse fields. Educational institutions should adapt curricula to equip future practitioners with the skills needed for a technology-driven world, emphasizing data literacy and ethical decision-making (Alford et al. 2025). Overall, OR/MS is expected to play a crucial role in navigating an increasingly interconnected and dynamic environment.

3. THE VALUE OF OR/MS

OR/MS tools and techniques are playing a pivotal role in generating value across a diverse array of organizations, from small startups to large multinational corporations, in corporate, public, and other sectors alike. The primary objective of implementing OR/MS tools is to optimize resource utilization, enabling organizations to achieve greater outcomes while minimizing the input of time, money, human resources, technology, and other resources. By employing sophisticated analytical techniques and methodologies, OR/MS tools facilitate informed efficacious decision-making, enhance operational efficiency, and drive strategic planning. Overall, the versatility and efficacy of OR/MS tools make them invaluable assets in the pursuit to do more with less, fostering innovation and efficiency across multiple domains.

In various sectors, organizations leverage these tools to solve complex problems, streamline processes, and improve overall performance. For instance, in the public sector, OR/MS applications can improve service delivery and resource allocation, while in military settings, they can optimize logistics and mission planning. In the realm of exploration—whether on Earth, like in mining or environmental assessments, or in space, such as with satellite missions—these tools help manage risks and maximize the effectiveness of every endeavor.

The availability of superior but less expensive computing provides greater opportunities to create value and efficacy through OR/MS. Such ubiquity of powerful computing helps provide better quantitative and business insights into complex problems. Furthermore, it improves performance in multiple ways including increasing power of analyses and recommendations about business processes, decreasing cost or investment, assessing the likely outcomes of decision alternatives, uncovering better/non-obvious alternatives, providing a superior basis for forecasting and planning, efficiently scheduling staff and equipment, improving productivity of organizations, increasing return on investment, improving quality, quantifying qualitative considerations, managing and reducing risk, achieving greater utilization from limited equipment, facilities, money, and personnel as well as supporting operational, tactical, and strategic planning (Saxena et al. 2021).

Nevertheless, the field of OR/MS has expanded so much that most OR/MS experts may only command expertise in a small section of it. Consequently, they are highly likely to tailor/model any problem situation to fit in a small number of OR/MS techniques of their own expertise. Nevertheless, the primary aim of OR/MS tools is to deliver valuable insights for informed decision-making. These tools are designed to analyze complex data sets, model various scenarios, and project potential outcomes, thereby equipping decision-makers with the necessary information to make well-founded choices. However, challenges often emerge when users misinterpret the outputs of OR/MS models as conclusive answers/decisions rather than simply decision aids. This leads to a range of issues like over-reliance on model outputs without considering the broader context, assumptions, or potential uncertainties inherent in the data. This may lead not only to a lack of critical thinking but also a diminished capacity to adapt to changing circumstances. Overlooking other important factors, such as qualitative insights or stakeholder perspectives, can ultimately lead to suboptimal choices that fail to address the intricacies of real-world situations. In addition, there is a considerable gap between theory and practice of problem-solving techniques in real-world settings.

To mitigate these risks, it is essential to cultivate a culture of critical analysis and continuous learning among users of OR/MS tools. Furthermore, there is a need to provide deep and rigorous treatment of OR/MS to practitioners without complicating mathematics. Training programs and workshops can help users understand the role of OR/MS tools in decision support, emphasizing that the outputs should serve as one of many inputs in a comprehensive decision-making framework. By fostering a more nuanced understanding of the capabilities and limitations of OR/MS models, organizations can enhance their decision-making processes and achieve better outcomes.

3.1. OR/MS and UN SDGs

It is important to note that with the prominence of United Nations (UN) Sustainable Development Goals (SDGs), there is renewed emphasis on the value of OR/MS (UN, 2015). These SDGs are designed to eradicate poverty, protect the planet, and ensure prosperity for all. OR/MS could help realize all 17 UN-SDGs. In the following, we discuss some of the ways OR/MS could/would support the realization of various specific UN-SDGs.

SDG-1 (No Poverty) can be supported by OR/MS by identifying the most effective interventions for poverty alleviation and optimizing resource distribution in social services/support programs. By analyzing data on socioeconomic factors, aid organizations can target support more specifically and effectively (Aiguobarueghian et al., 2024). Similarly, SDG-2 (Zero Hunger) can be supported by OR/MS via optimization of agricultural and food distribution supply chains, enhancement of food production processes, and minimization of waste (Helo & Hao, 2025; Prodromou et al., 2025). By modeling agricultural practices and logistics, stakeholders can improve food security and make better use of land and resources.

SDG-3 (Good Health & Well-Being) can be supported by OR/MS through, among others, optimal decision regarding patient flows, resource allocations, and treatment plans (Yachachín et al., 2025). Such advanced methods as simulation and queuing theory can help hospitals manage capacity and improve service delivery, ultimately resulting in sustainable and superior health outcomes (Harris et al., 2022; Kandwal et al., 2024; Sensor et al., 2021). Similarly, SDG-6 (Clean Water & Sanitation) can be supported by OR/MS via, for instance, improved demand forecasting, superior management of water resources, more appropriate design and optimization of water distribution systems, etc. (Aiguobarueghian et al., 2024).

SDG-4 (Quality Education) can be supported by OR/MS through, among others, inclusive and equitable quality education and skills training, promoting lifelong learning opportunities for all (Zhou & Alam, 2024). OR/MS techniques can help educational institutions allocate resources more effectively, ensuring that funding reaches areas of greatest need. OR/MS can assist in designing curricula that meet educational standards and address gaps in knowledge, ensuring that students receive quality education (Alford et al. 2025; Maqbool et al., 2024). OR/MS can identify at-risk students and track their progress, enabling timely interventions to improve retention and success rates (Abubakr et al., 2024). Indeed, quality education itself has an extremely critical role in achieving all the SDGs by creating SDG-centric awareness and culture (Cembranel et al., 2024; Contreras-Medina et al., 2025; Kleespies & Dierkes, 2022).

SDG-5 (Gender Equality) can be supported by OR/MS via workforce optimization through analysis of labor market data and identification of disparities in employment, promotions, and remunerations, thereby assisting organizations make informed decisions in implementing strategies to promote equality and achieve (Abubakr et al., 2024). By modeling resource allocation in educational programs and training initiatives, OR/MS can help ensure equitable access for women and marginalized groups (Zhou & Alam, 2024). Simulations and optimization models could help policymakers evaluate the impact of equality-focused policies on economic and social outcomes, allowing for informed, data-driven, and effective policy improvements (Maqbool et al., 2024).

SDG-7 (Affordable & Clean Energy) can be supported by OR/MS through analysis of energy systems and consumption patterns as well as through the development of sustainable energy solutions (Ramírez-Márquez et al., 2024). The optimal integration of renewable energy sources, improving energy efficiency, and reducing costs are all possible via judicious application of OR/MS. Similarly, SDG-11 (Sustainable Cities & Communities) can be supported by optimizing traffic flow, public transport routes, resource distribution in urban planning and transportation systems, etc. This leads to more efficient urban environments and improved quality of life for residents (Zhou et al., 2025). Likewise, SDG-13 (Climate Action) can be supported by OR/MS via modeling environmental impacts and optimizing mitigation strategies, helping institutions and regulators assess risks, implementing efficacious actions to ameliorate and counter climate change/degradation, etc. (Haugen et al., 2024).

SDG-9 (Industry, Innovation, & Infrastructure) can be supported by OR/MS through fostering innovation by modeling and optimizing industrial processes, streamlining operations, reducing waste, and enhancing productivity, all contributing to sustainable industrial growth. Similarly, SDG-12 (Responsible Consumption & Production) can be supported by facilitating sustainable supply chain management via analysis of lifecycle impacts, optimal resource usage, minimization of waste, etc. (Haugen et al., 2024; Helo & Hao, 2025; Heriyati et al., 2024; Ramírez-Márquez et al., 2024)

Above all, SDG-17 (Partnerships for the Goal) calls for actively and proactively pursuing collaboration among diverse stakeholders as essential ingredient for achieving the SDGs. OR/MS tools can enhance communication and cooperation by providing a common framework for data analysis, collaborative decision-making, fostering partnerships that drive progress, etc. (Heriyati et al., 2024). In short, the OR/MS discipline offers robust frameworks, tools, techniques, and values for analyzing complex systems, which make OR/MS invaluable in the pursuit of the UN SDGs. By leveraging on data-driven insights, organizations can make informed decisions that promote sustainability, equity, and resilience across various sectors.

4. TRADITIONAL OR/MS APPROACHES

The traditional OR/MS approaches are analytic and reductionist in nature (Jackson, 2016; Petropoulos et al., 2024). A hierarchical decomposition approach is followed to explore how a system works, leading to recommendations as to how things can be changed for better (Hirschheim, 1983; Daellenbach & McNickle, 2012). It often consists of hard and technical ways to tame the problem, embracing scientific methodologies and explicitly concerning the validation of results (Daellenbach & McNickle, 2012). An interesting analogy of OR/MS is that of "a doctor who, with little participation from the patient, indicates how health can be improved" (Keys 1991, pp. 2).

One simplifying tactic used by OR/MS practitioners is to focus on individual parts of the system and try to improve the performance of those parts (Jackson, 2016; Petropoulos et al., 2024). The assumption is that improving the performance of the parts of a system taken separately will improve the performance of the whole (Lawson, 2005). However, in reality, such an approach may even destroy the whole system (Ackoff, 2001; Warfield and Staley, 1996). Nevertheless, dividing large projects into manageable parts has remained a viable problem-solving strategy, which usually results in compartmentalization of information (Lawson, 2005).

From the early days of OR/MS, quantitative analyses were employed to provide some discernible and computable measures of effectiveness, enabling meaningful interpretations of the information generated. OR/MS provided useful information and insights to decision-makers for more informed and effective decision-making. Indeed scientific methods provide a basis for traditional OR/MS to relate theory to practice. However, OR/MS has traditionally assigned greater substance to its observance of scientific methods. In the next Section, we discuss some other important aspects of OR/MS as well as some problems associated with those.

5. ISSUES WITH TRADITIONAL OR/MS APPROACHES

In OR/MS, the usual starting point is a relatively well-structured and tightly defined problem with clearly articulated performance measures and alternative courses of action. It proceeds with formulation of a model for the problem situation, collecting necessary data, and solving and validating the model. A tendency to follow a mathematically accurate approach to problem solving and perceiving OR/MS as exact science turns out to be an impediment in effective problem solving.

Although such an exact approach is desirable in mathematical research, presuming it to be a necessary step in solving complex and composite real-world problems tends to mask the right perspective of the problem and prove a certain lemma becomes the focus of the exercise. Such an approach reinforces the view that the OR/MS academics largely solve academic problems quite removed from reality. This usual approach inherently contains several problems that may limit the efficacy of OR/MS tools, some of which are discussed below.

5.1. Deterministic / Hard Optimization Paradigm

An optimization model consists of an objective function and a set of constraints. The goal is to find a solution that gives the best value objective function while satisfying all the constraints included in the model. However, often the solution obtained is the solution to the model and not to the actual problem (Lawson, 2005). Although some techniques support stochastic processes, OR/MS models and tools still deal with abstractions and simplifications of reality more appropriate for deterministic systems (Ackoff, 2001).

In a typical OR/MS approach, mathematical modeling precedes the generation of alternative courses. Consequently, there is often a futile and time-consuming exercise of formulating a mathematical model that might not even produce a feasible solution keeping in view the plethora of qualitative constraints in socio-technical systems (Foulds and Thanchekary, 2001). However, once an OR/MS analyst has spent a significant amount of time in formulating a complex, albeit rigidly and myopically defined, mathematical model, such cognitive biases as anchoring may hinder the search for a viable course. These cognitive biases, as discussed further in this section later, might impede the much-needed leaps of faith and the need for breaking away from the course already followed (Greenberg, 2010; Tversky and Kahneman, 1974). Resource limitations further aggravate the problem by providing a rationale for not looking further.

The hard optimization and mathematical programming approach particularly fails when addressing complex problems with conflicting objectives. It often provides a unique optimal solution against a single, simple criterion (Ahmad 2008; Triantaphyllou *et al.*, 1996, 1998). The solution is often based on a model that fails to capture many of the realities of the practical situation and includes only hard constraints (Ahmad, 2005). Furthermore, it often cannot provide answers within reasonable computational time (Foulds and Thanchekary, 2001). Even though OR/MS practitioners may come up with an optimal solution to a certain problem given the objective situation, some highly subjective considerations often constitute the imperative and decisive role in the selection of the solution alternative (Churchman 1970, Foulds and Thanchekary, 2001).

5.2. Problem Complexity

Another issue with traditional OR/MS approaches is the problem of complexity. Complexity of a problem is often seen as the number of components, variables, and their combinations considered or needing to be considered in decision-making (Jackson, 2016; Warfield and Staley, 1996). It is commonly referred to as *combinatorial complexity* (Garey and Johnson, 1979). For instance, the scheduling of jobs on various machines available is a hard problem from a combinatorial perspective. The larger the number of machines or jobs at hand, the larger the number of possibilities that should be screened in search of an optimal solution (Imam *et al.*, 2015). However, the real-world systems also exhibit another subtle but critical sort of complexity – the *dynamic complexity* – discussed further in this section later. The dynamic complexity appears from complex interactions among various system variables and agents over time. In fact, many variables may change simultaneously, thus confounding the effects of any intervention into the system.

OR/MS still tends to focus on narrowly defined questions with more emphasis on breaking down problems into smaller and lower-level components that are more tractable to analysis and more immune to uncertainty (Drejer *et a*l. 2000). The tradition of reductionism in OR/MS can be traced back to F.W. Taylor and his Scientific Management approach to industrial engineering. Although the current practice of OR/MS employs tools boasting a much higher level of scientific sophistication and efficacy than early scientific management techniques, the scientific management approach from Taylor still provides a classic perspective to reductionism. Such reductionist approaches involved reducing a system to its simplest elements, analyzing them, and calculating how to improve each element – an antithesis of a much more plausible integrative approach discussed in Section 6.

The underlying notion in scientific management is that observation and measurement provide a secure ground for analysis (Taylor, 2018). However, such an approach, while influential in the development of modern management practices, has several debilitating limitations (Awofeso, 2019; Braverman, 1998; Gong et al., 2025; Wiggin, 2025). These limitations include, among others, oversimplification of work, ignorance of human and social factors, dehumanization of workers, ignorance of team dynamics, etc. In this regard, the systems thinking and participatory management are often suggested as alternatives to scientific management (Nyland & Bruce, 2025).

5.3. Social, Political, and Dynamic Complexity

The dynamic complexity may arise in seemingly very simple problem situations. For instance, most political or social problems seen from the so-called objective lens would appear as very simple problems that may be resolved by resorting to simple interventions (Drew, 2025). However, such interventions do not always result in keeping other variables constant and modeling the effects of variable of interest.

Furthermore, problems arising from social, behavioral, or political influences require an understanding of such influences before embarking on some OR/MS solution methodology (Drew, 2025). Due consideration of social and behavioral determinants of operational success might provide very high returns without resorting to rigid OR/MS techniques for optimizing some tangible aspects of the system. For instance, studies in Sun Microsystems™ have shown how spending about $20,000 in understanding and improving intangible determinants of product success yielded about $150 million dollars in returns (Rhodes, 2000).

A classic example in this regard is given about the waiting time for elevators in a building during the early days of OR/MS. Users of elevators constantly complained about the excessive perceived waiting time for elevators. Decision makers, and a series of OR/MS teams, embarked on a variety of futile interventions to reduce the waiting time of passengers. The solution finally came from an OR/MS professional who tried to break away from the usual approach of modeling a problem situation and then solving it optimally and tried to inspect the situation personally rather than relying on heaps of available data. The magic solution that followed the observation constituted installation of large mirrors in the elevator areas giving people a chance to appreciate appearances of themselves and others discreetly while waiting for the elevator (Taha, 2022). Interestingly, all complaints about excessive waiting times suddenly vanished.

Another example of disregarding social and behavioral aspects relates to Taylor's scientific management approach towards improving the efficiency of an individual and, purportedly, improving the workplace and business productivity. It disregarded many important dynamics of the whole system, such as potential compensation paid by businesses to workers suffering injuries in performing repetitive tasks in a rigid fashion. Union resistance and resulting social tensions were ignored as well as any other social aspect of socio-technical systems, as highlighted by the famous Hawthorne study (Greenberg, 2000). Consequently, even though the apparent efficiency of an individual worker during a short period may be improved, the overall business efficiency and productivity may be reduced and render the whole exercise of scientific micromanagement counter-productive.

5.4. Information Scarcity, Uncertainty, and Filtering

An OR/MS tool without data is like a person without a soul. However, the required data are often so scarce that it constrains meaningful application of quantitative and statistical tools available to us. Furthermore, the information we receive usually comes through various information filters and strong systematic assumptions (Sterman, 1994; Warfield and Staley, 1996). In a social or socio-technical system, these filters may involve propaganda and control of information. In other technical systems, where we receive data based on our measurements, these might involve distortions, errors, biases, uncertainty, ignorance, etc. (Sterman, 2017). In reality, measurements are based on selections – not only the selection of variables of interest but also the range, the format, the equipment, the environment, and the agent for measurements.

A classic example comes from the *Nimbus 7* project NASA launched in 1978, where the measurements of atmospheric ozone levels did not enable the detection of a gigantic ozone hole above Antarctica. It was only in 1985 when the first scientific evidence of this ozone hole was published, forcing NASA scientists to investigate the reasons of this failure. The cause was the computer program that was based on the belief that very low readings of ozone may only come from instrumental malfunctioning. Ironically, a revisit to the original data collected by Nimbus 7, before the filtering, showed that this ozone hole, which has potentially disastrous consequences for our planet, could have been detected in 1978 (Meadows *et al.*, 1992; Sterman, 2017). In short, a measurement system was created that simply filtered important pertinent information.

Interestingly, such filtering is not only present in information procurement but also in the selection of the problem and its potential solution. The famous *garbage can model* illustrates how a problem is pseudo-randomly selected from various existing problems and then a pseudo-randomly selected solution approach is pursued (Greenberg, 2010). Indeed, solving the wrong problem, facetiously referred to as *type III error*, is not rare. Again, a classic example comes from the very early days of the OR/MS discipline when scientists in the Royal Air Force's OR group started working on optimally fortifying fighter planes. The approach involved meticulous measurement and data collection regarding most hit or damaged parts of the warplanes returned from war missions. The underlying idea was to fortify the more vulnerable parts of the plane. The fallacy was that a wrong conception of vulnerability was employed. This fallacy was recognized when a senior air force officer pointed out that they were basing their analyses on those planes that have been able to return from combat missions.

Apart from scarcity and filtering of the problem-pertinent data, the available data often support divergent and competing interpretations. In an attempt to draw inferences from the data, we must systematically resort to some simplifying assumptions. Nevertheless, our mental and mathematical model formulations are constrained by what we resolved to define, observe, and measure and other 'externalities' are usually disregarded/unnoticed in problem solving and decision-making process (Tversky and Kahneman, 1974). Consequently, such information unavailability, filtering, and uncertainty often result in false pretenses as well as unintentional *anchoring bias*, which refers to a cognitive bias that describes the common human tendency to rely too heavily on the first piece of information offered when making decisions (Greenberg, 2010).

Subjectivity and uncertainty in domain knowledge have been classified by some researchers as incomplete, inconsistent, imprecise, and vague knowledge (Ahmad *et al.* 2005). The term *incompleteness* implies the unavailability of some of the information, requiring the employment of heuristics and approximate reasoning. *Inconsistency* suggests a disparity or conflict in the knowledge, causing problems in using the available information. *Imprecision* refers to values that are loosely defined or measured inaccurately. *Vagueness* implies subjectivity in the estimate about some value or fact, causing difficulty in aptly understanding the information (Ahmad *et al.*, 2004b).

Often, the preferences and constraints faced in solving problems are intrinsically incomplete, imprecise, inconsistent, and vague (Ahmad, 2004b, 2005, 2008). In addition, the uncertainty tackled in operations research literature is largely limited to the uncertainty in the occurrence of a well-defined event such as uncertainty in demand forecasts. Nevertheless, many problems also involve uncertainty in the event itself due to incomplete, imprecise, inconsistent, and vague information. However, even under a relatively deterministic environment, the cost of procuring exact and complete information could often be prohibitive as well as implicitly involve systematic judgment (Ahmad *et al.*, 2004b; Sterman, 2017).

The limited, imperfect, and uncertain nature of information affects our decisions. The ambiguity arising from the confounded nature of interaction among system variables and multiple interpretations of linguistic terms significantly alters the human ability to identify the structure and parameters of a system from the examination of the system (Ahmad et al., 2015; Jackson, 2016; Warfield and Staley, 1996). In reality, people often reason in terms of vague and context dependent concepts in dealing with situations where they encounter uncertainty (Turban and Aronson, 2005).

Such a multitude of tangible and intangible considerations in real world applications underscores the need for a composite/multi-criteria fitness evaluation regime combining various desirable characteristics (Triantaphyllou *et al.*, 1998). One reason is that different fitness criteria represent different perspectives and are very much likely to be competing and conflicting. However, the traditional approach in OR/MS to decision-making with multiple objectives is a fundamentally ad hoc methodology. Generally, it works with optimizing the most important objective, while meeting minimal aspiration levels on all other objectives. In short, arbitrary boundary choices are used with little justification for it, involving a high degree of subjectivity. Such an ad hoc approach evades many fundamental issues in MCDM (Triantaphyllou *et al.*, 1998). A situation with more than one decision-maker further aggravates the problem.

Indeed, the uncertainty in the environment calls for flexibility in order to respond to changes quickly. However, most OR/MS tools are designed to achieve a rigidly defined optimality, undermining the very notion of flexibility and adaptability.

5.5. Dynamic Reality and Rationality

Another salient issue is the dynamic nature of reality and the problem situation. It is to be noted that decision-making is not a static process. It usually involves a planning horizon during which the environment encompassing the problem may change significantly. Irrespective of the degree to which we agree with such assertions, there is no denying the fact that most problem situations are dynamic in nature, as new and unintended dynamics may emerge through interactions among various components in a system or its ecology (Daellenbach and McNickle, 2012). Such a dynamic and evolutionary character is also true for client's requirements and preferences, pronouncing the need for the OR/MS practitioners to adapt their problem-solving approach quite often. In fact, such dynamism in preferences frequently results from client's interaction with existing or intermediate solutions, a phenomenon often referred to as *dynamic rationality* (Bouyssou & Vincke, 1998; Sterman, 2017).

This dynamic rationality challenges traditional OR/MS approaches by emphasizing the importance of adaptability and real-time decision-making in uncertain environments. While the traditional OR/MS often relies on static assumptions, dynamic rationality, however, recognizes that conditions can change rapidly, requiring models that can adapt to new information. Similarly, while traditional OR/MS typically requires comprehensive datasets to inform decision-making, dynamic rationality suggests that in rapidly changing situations, having real-time data is more crucial than extensive historical data, which may become outdated. Moreover, traditional approaches may use deterministic models that do not account for uncertainty, dynamic rationality incorporates probabilistic models that can adjust as new data emerges, allowing for more robust decision-making.

5.6. Bounded Rationality

A decision strategy that is completely informed, perfectly logical, and oriented towards economic reward is referred to as *perfect rationality* (Simon, 1955). However, in practice, decision makers often resort to *bounded-rationality* reflecting an inadequacy of tangible and intangible resources – such as limited information and information processing cognitive abilities (Greenberg, 2010; Simon, 1957). The most taxing issue OR/MS practitioners face is the *ill-structured* nature of problems that tend to be complex, relatively novel, subjective, and uncertain (Ignizio & Cavalier, 2019). Such undertakings require a high degree of creativity and expertise (Ignizio, 2019).

Besides all limitations, ambiguity, and complexity of information, humans may still interpret available information in ways that might be far from a rational approach. The search for an objective measure for evaluation and comparison of alternatives adds to other constraints that hamper effectual decision-making (Simon, 1955, 1957). Consequently, decision makers often resort to *satisficing* tactics instead of rational optimizing methodologies (March & Simon, 1958). The term refers to the process of formulating an adequate degree of acceptability for a solution to a problem and screening solutions until one that surpasses this benchmark is found (Greenberg, 2010).

5.7. Flawed Judgments and Cognitive Biases

As mentioned, there are various biases involved in decision-making. For instance, decision-makers often resort to making presumptions regarding some aspects of the available information, an affinity referred to as *framing* error (Greenberg, 2010). Furthermore, even when objective information remains unchanged, people's judgment is strongly affected by the frame in which information is presented. Moreover, cognitive biases for procuring and dealing with information in an error-prone fashion may emerge in unsuspecting manners. For instance, there is an instinctive human tendency to rely on more recent or readily available information, a phenomenon commonly referred to as the *availability heuristic* (George & Jones, 2024). Indeed, memory is distorted by hindsight, the availability and salience of examples, and the desirability of outcome. In addition, there is often a propensity to expose oneself only to the information that conforms to one's own analysis and belief of the situation, referred to as the *confirmation bias* (Greenberg, 2010). Sometimes, such biases are a result of the *information overload*, where more information is acquired or available than is necessary to make effective decisions (George and Jones, 2024).

Research provides cogent evidence that even seasoned scientists and experienced professionals suffer from many judgmental and cognitive biases (Kahneman *et al.*, 1982). Consequently, despite all virtues of an expert's judgment, judgment alone may not be an adequate assessment tool (Jackson, 2016; Lawson, 2005; Petropoulos et al., 2024). Nevertheless, failure of foresight and judgmental flaws often make people blame the technology or tool for the failure.

5.8. Flawed Model Development Process

A mathematical model expresses the relationships among various components of the system in quantitative terms. Indeed, mathematical models are relatively easy to manipulate and facilitate exploration of the effects of changes in inputs (Daellenbach and McNickle, 2012). Nevertheless, a model should be developed with the purpose of solving the problem with a clear purpose, which is to solve the problem at hand (Churchman 1970; Petropoulos et al., 2024).

To address complex problems arising in industry, many OR/MS practitioners fall back on constructing complex models and try to solve them using optimization techniques (Ackoff, 2001; Paola, 2021; Foulds and Thanchekary, 2001; Sterman 2017). However, researchers have increasingly asked whether such an approach can deal successfully with many of the complex, ill-defined, difficult-to-model issues OR/MS practitioners are required to solve giving rise to other approaches and such soft systems methodologies as decision support systems, expert systems, soft computing, etc. (Foulds and Thanchekary, 2001; Konar, 2018).

In practice, there are many difficult problems for which an exact/optimal solution approach would take forever, even with all the computing resources of the world at the disposal of problem solvers, just to get a feasible or workable solution (Garey and Johnsons, 1979). However, many OR/MS practitioners assume that implementing sophisticated techniques will solve all problems (Ackoff, 2001; Paola 2021). Furthermore, many OR/MS proponents stress the need for adopting a 'mathematically correct' solutions approach. In reality, many reasoning mechanisms not based on any sound theoretical or mathematical basis have been very successful in practice (Karray and de Silva, 2004; Negnevitsky, 2019; Shafer, 1976; Taguchi, 1989; Turban and Aaronson, 2005).

5.9. Inefficacious Model Testing

In OR/MS, model testing and validation are presented as essential to any problem-solving efforts. Indeed, such a process significantly contributes to establishing the credibility of the model itself. Such efforts require both internal validity (i.e. checking that the model is logically and mathematically correct) as well as external validity (i.e. checking that the model actually a sufficient representation of reality). However, in practice, more efforts are expended in validating the model and establishing the model as the right or appropriate model in comparison to the efforts invested in testing the model. This approach erodes the utility of the model and, more so, the credibility of the modeler and the OR/MS discipline. Model testing should instead be designed to uncover errors in the approach, understand limitations of the model, enhance the model efficacy, and seek a more appropriate model for effective decision-making.

This exercise should also be directed at testing the validity of assumptions underlying the model formulation, its robustness, and its sensitivity to various factors. Ironically, models may not essentially be verified or validated as, by definition, all models are abstraction/simplification of reality (Jackson, 2016; Petropoulos et al., 2024). Furthermore, a parsimonious or elegant model, one that explains a large part of the variability in the system and its behavior by employing the smallest number of system variables, is deemed a desirable outcome (Daellenbach and McNickle, 2012). Nevertheless, such a desirability of parsimony should not deter, or digress, analysts from critically assessing the viability of the model as well as the underlying simplifying assumptions (Churchman 1970). Furthermore, the model should be tested for an appropriate time horizon, as an unrealistically shorter or longer time horizon may distort the conclusions emanating from the whole exercise.

5.10. Erroneous Inferencing

Generally, the success or failure of some implementation of the OR/MS methodology is gauged using performance metrics that have limited capability to capture the essence of success or failure. Once again, such performance metrics disregard the importance, sometimes even the existence, of organizational, behavioral, political, or socio-technical factors in success or failure of any solution methodology (Drew, 2025).

As mentioned, the multitude of tangible and intangible considerations in real world applications advocated the use of MCDM regime. Nevertheless, the subjectivity, uncertainty, and incommensurable units render MCDM paradigm inherently difficult to realize in many applications (Ahmad, 2008; Ficko et al. 2021). In addition, the tendency of people relying on a pre-existing perspective of their skills in judging their performance, or potential for performing well, in present and future assignments results in overconfidence (Daellenbach and McNickle, 2012).

6. OPPORTUNITIES FOR INCREASING EFFICACY OF OR/MS TOOLS

Despite an apparently long list of impediments in success and adoption of OR/MS tools, these tools have been very successful in solving a variety of problems and this is expected to continue to hold in future (Pidd 2004). However, we believe that the OR/MS techniques can be more effectively employed when embedded within decision support and soft OR/MS paradigms. Nevertheless, most OR/MS texts fail to provide readers exposure to such approaches and confine themselves to gory computational details of some quantitative techniques. However, it is interesting to note that the very first organized textbook on OR/MS that appeared in 1957, written by Churchman, Ackoff, and Arnoff, clearly stressed the comprehensiveness of OR/MS as a systems approach (Churchman 1957). Unsurprisingly, the first two authors were the earliest and most vocal critics of the hard optimization approach that was fast becoming a norm in the OR/MS knowledge domain at that time (Jackson, 2016).

We believe that teachings in OR/MS discipline should be complemented by its role in organizations providing some perspective of organizational dynamics and business realities. Fortunately, many academic programs in forward-looking higher education institutions realize this requirement and provide students with an essentially interdisciplinary outlook (Alford et al. 2025). In this Section, we discuss some interdisciplinary solution paradigms that offer promise to increase efficacy of OR/MS. By recognizing the limitations of traditional OR/MS and complementing it with various interdisciplinary approaches, we create more dynamic, responsive, and human-centered environments that promote both efficiency, efficacy, and well-being.

6.1. Systems Perspective

A systems perspective promotes the notion that real-world problems are part of a complex system and should be solved with an integrated perspective (Daellenbach and McNickle, 2012; Jackson, 2016). Such an integrated and holistic perspective provides the ability to appreciate that doing one thing does not essentially keep all other aspects of the system intact and that everything is linked to everything else. Although there are many schools of thought in systems thinking itself, all seems to agree that a systems and integrated view of the world, though rare, is desirable (Carrubbo et al., 2025; Doyle, 1997; Lane, 1994; Richardson, 1996). Just one decision or intervention may provoke reaction from others, the Counterintuitive Behavior of the social systems (Forrester, 1971; Daellenbach and McNickle, 2012). Even a single or simple change in one of the factors under consideration may result in unpredictable changes in various other factors influencing the system (Doyle, 1997).

The systems approach typically starts with problem definition and proceeds with generation of alternatives, model formulation, and analysis/selection of alternative. Notably, model formulation follows generation of alternatives where we identify alternative courses, albeit vague ones, that satisfy the project objective. The organizational, behavioral, political, and environmental constraints may dictate only a handful of different courses of actions (Drew, 2025; Sterman, 1994). Gathering information on such alternative courses helps in formulating mathematical models that accurately simulate the performance of each alternative.

The systems approach does not necessarily call for an analytical solution (Keys, 1991). A key principle is to scrutinize a problem from multiple perspectives and mull over long-term consequences of decisions, including behavioral, organizational, political, and moral implications (Drew, 2025). A systems perspective often requires interdisciplinary teams for problem solving. Indeed, it has been frequently argued by many researchers that the traditional rational/scientific thinking in OR/MS is largely based on reductionist and cause-and-effect modes and may not be able to cope with the complexity and ill-structured nature of most real-world problems (Ossenbruggen, 1994).

The systems approach also employs qualitative notions of synthesis and emergence that can be used to tackle ill-structured problems giving more weight to efficacy than validation. Such approaches lead to explanations of why phenomena are, rather than just how those operate (Keys, 1991). Such an approach advocates both subjective evaluation of qualitative factors and an objective analysis of quantitative factors in decision-making. Nevertheless, a qualitative analysis often proves to be critical in problem solving. In essence, systems analysis is more of a process of discovery and learning than an analytical exercise in optimization (Ossenbruggen, 1994).

The systems approach is not as closely associated with science as tools used in OR/MS approaches. The methodologies employed in a systems approach are rigorous but differ in fundamental ways from those of science. For instance, a systems approach may not instill the same reliance on quantitative/empirical data as OR/MS does. Furthermore, model validation and manipulation offer difficulties for systems approaches. However, these approaches provide ways to work when there are limits on effective use of OR/MS.

The relationship or interrelationship between the OR/MS and the systems thinking has been discussed at length by many researchers (e.g. Ackoff, 2001; Churchman 1957, Daellenbach and McNickle, 2012; Jackson, 2016; Jackson and Keys, 1985; Keys, 1991; Ossenbruggen, 1994; Paola, 2021; Silver and Meal, 1998; Sterman, 1994; Stainton, 1984). Some even argue that the OR/MS is only a strand in different version of the systems approach (Jackson, 2016).

6.2. Soft OR/MS Paradigm

As already mentioned, researchers have been advocating soft OR/MS approaches for several decades (Daellenbach & McNickle, 2012). They aptly indicate that an overwhelming majority of real-world problem scenarios violate many, if not most, of the assumptions underlying the hard OR/MS approaches. Indeed, behavioral, social, and political facets are largely ignored in hard OR/MS approaches (Drew, 2025). Soft OR/MS approaches were designed to tackle situations that are ill-structured, ill-defined, and involve different stakeholders with differing and possibly conflicting opinions.

Some salient characteristics of such soft OR/MS approaches include structuring the issues in problem situations instead of narrowly focused problem solving, facilitating dialogue among stakeholders and eliciting resolution of the problem from stakeholders themselves, instead of imposing a solution from the analyst, focusing the 'what' questions before taking up the 'how' questions, etc. The hard OR/MS deals mainly with the 'content' of the problem and gives little weight to the 'process'. In contrast, soft OR/MS emphasizes both content and process starting with an attempt to obtain a reasonably comprehensive perspective and shared understanding of the pertinent issues (Karray and DeSilva, 2004).

There is a range of soft OR/MS methodologies starting from Game Theory to Theory of Constraints. A detailed discussion of these various soft OR/MS approaches is beyond the scope of this paper and inquisitive readers may find good reviews in literature (e.g. Daellenbach and McNickle, 2012; Jackson, 1991; Ossenbruggen, 1994; Petropoulos et al., 2024). However, all such approaches involve active participation of stakeholders and, unlike hard OR/MS, these do not constitute some sequence of systems/phases whose output becomes the input to the next system. All these approaches are recognized as learning systems used iteratively and flexibly, drawing from philosophy, psychology, sociology, and organizational theory (Ossenbruggen, 1994).

Several researchers have argued that the potential of various soft computing technologies is limited only by the imagination of their users (Negnevitsky, 2019; Zha, 2003). However, there is no dearth of OR/MS practitioners who look down upon these soft approaches and advocate/practice less-pertinent hard optimization techniques even in highly subjective and uncertain problem domains (Ahmad, 2008; Ficko et al., 2021). Nevertheless, OR/MS practitioners should realize that their proposed solutions would be implemented in the wake of numerous organizational and political considerations and soft thinking might provide effective decision alternatives. Disregarding such issues may greatly diminish the chances of their success (Drew, 2025).

6.3. Decision Support Paradigm

The Decision Support System (DSS) paradigm to problem-solving provides a means for assisting decision makers in retrieving, summarizing, and analyzing data relevant to the decision. Consequently, it results in a reduction in the cognitive overload faced by the decision maker(s). Research has shown that DSS techniques are useful in generating and evaluating a large number of alternative solutions and effectively helping decision-makers in arriving at better decisions (Ghorbian &

Ghorbian, 2025; Turban and Aronson, 2005; Greenberg, 2010). However, it should be emphasized that a DSS neither automates the decision process nor imposes solutions. It simply provides analytical and information processing support in an interactive environment.

Indeed, this soft decision-making paradigm has successfully been applied in a variety of complex and subjective task domains and power of the OR/MS tools can be enhanced by incorporating its models and methods within a decision support system (Ayyub, 2001; Foulds and Thanchekary, 2001; Handoko et al., 2025; Konar, 2018; Marakas 2015; Turban and Aronson, 2005). Our personal experience of successfully using Decision Support and Expert Systems paradigms and Soft Computing tools for solving highly complex, non-linear, subjective, and uncertain problems of layout optimization have sensitized us to the potential of such tools coupled with OR/MS methodologies (Ahmad, 2005).

Indeed, it is irrational to expect that a specific solution would surpass all others for every evaluation objective under consideration. Furthermore, the objectives could often be fuzzy and dynamic in nature. Consequently, the generation of superior alternatives in a flexible and automated manner through some decision support mechanism is critical to many planning processes. Decision Support Systems (DSS) represent a class of computerized information systems that utilize the knowledge about a specific application domain to assist decision makers by recommending appropriate actions and strategies (Ahmad et al., 2008; Turban and Aronson, 2005). The DSS problem-solving paradigm provides a means for assisting decision makers in retrieving, summarizing, and analyzing relevant data. Consequently, it results in a reduction in the cognitive overload faced by the decision maker(s). Research has shown that DSS techniques are useful in generating and evaluating a large number of alternative solutions and effectively helping decision-makers in arriving at better decisions (Ahmad, 2002; Turban and Aronson, 2005; Greenberg, 2010).

However, it should be emphasized that a DSS neither automates the decision process nor imposes solutions. It simply provides analytical and information processing support in an interactive environment. Furthermore, a DSS should be much more than just a mere collection of models and solution techniques (Sharda et al., 2020). The essence of the DSS is the user-system interface that allows the planners to interactively conduct experiments, input decision rules and criteria, tackle issues with unstructured and semi-structured scenarios, and allow for multiple criteria decision-making as well as adapt soft constraints while providing flexibility, user friendliness, and convenience.

There is a need to develop decision systems that facilitate the implementation of more sophisticated OR/MS tools giving people systems with which they can creatively solve their own problems. This need is more pronounced by the fact that the mantra in the industrial and public sector organizations is to empower employees.

OR/MS can provide them with the means to monitor, comprehend, intervene, and continuously improve their own performance. It is our belief that judicious institution of the decision support and expert systems paradigms may nicely bring together the scientific methods of OR/MS, the soft thinking of systems approach, and the learning from social/behavioral sciences to engender a more effective problem-solving environment. We perceive these as complementary disciplines.

6.4. Simulation

Understanding the behavior of a system requires knowledge about how the state of the system changes over time. Collecting the relevant data for understanding the behavior might not be possible because of various limitations such as high observation cost, long observation time, or even the non-existence of the system itself. In such scenarios, simulation or systematic imitation of the behavior of the system over time is a general approach and constitutes one of the most pertinent tools for an OR/MS analyst (Ahmad, 2005).

The typical absence of some encompassing, closed-form, and analytical fitness functions also render computer simulations a useful alternative (Ahmad, 2005). Such an approach would typically provide detailed analysis, modeling, and evaluation of complex problems (Bookbinder and Higginson, 1986). However, simulation models are not easily amenable to optimization and make procurement of a superior alternative difficult to achieve (Chan *et al.*, 1995).

6.5. Artificial Intelligence (AI)

AI can significantly enhance or even replace certain aspects of OR/MS through a multitude of manners (Papadopoulos et al., 2022). We believe, this wealth of opportunities offered by AI-augmented OR/MS warrants a separate section on it. Consequently, we discuss these opportunities in the subsequent section.

7. AI-AUGMENTED OR/MS

AI has the potential to significantly enhance or even replace certain aspects of OR/MS in various ways (Fan et al. 2024). For instance, AI can analyze large datasets much faster and more accurately than traditional methods, enabling organizations to identify patterns and trends that may have previously gone unnoticed. Moreover, techniques such as machine learning can improve predictive modeling. Additionally, AI can automate routine optimization tasks, freeing up human analysts to focus on more strategic, complex problems that require nuanced thinking. Furthermore,

AI-driven simulations can create models that adjust in real time to changing conditions. By integrating natural language processing, AI can also make OR/MS tools more accessible to decision-makers, allowing them to interact with these systems without needing extensive technical expertise (Papadopoulos et al., 2022). In the following we elaborate on some of such opportunities offered by AI-augmented OR/MS (Kobbacy et al., 2017).

First, AI can promote data-driven decision-making via machine learning and real-time analytics (Fan et al. 2024). AI can analyze vast datasets to identify patterns and make predictions using machine learning models, often outperforming traditional statistical methods used in OR/MS. Furthermore, AI systems can process and analyze data in real-time, allowing for quicker decision-making compared to static OR/MS models (Papadopoulos et al., 2022). For instance, integrating AI with the Internet of Things (IoT) provides real-time data that can enhance operational efficiency and decision-making (Xiao & Zhang, 2024). Moreover, using AI in conjunction with blockchain technologies can improve transparency and various other metrics.

Second, AI can use advanced soft computing techniques and algorithms like fuzzy logic, genetic algorithms, neural networks, and reinforcement learning for solving complex optimization problems that are intractable with traditional OR/MS methods. Indeed, AI could furnish dynamic optimization capabilities by adaptively optimizing solutions as conditions change, making AI particularly useful in dynamic fields like sustainable supply chain management (Helo & Hao, 2025). In contrast, traditional OR/MS often relies on static models (Fan et al. 2024).

Third, AI can create more sophisticated simulations that better reflect real-world complexities, improving the accuracy of predictions and insights (Gupta et al. 2022). Indeed, the ability of AI to handle multiple scenarios simultaneously and provide insights into various potential outcomes more efficiently. In contrast, the traditional simulation models are not only overly complex, time-consuming, and costly to build but also relatively static.

Fourth, AI can augment OR/MS by adding sentiment analysis, the process of interpreting and classifying emotions, into the decision-making and forecasting by analyzing social media and other unstructured data sources to gauge public sentiment and its potential impact on operational, tactical, and strategic decisions (Farooq et al., 2025a; Farooq et al., 2025b). Indeed, both OR/MS and AI excel predictive analyses like forecasting demand, sustainable supply chain needs, and other key areas by utilizing historical data and recognizing trends (Gupta et al. 2022; Helo & Hao, 2025). Nevertheless, traditional OR/MS tools are not directly amenable to incorporating unstructured data like emotions in predictive analytics. By integrating sentiment analysis into OR/MS, organizations can make their decision-making processes more responsive, customer-centric, and data-driven, ultimately improving the overall performance and competitiveness.

Fifth, AI-based tools often come with intuitive and user-friendly interfaces like chatbots that can democratize access to complex but powerful analytics, allowing non-experts to utilize powerful analytical capabilities. AI-driven platforms often feature visually appealing dashboards that present data and insights in an easily digestible format. Users can quickly grasp complex information through charts, graphs, and interactive elements. Many AI tools allow users to manipulate data and models through simple drag-and-drop actions, reducing the learning curve and making the technology accessible to non-experts. AI tools that utilize Natural Language Programming (NLP) to provide conversational and voice command interfaces enabling users to interact with the system using natural language queries (Wickman et al., 2025). This means users can ask questions and receive answers without needing to understand complex programming or technical jargon, simplifying the user experience.

Sixth, as an extension of intuitive and user-friendly interfaces, users can customize and personalize their interfaces to show the metrics and data most relevant to their roles, enhancing usability and focus. Furthermore, some AI systems learn from user interactions and adapt their recommendations and interfaces over time, becoming more personalized and efficient for individual users. Moreover, many AI tools can be integrated easily with existing software, such as ERP systems or CRMs, enabling seamless integration and permitting users to work within familiar environments while benefiting from advanced analytics. Indeed, advanced users can leverage Application Programming Interfaces (APIs) to further customize their workflows and access AI capabilities directly within their preferred applications.

While AI can enhance many aspects of OR/MS, it may not entirely replace the need for OR/MS professionals, especially in areas requiring human judgment, ethical considerations, and contextual understanding (Ghorbian & Ghorbian, 2025; Gupta et al. 2022). Nevertheless, the integration of AI and OR can lead to more effective decision-making and optimized operations (Kobbacy et al., 2017).

8. SUMMARY

We argue that the objective of OR/MS is not to provide 'optimal' solutions, as most traditional texts and practitioners promise, but to provide a basis for more informed and insightful decision-making. However, the usual reductionist approach adopted in many OR/MS techniques only permits attention to details of each component of the system under study. In contrast, systems thinking permits attention to the systematic and interactive roles of components as an integrated whole. Often both modes of thinking are required to procure wider comprehension of a system. A soft OR/MS or a system thinking approach, though no panacea in efficacious problem solving, is a nice complement to traditional OR/MS. Its inherent nature of attempting to take an integrated view and permissibility of soft/violable system boundaries is likely to lead to more encompassing, more insightful, and more effective decision-making.

At a meta-level, OR/MS, Systems approach, and Decision Support paradigms may be conceived as sub-disciplines of a wider faculty. We believe that the limits of efficacious use of OR/MS approaches can be overcome by using systems approaches. We believe that the limits of efficacious use of OR/MS approaches can be overcome by using systems approaches. Indeed, OR/MS approaches may be perfectly effective in some cases while systems approaches may be deployed with the same effect in other cases. Nevertheless, both approaches can make unique contributions towards effective decision-making in tandem with each other. We further believe that it is in the benefit of the scientific community and the humanity that some effective meta-level framework guiding the effectual employment of both approaches be formulated. Such a framework would facilitate an environment in which OR/MS (alternatively, Systems) disciplines have a positive view of relative strengths, an appreciation of the unique contributions of the other disciplines, and the potential learning from both to support effective decision-making and problem solving.

Despite all arguments and counter-arguments in favor of one philosophy or the other regarding applicability of OR/MS, one aspect stands out: OR/MS practitioners should adopt an integrated systems methodology, whichever manner we define it, for solving real world problems plagued by behavioral, organizational, political, and environmental complexities, uncertainties, and constraints. Though AI can be deployed to augment numerous aspects of OR/MS, AI may not become a substitute for employing OR/MS professionals. This is truer in domains requiring human reasoning, contextual insights, ethical judgements, and contextual understanding. Nevertheless, the seamless integration of AI and OR/MS is expected to result in more efficacious decision-making, improved operations, and sustainable outcomes by fostering data-driven decision-making and leveraging powerful soft-computing tools in incorporating unstructured data sources.

This chapter is expected to further the debate in this direction. Nevertheless, this article is limited in many ways. For instance, we have included a very generalized discussion of tools, without detailed elaboration of such tools, due to space and scope constraints. Furthermore, we have not expended much space on past efforts made in making OR/MS tools more effective and useful through incorporation of systems thinking in a wider problem-solving approach. However, we believe there is still a relative dearth in such efforts and some formulation of a wider framework for facilitating complementary utilization of OR/MS and other AI and soft systems approaches in effective decision-making. Consequently, more efforts in this direction are required.

ACKNOWLEDGMENT

This work was supported by the Deanship of Scientific Research, Vice Presidency for Graduate Studies and Scientific Research, King Faisal University, Saudi Arabia (Grant No. KFU251816).

REFERENCES

Abubakar, A. A., Al-Mamary, Y. H., Singh, H. P., Singh, A., Alam, F., & Agrawal, V. (2024). Exploring factors influencing sustainable human capital development: Insights from Saudi Arabia. *Heliyon, 10*(16), e35676. DOI: 10.1016/j.heliyon.2024.e35676 PMID: 39229536

Ackoff, R. L. (2001). OR: After the post mortem. *System Dynamics Review, 17*(4), 341–346. DOI: 10.1002/sdr.222

Ahmad, A. R. (2002). A Research Framework for Intelligent Decision Support in the Web Page Layout Design and Presentation Planning, *M.S. Thesis*. Management Sciences, Faculty of Engineering, University of Waterloo.

Ahmad, A. R. (2005). An Intelligent Expert System for Decision Analysis and Support in Multi-Attribute Layout Design, *Ph.D. Thesis*. Systems Design Engineering, Faculty of Engineering, University of Waterloo.

Ahmad, A. R., Basir, O. A., & Hassanein, K. (2004, July). Intelligent Decision Support System for Layout Design. In *Proceedings of the 9th Asia-Pacific Decision Sciences Institute Conference (APDSI'04)*.

Ahmad, A. R., Basir, O. A., Hassanein, K., & Imam, M. H. (2004b, October). Decision preferences, constraints, and evaluation objectives in layout design: a review of modeling techniques. In Proc. of the 5th Int'l Conf. on Op. and Quan. Manag (ICOQM-V).

Ahmad, A. R., Basir, O. A., & Hassanein, K. (2008). "An Intelligent Expert Systems' Approach to Layout Decision Analysis and Support under Uncertainty", Intelligent Decision Making: An AI-Based Approach. *Studies in Computational Intelligence, 97*, 321–364.

Ahmad, A. R., Tasadduq, I. A., Imam, M. H., & Shaban, K. B. (2015). Automated Discovery and Utilization of Tacit Knowledge in Facility Layout Planning and Optimization. *Journal of Software and Systems Development, 2015*, 27–39. DOI: 10.5171/2015369029

Aiguobarueghian, I., Adanma, U. M., Ogunbiyi, E. O., & Solomon, N. O. (2024). An overview of initiatives and best practices in resource management and sustainability. *World Journal of Advanced Research and Reviews*, *22*(2), 1734–1745. DOI: 10.30574/wjarr.2024.22.2.1519

Alford, K. R., Stedman, N. L., Bunch, J., Baker, S., & Roberts, T. G. (2025). Real-World Experiences in Higher Education: Contributing to the Developing a Systems Thinking Paradigm. *Journal of Experiential Education*, *48*(1), 169–188. DOI: 10.1177/10538259241259626

Awofeso, O. (2019). Managing formal organizations in the 21st century: A critique of Fredrick Taylor's scientific management theory. *Journal of Public Management Research*, *5*(2), 1–11. DOI: 10.5296/jpmr.v5i2.15970

Ayyub, B. M. (2001). *Elicitation of Expert Opinions for Uncertainty and Risks*. CRC Press. DOI: 10.1201/9781420040906

Bouyssou, D., & Vincke, P. (1998). Preference Modeling. *Annals of Operations Research*, *80*, 147–180. DOI: 10.1023/A:1018988520000

Braverman, H. (1998). *Labor and monopoly capital: The degradation of work in the twentieth century*. NYU Press.

Carrubbo, L., Cosimato, S., & Gagliardi, A. R. (2025). Towards dynamic decision-making in government as service organization: insights from systems thinking. Transforming Government: People. *Process and Policy*, *19*(1), 108–129.

Cembranel, P., Gewehr, L., Dal Moro, L., Fuchs, P. G., Birch, R. S., & Andrade Guerra, J. B. S. O. D. A. (2024). The pivotal role of higher education institutions in cultivating a sustainable development goals-centric culture. *International Journal of Sustainability in Higher Education*, *25*(7), 1385–1411. DOI: 10.1108/IJSHE-01-2024-0057

Chen, C. (2020, December 23). A Glimpse of the First Eight Months of the COVID-19 Literature on Microsoft Academic Graph: Themes, Citation Contexts, and Uncertainties. *Frontiers in Research Metrics and Analytics*, *5*, 607286. DOI: 10.3389/frma.2020.607286 PMID: 33870064

Churchman, C. W. (1970). Operations Research as a Profession. *Management Science*, *17*(2), B-37–B53. DOI: 10.1287/mnsc.17.2.B37

Churchman, C. W., Ackoff, R. L., & Arnoff, E. L. (1957). *Introduction to Operations Research*. John Wiley& Sons.

Contreras-Medina, D. I., Nieto, E. S. D., Medina-Cuéllar, S. E., & Aguilar-Rivera, N. (2025). Foundational concepts: An introduction to ecocentrism, knowledge management and sustainability. In *Ecocentrism for Knowledge Management and Sustainability* (pp. 1–8). Routledge.

Daellenbach, H., & McNickle, D. (2012). *Management Science: Decision Making Through Systems Thinking* (2nd ed.). Palgrave Mcmillan. DOI: 10.1007/978-1-137-07512-3

Doyle, J. (1997). The Cognitive Psychology of Systems Thinking. *System Dynamics Review*, *13*(3), 253–265. DOI: 10.1002/(SICI)1099-1727(199723)13:3<253::AID-SDR129>3.0.CO;2-H

Drews, S. (2025). Towards systems-oriented behavioral ecological economics. *Ecological Economics*, *227*, 108402. DOI: 10.1016/j.ecolecon.2024.108402

Faheem, M. A., Anwer, S., Rayhan, Z., Ullah, M. A., Paudel, R., Ahmed, M. F., & Khan, H. (2024). AI-Driven Innovation In HRM And Its Impact On Business Management: An In-Depth Study Of Technology Advancement And Strategic Implementation. *Nanotechnology Perceptions*, *20*, 1174–1204.

Fan, Z., Ghaddar, B., Wang, X., Xing, L., Zhang, Y., & Zhou, Z. (2024). Artificial intelligence for operations research: Revolutionizing the operations research process. arXiv preprint arXiv:2401.03244.

Farooq, M., Ramzan, M., & Yen, Y. (2025a). *AI in Consumer Behavior Management*. IGI Global., DOI: 10.4018/979-8-3693-4322-7.ch011

Farooq, M., Ramzan, M., & Yen, Y. (2025b). *Exploring the Impact of Artificial Intelligence on Brand Management: A Multi-Dimensional Analysis*. IGI Global., DOI: 10.4018/979-8-3693-4322-7.ch006

Feroz, A. K., & Stines, A. (2024). Utilization of AI, big data, and internet of things (IoT) in sustainable innovation and their role in the emergence of novel business models. *Issues in Information Systems*, *25*(1).

Ficko, M., Berus, L., Palčič, I., & Klančnik, S. (2021). Design of Facility Layout for Industry 4.0. In *Research Anthology on Cross-Industry Challenges of Industry 4.0* (pp. 219–244). IGI Global. DOI: 10.4018/978-1-7998-8548-1.ch012

Forrester, J. W. (1971). The counterintuitive behavior of social systems. *Technology Review*, *73*(3), 52–68.

Foulds, L., & Thachenkary, C. (2001). Empower to the people–how decision support systems and IT can aid the OR analyst. OR/MS Today, June.

Garey, M. R., & Johnson, D. S. (1979). *Computers and Intractability*. W.H. Freeman Press.

Gass, S. I., & Assad, A. A. (2005). *An annotated timeline of operations research: An informal history* (Vol. 75). Springer Science & Business Media.

George, J. M., & Jones, G. R. (2024). *ISE Contemporary Management* (13th ed.). McGraw-Hill.

Ghorbian, M., & Ghorbian, S. (2025). Empowering Clinical Decision-Making with Generative AI in Intelligent Decision Support Systems. In *Generative Artificial Intelligence (AI) Approaches for Industrial Applications* (pp. 241–259). Springer Nature Switzerland. DOI: 10.1007/978-3-031-76710-4_12

Gong, Q., Fan, D., & Bartram, T. (2025). Integrating artificial intelligence and human resource management: A review and future research agenda. *International Journal of Human Resource Management*, *36*(1), 103–141. DOI: 10.1080/09585192.2024.2440065

Greenberg, J. (2010). *Behavior in Organizations* (10th ed.). Pearson Education.

Gupta, S., Modgil, S., Bhattacharyya, S., & Bose, I. (2022). Artificial intelligence for decision support systems in the field of operations research: Review and future scope of research. *Annals of Operations Research*, *308*(1), 215–274. DOI: 10.1007/s10479-020-03856-6

Handoko, B., Franty, Y. K., & Indrayatna, F. (2025). Quantifying Uncertainty of Insurance Claims Based on Expert Judgments. *Mathematics*, *13*(2), 245. DOI: 10.3390/math13020245

Harris, O. O., Bialous, S. A., Muench, U., Chapman, S., & Dawson-Rose, C. (2022). Climate change, public health, health policy, and nurses training. *American Journal of Public Health*, *112*(S3), S321–S327. DOI: 10.2105/AJPH.2022.306826 PMID: 35679559

Hassan, S., Dhali, M., Zaman, F., & Tanveer, M. (2021). Big data and predictive analytics in healthcare in Bangladesh: Regulatory challenges. *Heliyon*, *7*(6), e07179. DOI: 10.1016/j.heliyon.2021.e07179 PMID: 34141936

Haugen, J. B., Link, J. S., Cribari, K., Bundy, A., Dickey-Collas, M., Leslie, H. M., Hall, J., Fulton, E. A., Levenson, J. J., Parsons, D. M., Hassellöv, I.-M., Olsen, E., DePiper, G. S., Gentry, R. R., Clark, D. E., Brainard, R. E., Mateos-Molina, D., Borja, A., Gelcich, S., & Agnalt, A. L. (2024). Marine ecosystem-based management: Challenges remain, yet solutions exist, and progress is occurring. *Npj Ocean Sustainability*, *3*(1), 7. DOI: 10.1038/s44183-024-00041-1

Helo, P., & Hao, Y. (2022). Artificial intelligence in operations management and supply chain management: An exploratory case study. *Production Planning and Control, 33*(16), 1573–1590. DOI: 10.1080/09537287.2021.1882690

Heriyati, P., Yadav, N., & Tamara, D. (2024). Accomplishing sustainable development goals through international management system standards and multinational supply chains. *Business Strategy and the Environment, 33*(6), 5214–5231. DOI: 10.1002/bse.3752

Ignizio, J. P., & Cavalier, T. M. (2019). *Linear Programming*. Pearson.

Imam, M. H., Tasadduq, I. A., & Ahmad, A. R. (2015). A Hybrid Algorithm for Optimising Facility Layout. *South African Journal of Industrial Engineering, 26*(1).

INFORMS. (2025). Institute for Operations Research and the Management Sciences, https://www.informs.org/, accessed on Feb. 25, 2025.

Jackson, M. C. (2016). *Systems thinking: Creative holism for managers*. John Wiley & Sons, Inc.

Jackson, M. C., & Keys, P. (1985). Introduction. *The Journal of the Operational Research Society, 36*(9), 753–755. DOI: 10.1057/jors.1985.140

Kahneman, D., Slovi, P., & Tversky, A. (1982). *Judgment Under Uncertainty: Heuristics and Biases*. Cambridge University Press. DOI: 10.1017/CBO9780511809477

Kandwal, N. P., Nautiyal, A., & Juyal, S. A. (2024, December). Promoting Health and Well-being through Integrative Wellness Practices: Pathways to Sustainable Development Goal 3. In *5th International Conference on the Role of Innovation, Entrepreneurship and Management for Sustainable Development (ICRIEMSD-2024)* (pp. 362-375). Atlantis Press. DOI: 10.2991/978-94-6463-612-3_24

Karray, F., & De Silva, C. (2004). *"Soft Computing and Intelligent Systems Design: Theory, Tools and Applications", Readings*. Addison-Wesley.

Kleespies, M. W., & Dierkes, P. W. (2022). The importance of the Sustainable Development Goals to students of environmental and sustainability studies—A global survey in 41 countries. *Humanities & Social Sciences Communications, 9*(1), 218. DOI: 10.1057/s41599-022-01242-0

Kobbacy, K. A., Vadera, S., & Rasmy, M. H. (2007). AI and OR in management of operations: History and trends. *The Journal of the Operational Research Society, 58*(1), 10–28.

Konar, A. (2018). *"Artificial Intelligence and Soft Computing: Behavioral and Cognitive Modeling of the Human Brain", Readings*. CRC Press.

Lane, D. (1994). With Little Efforts from our Friends: How System Dynamics and "Soft" O.R. can learn from each other. *System Dynamics Review, 10*(2-3), 101–134. DOI: 10.1002/sdr.4260100203

Lawson, D. (2005). *Engineering Disasters – Lessons to be learned, Readings.* ASME Press.

Maqbool, S., Zafeer, H. M. I., Zeng, P., Maqbool, S., Draissi, Z., & Javed, S. (2024). Inventive leadership styles and their impact for achieving sustainable development goals in education at secondary schools: A case study from Multan, Pakistan. *Humanities & Social Sciences Communications, 11*(1), 1–11. DOI: 10.1057/s41599-024-03086-2

Marakas, G. M. (2015). *Decision Support Systems.* Pearson India.

March, J. G., & Simon, H. A. (1958). *Organizations.* Wiley.

Meadows, D. H., Meadows, D. L., & Randers, J. (1992). Beyond the limits. [Human populations and resource depletion]. World Watch;(United States), 6(1).

Negnevitsky, M. (2019). *Artificial Intelligence: A Guide to Intelligent Systems* (3rd ed.). Pearson India.

Norman, D. A. (1993). *Thing That Make Us Smart: Defending Human Attributes in the Age of Machines.* Persius Books Group.

Nyland, C., & Bruce, K. (2025). Taylorism, trade unions and participatory management. *Journal of Management History.* Advance online publication. DOI: 10.1108/JMH-09-2024-0136

Science of Better. (2015). Operations Research: The Science for Better, http://www.scienceofbetter.org

Ossenbruggen, P. J. (1994). *Fundamental Principles of Systems Analysis and Decision-Making.* John Wiley.

Adinolfi, P. (2021). A journey around decision-making: Searching for the "big picture" across disciplines. *European Management Journal, 39*(1), 9–21.

Papadopoulos, T., Sivarajah, U., Spanaki, K., Despoudi, S., & Gunasekaran, A. (2022). Artificial Intelligence (AI) and data sharing in manufacturing, production and operations management research. *International Journal of Production Research, 60*(14), 4361–4364. DOI: 10.1080/00207543.2021.2010979

Petropoulos, F., Laporte, G., Aktas, E., Alumur, S. A., Archetti, C., Ayhan, H., Battarra, M., Bennell, J. A., Bourjolly, J.-M., Boylan, J. E., Breton, M., Canca, D., Charlin, L., Chen, B., Cicek, C. T., Cox, L. A.Jr, Currie, C. S. M., Demeulemeester, E., Ding, L., & Zhao, X. (2024). Operational Research: Methods and applications. *The Journal of the Operational Research Society*, *75*(3), 423–617. DOI: 10.1080/01605682.2023.2253852

Pidd, M. (2001). The Futures of O.R. *The Journal of the Operational Research Society*, *52*(11), 1181–1190. DOI: 10.1057/palgrave.jors.2601268

Pidd, M. (2004). The Futures of O.R. *The Journal of the Operational Research Society*, *55*, 791–800. DOI: 10.1057/palgrave.jors.2601748

Prodromou, A., Papademetriou, C., & Savvides, D. (2025). The Strategic Role of Green HRM in Hospitality Industry for Reaching Sustainable Development Goals. In Sustainability Through Green HRM and Performance Integration (pp. 1-38). IGI Global.

Ramírez-Márquez, C., Posadas-Paredes, T., Raya-Tapia, A. Y., & Ponce-Ortega, J. M. (2024). Natural resource optimization and sustainability in society 5.0: A comprehensive review. *Resources*, *13*(2), 19. DOI: 10.3390/resources13020019

Rhodes, J. S. (2000). "Usability can save you money", http://www.webword.com/moving/savecompany.html, accessed on Nov. 02 2004, December 2000.

Rand, G. K. (1998). Re-Engineering the Manufacturing System: Applying the Theory of Constraints. *The Journal of the Operational Research Society*, *49*(7), 772–773.

Saxena, A., Ramaswamy, M., Beale, J., Marciniuk, D., & Smith, P. (2021). Striving for the United Nations (UN) Sustainable Development Goals (SDGs): What will it take? *Discover Sustainability*, *2*(1), 20. DOI: 10.1007/s43621-021-00029-8

Sensor, C. S., Branden, P. S., Clary-Muronda, V., Hawkins, J. E., Fitzgerald, D., Shimek, A. M., Al-Itani, D., Madigan, E. A., & Rosa, W. E. (2021). Nurses achieving sustainable development goals: The United Nations and Sigma. *The American Journal of Nursing*, *121*(4), 65–68. DOI: 10.1097/01.NAJ.0000742544.07615.db PMID: 33755634

Shafer, G. (1976). *A Mathematical Theory of Evidence*. Princeton University Press. DOI: 10.1515/9780691214696

Sharda, R., Delen, D., & Turban, E. (2020). *Analytics, Data Science, & Artificial Intelligence: Systems for Decision Support*. Pearson.

Simon, H. A. (1955). A Behavioral Model of Rational Choice. *The Quarterly Journal of Economics, 69*(1), 99–118. DOI: 10.2307/1884852

Simon, H. A. (1957). *Administrative behavior*. Simon and Schuster.

Stainton, R. S. (1984). Applicable Systems Thinking. *European Journal of Operational Research, 18*(2), 145–154. DOI: 10.1016/0377-2217(84)90180-2

Sterman, J. D. (1994). Judgment under Uncertainty: Heuristics and Biases. *System Dynamics Review, 10*(2-3), 291–330. DOI: 10.1002/sdr.4260100214

Sterman, J. D. (2017). *Business Dynamics: An Introduction*. Mc-Graw Hill.

Taguchi, G., Elsayed, E. A., & Hsiang, T. (1989). *Quality Engineering in Production Systems*. McGraw Hill.

Taha, H. A. (2022). Operations Research: An Introduction, 11th Kindle Ed., Pearson.

Taylor, B. W. (2018). *Introduction to Management Science* (13th ed.). Pearson.

Thomas, W. (2015). *Rational action: The sciences of policy in Britain and America, 1940-1960*. MIT Press. DOI: 10.7551/mitpress/9997.001.0001

Triantaphyllou, E., & Lin, C. (1996). Development and Evaluation of Five Fuzzy Multi-Attribute Decision-Making Methods. *International Journal of Approximate Reasoning, 14*(4), 281–310. DOI: 10.1016/0888-613X(95)00119-2

Triantaphyllou, E., Shu, B., Nieto Sanchez, S., & Ray, T. (1998). Multi-Criteria Decision Making: An Operations Research Approach. In Webster, J. D. (Ed.), *Encyclopedia of Electrical and Electronics Engineering*. John Wiley.

Turban, E., & Aronson, J. E. (2005). *Decision Support Systems and Intelligent Systems*. Prentice Hall Inc.

Tversky, A., & Kahneman, D. (1974). Judgment under Uncertainty: Heuristics and Biases. *Science, 185*(4157), 1124–1131. DOI: 10.1126/science.185.4157.1124 PMID: 17835457

United Nations. (2015). *Transforming our world: The 2030 agenda for sustainable development*.

Warfield, J. N., & Staley, S. M. (1996). Structural Thinking: Organizing Complexity Through Disciplined Activity. *Systems Research, 13*(1), 47–67. DOI: 10.1002/(SICI)1099-1735(199603)13:1<47::AID-SRES27>3.0.CO;2-A

Wiggin, T. (2025). Weaponizing the Workplace: How Algorithmic Management Shaped Amazon's Antiunion Campaign in Bessemer, Alabama. *Socius: Sociological Research for a Dynamic World*, *11*, 23780231251318389. DOI: 10.1177/23780231251318389

Xiao, Z., & Zhang, Y. (2024). The influence of information sharing on supply chain performance in Internet of Things. Journal of Computational Methods in Sciences and Engineering, 14727978241300832.

Yachachín, O. M. L., Ocoruro, E. A. V., & La Riva, M. M. (2025). Expectation and Perception of Quality of Care from the Perspective of User Satisfaction in a Private Clinic: Wellness and Health. *Journal of Lifestyle and SDGs Review*, *5*(2), e04249–e04249. DOI: 10.47172/2965-730X.SDGsReview.v5.n02.pe04249

Zha, X. F. (2003). Soft Computing Framework for Intelligent Human–Machine System Design, Simulation and Optimization. *Soft Computing*, *7*(3), 184–198. DOI: 10.1007/s00500-002-0196-4

Zhou, L., & Alam, G. M. (2024). Commercial higher education strategies for recruiting international students in China: A catalyst or obstacle for sustainable education and learning? *Discover Sustainability*, *5*(1), 33. DOI: 10.1007/s43621-024-00216-3

Zhou, Z., Syamsunur, D., Muhamad, N. S., Ghazalli, A. Z., Azahar, M. A., & Jusoh, M. N. H. B. (2025). A Tentative Analysis of the Interactions Between the Building Sector and SDGs. *Journal of Lifestyle and SDGs Review*, *5*(2), e03212–e03212. DOI: 10.47172/2965-730X.SDGsReview.v5.n02.pe03212

Chapter 3
Green HRM Bridging Sustainability and Workforce Management

Mohi ud Din
Beaconhouse International College, Pakistan

Muhammad Faizan Khan
https://orcid.org/0009-0009-3676-7311
Kohsar University, Murree, Pakistan

ABSTRACT

The growing global emphasis on sustainability and environmental protection has led to the integration of eco-friendly initiatives across industries, resulting in the emergence of green human resource management (Green HRM). "Green HRM" is an extension of sustainable HRM, focused on integrating environmental management into human resource policies. It involves the use of eco-friendly HR practices that support the organization's environmental sustainability agenda. The primary goal of Green HRM is to promote a culture of environmental stewardship and sustainability within the organization, helping to reduce carbon footprints, minimize waste, and encourage responsible use of resources. At its core, Green HRM aligns organizational goals with environmental strategies. It involves practices such as "green recruitment," "green training," "green performance management," and "green rewards."

DOI: 10.4018/979-8-3373-0139-6.ch003

1. INTRODUCTION

The growing global emphasis on sustainability and environmental protection has led to the integration of eco-friendly initiatives across industries, resulting in the emergence of Green Human Resource Management (Green HRM). Green HRM refers to the policies, practices, and strategies that organizations implement to promote sustainable use of resources, reduce environmental impact, and align business operations with environmental goals. The concept goes beyond traditional HRM by embedding environmental sustainability into core HR functions, from recruitment and training to performance management and employee engagement. In this chapter, we will explore the principles, practices, and benefits of Green HRM, along with its evolution and future prospects (Cooke et al., 2020).

Green HRM is significant because it is very important to change employee behavior as well as company structure and strategy to their long-term business approach. Integrating this in the HR practices of organizations, leads to their practice where they can bring substantial transformation on strategies to mitigate waste, conserve energy and promote sustainable resource utilization. Green HRM principles go beyond corporate social responsibility (CSR) philosophy, making sustainability a composite of workforce management and ethics (Ahmed et al., 2021).

Sustainable talent management, one of the tenets of Green HRM, emphasizes the hiring, attracting, and retaining of employees whose values are consistent with an organization's environmental orientation. This includes integrating new eco-centric criteria into job descriptions, recruitment and selection measures, so that new hires are environmentally conscious. Organizations also run training to help employees understand environmental challenges and provide them with the knowledge and skills they need to integrate sustainability into the workplace (Alam & Niu, 2021).

Green HRM also includes the performance management of employees by checking how much an employee is contributing to environmental sustainability in their KPIs. This can include things like resource conservation efforts, energy efficiency initiatives, sustainability programs, and similar metrics. In this aspect, having green goals in performance evaluation and reward systems pushes the employees of the company to adopt ethical behaviors towards the environment (Al-Hawari et al., 2021).

Green HRM is ultimately an exercise in employee engagement. Many organizations adopt eco-friendly workplace policies, such as waste-reduction programs, paperless work environments, and sustainable commuting options. Companies may also form green committees or teams to promote employee involvement in environmental initiatives. This strengthens the sense of responsibility and organizational commitment towards sustainability resulting in an environmentally sustainable organizational culture (ud Din et al., 2025).

Green HRM has emerged as a response to the increasing realization of environmental significance and the pressure on businesses from stakeholders, governments, and regulatory agents to take up sustainable enterprise practices. But while organizations once regarded sustainability as a box to check, it has increasingly become a strategic imperative. With the advancements in digital technologies, artificial intelligence, and data analytics, more solutions are provided through the implementation of Green HRM as it is beneficial to track and measure the environmental performance and hence improvement can be made. In addition, organizations are paying greater attention to circular economy paradigms, in which HRM has an essential role in responsible consumption, minimization of waste and optimization of resource usage (Alam & Niu, 2021).

The Green HRM is a new paradigm of managing human resources that can aid the organization to manage the multifaceted pragmatic issue of deriving the greatest utility from the recruitment and maintenance of talent in realizing the sustainability vision of management as a whole. As green policies become ingrained into recruitment, training, performance management, and employee engagement processes, organizations will embed environmental responsibility into organizational culture while gaining sustainable business benefits. Sustainable practices in HR will become the very essence of the future work and organizations moving towards a sustainable and virtuous business landscape will heavily rely on Green HRM (Anwar et al., 2020).

1.2 What is Green HRM?

According to ud Din et al. (2025), "Green HRM" is an extension of sustainable HRM, focused on integrating environmental management into human resource policies. It involves the use of eco-friendly HR practices that support the organization's environmental sustainability agenda. The primary goal of Green HRM is to promote a culture of environmental stewardship and sustainability within the organization, helping to reduce carbon footprints, minimize waste, and encourage responsible use of resources. At its core, Green HRM aligns organizational goals with environmental strategies. It involves practices such as "green recruitment", "green training", "green performance management", and "green rewards". These practices help employees become environmentally conscious and contribute positively to the organization's sustainability goals.

1.3 Origin of Green HRM

The concept of Green HRM can be traced back to the increasing awareness of environmental issues in the 1980s and 1990s, particularly in response to climate change and the depletion of natural resources. Corporate Social Responsibility (CSR) and environmental management systems (EMS) began influencing HR practices, creating a platform for sustainability-focused HR policies. Early adopters recognized that the HR function could contribute to environmental management by embedding eco-friendly practices in recruitment, training, performance management, and employee engagement (Jackson et al., 2011; Renwick et al., 2013).

1.4 Evolution of Green HRM

The idea of Green HRM emerged as a response to the growing pressure on organizations to reduce their environmental impact. While traditional HRM focused on maximizing employee productivity and organizational efficiency, the rise of corporate social responsibility (CSR) and sustainability in the 1990s highlighted the need for HR to contribute to broader environmental objectives. This shift led to the development of HR strategies that align employee behavior with environmental management systems (EMS). By the early 2000s, environmental management became a key consideration in HR practices. This led to the formal introduction of "Green HRM" as a distinct field within HRM. Since then, it has evolved from being a niche concept to becoming a mainstream practice in many organizations, particularly those committed to sustainability and social responsibility (Milliman & Clair, 2017; Wehrmeyer, 2017).

Key Milestones:

- 1987: Brundtland Report introduces sustainable development.
- 1990s: Rise of CSR integrating sustainability into business practices.
- Early 2000s: Development of environmental policies within HR functions. (Bansal & Roth, 2000; Brundtland & Mansour, 2010; Carroll, 1999)

The evolution of Green HRM parallels the global shift towards sustainability, fueled by environmental legislation, societal demands for corporate accountability, and international frameworks like the United Nations Sustainable Development Goals (SDGs).

Stages of Evolution:

- **Reactive Stage:** HR practices included basic compliance with environmental regulations.
- **Proactive Stage:** HR practices started incorporating eco-friendly policies.

- **Strategic Stage:** Environmental sustainability became embedded in HR practices. (Bansal & Roth, 2000; Jabbour & Santos, 2008; Renwick et al., 2013)

1.5 Meaning and Definitions of Green HRM

Green HRM refers to HR practices that promote sustainable use of resources, minimize environmental degradation, and foster a culture of environmental responsibility. Various definitions highlight Green HRM as an approach that aligns employee behaviors and HR policies with ecological objectives, promoting sustainable workplace practices and enhancing the organization's environmental impact.

Key Definitions:

- **Definition 1**: "Green HRM involves HR practices focused on enhancing environmental sustainability" (Renwick et al., 2013).
- **Definition 2**: "The integration of sustainable environmental goals into HR policies and practices" (Jabbour & Santos, 2008).
- **Definition 3**: Green HRM can be described as the integration of environmentally conscious practices within HR functions. It includes actions such as green recruitment, training, rewards, and empowerment. These practices aim to foster a pro-environmental behavior among employees, ultimately contributing to an organization's environmental sustainability goals (ud Din, M., & Ahmad, A., 2023).

2. KEY COMPONENTS OF GREEN HRM

Green HRM can be broken down into several key components, each addressing a specific area of HR management with an environmental focus:

2.1. Green Recruitment and Selection

Green recruitment and selection play a pivotal role in creating an environmentally responsible workforce that actively supports the organization's sustainability objectives. By embedding eco-conscious practices into hiring, organizations not only attract individuals with a strong commitment to sustainability but also reinforce their own dedication to environmental goals. As sustainability becomes a central component of corporate strategy, green recruitment and selection will remain essential for organizations aiming to lead in eco-friendly practices.

Green recruitment and selection, a crucial element of Green Human Resource Management (Green HRM), focus on hiring candidates who align with an organization's environmental goals and are committed to sustainable practices. As companies strive for sustainability, green recruitment and selection processes have become essential to building a workforce that not only meets operational goals but also contributes to the organization's eco-friendly mission.

2.2 The Concept of Green Recruitment and Selection

Green recruitment and selection are the processes of attracting, selecting, and hiring individuals who demonstrate environmental awareness and a commitment to sustainable practices. This approach seeks candidates who will actively participate in, support, and expand the organization's environmental objectives. By incorporating green criteria into job descriptions, interview questions, and selection metrics, HR departments aim to hire individuals who not only meet the technical requirements of their roles but also resonate with the company's sustainability-focused culture.

2.3 Strategies for Green Recruitment

Green recruitment incorporates several strategies to ensure that the hiring process aligns with sustainability principles:

2.4 Eco-Friendly Job Advertisements

Companies are increasingly adopting eco-friendly methods for advertising open positions. Instead of relying on traditional paper-based advertisements, organizations use digital job boards, company websites, and social media platforms to minimize paper consumption and reduce their carbon footprint. Job postings can include language that emphasizes the company's commitment to sustainability, which attracts candidates who share similar values.

2.5 Promoting Environmental Culture in Branding

Employer branding that highlights the organization's environmental goals can attract like-minded candidates. Showcasing achievements, sustainability certifications, and environmental initiatives on the company's website and social media platforms appeals to candidates who value eco-friendly workplaces. For example, companies like Patagonia, which focuses on eco-friendly products and sustainable practices, attract employees who are passionate about environmental causes.

2.6 Remote Hiring Practices

To reduce emissions associated with travel, companies can adopt virtual interviews and remote hiring practices. This practice is beneficial for both sustainability and cost-efficiency, as it eliminates the need for travel while allowing companies to access a wider talent pool.

2.7 Environmental Criteria in Job Descriptions

Including environmental responsibilities in job descriptions conveys that sustainability is a priority within the organization. For example, job roles might include responsibilities related to waste reduction, energy conservation, or sustainable sourcing practices. Emphasizing these criteria helps in attracting candidates who are not only qualified for the position but also committed to green initiatives.

2.8 Green Selection Practices

The selection process in Green HRM includes evaluating candidates for their commitment to environmental practices. Organizations integrate environmental criteria into traditional selection tools to assess candidates' potential for supporting green initiatives:

3. ECO-CONSCIOUS INTERVIEW QUESTIONS

Interview questions can be designed to assess candidates' environmental awareness and personal commitment to sustainability. For instance, asking candidates about their experience with eco-friendly practices in previous roles can help identify those with a genuine interest in environmental sustainability.

3.1 Evaluating Environmental Fit

Beyond assessing technical and interpersonal fit, companies evaluate a candidate's environmental fit—how well the candidate aligns with the company's sustainability values. This can involve assessing personal values and prior experience with green initiatives.

3.2 Green Onboarding

Once candidates are selected, a green onboarding process ensures that they quickly integrate into the organization's eco-friendly culture. Onboarding sessions can introduce new hires to the company's sustainability practices, from waste reduction programs to energy-saving policies.

3.3 Green Training and Development

Green Training and Development is a cornerstone of Green Human Resource Management (Green HRM), focusing on educating and empowering employees to adopt environmentally sustainable practices within the organization. As businesses increasingly integrate sustainability into their core strategies, training employees to understand and contribute to environmental goals is essential for fostering a culture of ecological responsibility. Green Training and Development aims not only to equip employees with eco-friendly practices but also to enhance their environmental awareness, driving long-term organizational sustainability.

Green Training and Development is a vital component of an organization's sustainability strategy, embedding environmental responsibility within its workforce. By educating employees on eco-friendly practices and encouraging sustainable behaviors, organizations create a culture that prioritizes environmental well-being alongside business performance. As more companies recognize the importance of Green HRM, green training will continue to play a critical role in preparing a workforce that is not only skilled but also dedicated to fostering a sustainable future. Through structured and purposeful green training, organizations can achieve their environmental objectives and contribute to a more sustainable world.

Green Training and Development refers to structured programs and initiatives that aim to instill a deep understanding of environmental sustainability in employees and equip them with the skills necessary to implement eco-friendly practices. Unlike traditional training, which focuses primarily on job-specific skills and career growth, green training emphasizes the ecological impact of everyday business operations and encourages employees to incorporate sustainable behaviors into their work routines. By prioritizing green training, organizations contribute to a shift in employee mindsets, fostering an enduring commitment to sustainability that transcends the workplace.

Key Components of Green Training

Green Training and Development typically include several components designed to promote environmental practices:

3.4 Environmental Awareness Programs

Environmental awareness is the foundation of green training. Awareness programs introduce employees to the organization's sustainability goals, eco-friendly policies, and the importance of environmental responsibility. Through workshops, seminars, or online modules, employees learn about global environmental challenges, such as climate change, waste management, and resource conservation, and understand how the organization's initiatives address these issues. Such programs provide a context for why sustainable practices matter, helping employees align with the organization's green objectives.

3.5 Skill-Based Green Training

Skill-based training focuses on equipping employees with specific skills that contribute to environmental sustainability. For example, employees may receive training on energy-saving techniques, resource conservation, waste reduction practices, or environmentally responsible procurement processes. This targeted training is especially valuable in industries where daily operations significantly impact the environment, such as manufacturing, logistics, and retail. By imparting practical skills, organizations empower employees to make sustainable choices that reduce the company's ecological footprint.

3.6 Green Leadership Development

Developing green leaders within an organization is crucial for sustaining long-term environmental goals. Green leadership development programs train managers and supervisors to lead sustainable practices in their teams, encouraging a trickle-down effect throughout the organization. Leaders are educated on sustainable decision-making, eco-friendly project management, and effective ways to promote green values among their team members. By fostering green leadership, organizations ensure that sustainability remains a top priority across all levels of the organization.

3.7 Encouraging Personal Responsibility

Beyond specific skills, green training also emphasizes the importance of personal accountability in achieving sustainability. Employees are encouraged to adopt eco-friendly habits, such as minimizing paper use, recycling, reducing energy consumption, and promoting sustainable commuting options. Training programs may include activities that help employees assess their ecological footprint and set personal sustainability goals, making environmental responsibility an integral part of both professional and personal life.

3.8 Benefits of Green Training and Development

Green Training and Development offer multiple benefits for organizations and their employees. For businesses, green training contributes to cost savings by promoting resource efficiency, reducing waste, and lowering energy consumption. It also strengthens the organization's reputation as an environmentally responsible employer, attracting talent that values sustainability. For employees, green training enhances job satisfaction by allowing them to contribute to meaningful environmental goals, instilling a sense of purpose in their work. Employees trained in green practices also develop valuable skills that are increasingly sought after as organizations worldwide adopt sustainable practices.

4. GREEN PERFORMANCE MANAGEMENT

Green Performance Management is a critical dimension of Green Human Resource Management (Green HRM), focusing on assessing and rewarding employees based on their contributions to organizational environmental goals. By incorporating sustainability metrics into performance evaluations, Green Performance Management encourages employees to adopt eco-friendly practices, thereby embedding a culture of environmental responsibility within the organization. As businesses worldwide emphasize corporate social responsibility, green performance management has emerged as an essential tool for promoting sustainability across all levels of an organization.

Green Performance Management goes beyond traditional evaluation metrics, which typically focus on productivity, efficiency, and skill proficiency. Instead, it integrates environmental objectives into performance appraisals, making sustainable behaviors a key criterion in evaluating employee contributions. By aligning individual performance goals with the organization's green agenda, this approach motivates employees to prioritize environmentally conscious actions within their roles. This alignment fosters a sense of accountability, as employees become aware that their environmental impact will influence their performance evaluations and career progression.

Green Performance Management is a transformative approach that aligns employee evaluation with sustainability objectives, fostering a workforce committed to environmental stewardship. By setting green goals, incorporating environmental metrics, and recognizing sustainable practices, organizations can create a performance management system that promotes both productivity and eco-responsibility. As sustainability becomes an increasingly critical aspect of business strategy, Green Performance Management will play a central role in helping organizations meet their environmental objectives and build a more sustainable future.

4.1 Key Elements of Green Performance Management

Setting Green Objectives and Goals

Green Performance Management begins with setting clear, measurable sustainability objectives for employees. These objectives may vary based on job roles and departments but are aimed at reducing the organization's ecological footprint. For instance, employees may be tasked with reducing energy usage, minimizing paper consumption, promoting waste recycling, or improving resource efficiency in their specific functions. By establishing green goals alongside traditional performance targets, organizations ensure that sustainability remains a priority throughout the evaluation period.

Incorporating Environmental Metrics in Appraisals

Integrating environmental metrics into performance appraisals is central to Green Performance Management. These metrics may include the amount of resources saved, waste reduction rates, compliance with eco-friendly protocols, and active participation in sustainability initiatives. For instance, in departments like procurement, evaluation metrics might assess the extent to which eco-friendly suppliers are selected. By incorporating these green metrics, performance appraisals reflect a holistic view of an employee's contributions, rewarding both professional and environmental achievements.

Recognition and Rewards for Sustainable Practices

Recognizing and rewarding employees for meeting green objectives is essential to reinforce sustainable behaviors. Green rewards can be financial, such as bonuses for achieving environmental targets, or non-financial, such as public recognition, eco-friendly incentives (e.g., reusable products or extra paid leave for volunteering in environmental projects). This acknowledgment of sustainable practices not only motivates employees to maintain eco-friendly habits but also signals the organization's commitment to its environmental goals. When employees know their green contributions are valued, they are more likely to engage in and advocate for sustainable practices.

Continuous Monitoring and Feedback

Effective Green Performance Management requires continuous monitoring and feedback to help employees understand their progress toward sustainability goals. Regular check-ins allow managers to offer guidance, suggest areas for improvement, and keep employees engaged with the organization's green agenda. For example, quarterly reviews may assess how employees are meeting their green objectives and adjust them if necessary. Ongoing feedback ensures that green performance management is dynamic and responsive to evolving environmental challenges and opportunities within the organization.

4.2 Benefits of Green Performance Management

Green Performance Management offers numerous benefits. For organizations, it ensures alignment between corporate sustainability goals and individual performance, helping to reduce the overall environmental impact. By fostering a culture where eco-friendly behaviors are acknowledged and rewarded, organizations enhance their reputation as socially responsible employers, attracting talent with similar values. For employees, green performance management creates a sense of purpose and engagement, as they see their work contribute to broader environmental goals. Furthermore, it enhances professional development by teaching employees the value of sustainable practices in their daily work.

4.3 Green Rewards and Compensation

Green Reward and Compensation refer to a system where employees are rewarded not only for their traditional work performance but also for their eco-friendly actions and contributions to sustainability. Traditional rewards, typically based on productivity and job performance, are expanded to include incentives for environmental initiatives, such as energy conservation, waste reduction, and sustainable resource management. By introducing green criteria into reward and compensation systems, organizations can align individual goals with corporate sustainability efforts, ensuring that employees recognize the importance of environmental responsibility in their roles.

Green Reward and Compensation is an integral aspect of Green Human Resource Management (Green HRM) that focuses on recognizing and incentivizing employees' contributions to an organization's environmental goals. As organizations strive to integrate sustainability into their operations, Green Reward and Compensation emerges as a key mechanism to encourage employees' commitment to eco-friendly behaviors. By aligning rewards with sustainability objectives, organizations can motivate employees to actively participate in initiatives that reduce their environmental footprint, thus fostering a workplace culture that prioritizes sustainable development.

Green Reward and Compensation is an essential component of Green HRM that aligns employee incentives with the organization's environmental objectives. By offering financial incentives, non-financial rewards, and green benefits, organizations can foster a culture of sustainability and encourage eco-friendly practices at all levels of the workforce. As environmental sustainability continues to gain importance in business, Green Reward and Compensation will play a pivotal role in motivating employees to take active responsibility for reducing their environmental impact, ultimately supporting a more sustainable future for both organizations and society.

4.4 Types of Green Rewards and Compensation

Green Reward and Compensation can be structured in various ways to recognize employee contributions to sustainability (Amjad et al., 2021):

Financial Incentives

Financial rewards are a powerful motivator and can effectively encourage employees to adopt eco-friendly practices. These incentives may include bonuses for achieving specific environmental targets, such as reducing office energy usage or meeting recycling goals. Some organizations provide salary increments or special allowances to employees who demonstrate exceptional commitment to green initiatives. For example, employees in roles focused on environmental management may receive performance bonuses tied to the company's sustainability metrics, thus reinforcing the importance of eco-friendly actions.

Non-Financial Rewards

Non-financial rewards, which are often just as effective as monetary incentives, recognize employees' green contributions in more symbolic ways. Public recognition, awards, and certificates acknowledging employees' efforts in sustainability initiatives are common examples. Additionally, eco-friendly gifts, such as reusable products, sustainable office supplies, or even paid time off for volunteering in environmental programs, are valuable non-financial incentives that signal the organization's commitment to sustainability. Such rewards not only encourage employees to continue eco-friendly practices but also increase their sense of connection to the company's environmental values.

Green Benefits and Allowances

Green Reward and Compensation may also include green benefits, such as subsidies for public transportation, carpool programs, or bike-to-work allowances. These benefits support employees in reducing their personal carbon footprint while commuting, making sustainable choices more accessible. Some companies also provide grants for employees who wish to pursue environmentally-focused training, certifications, or participate in external sustainability initiatives. By offering green benefits, organizations enable employees to integrate sustainability into their lifestyle, thus supporting a holistic approach to environmental responsibility.

4.5 Advantages of Green Reward and Compensation

Green Reward and Compensation bring substantial advantages to both the organization and its employees. For the organization, these rewards promote a culture of sustainability, making eco-friendly practices a core part of the organizational identity. By aligning rewards with green initiatives, companies can attract environmentally conscious talent, enhancing their employer brand and supporting their corporate social responsibility (CSR) efforts. Furthermore, these practices encourage employees to become active participants in sustainability, which can lead to cost savings from reduced energy usage, lower waste management costs, and resource conservation. For employees, green rewards enhance job satisfaction and motivation, as they feel recognized for contributing to meaningful environmental goals. These rewards provide a sense of purpose and fulfillment, particularly for employees passionate about sustainability. Additionally, green compensation plans offer unique benefits that support employees' personal and professional growth, while reinforcing eco-conscious behaviors that can positively impact their lifestyle choices.

4.6 Green Employee Engagement

According to Ali Ababneh et al. (2021), green employee engagement is a vital aspect of Green Human Resource Management (Green HRM), focusing on the active involvement of employees in an organization's environmental sustainability initiatives. As companies increasingly commit to reducing their ecological footprint, engaging employees in green practices becomes essential to achieving these goals. Green employee engagement involves creating a culture where employees are not only aware of the organization's environmental objectives but are also motivated and empowered to contribute actively to these efforts. Through strategies that emphasize shared environmental responsibility, organizations can foster a workforce that is both environmentally conscious and actively engaged in promoting sustainability.

Green Employee Engagement goes beyond traditional employee involvement, which typically focuses on job satisfaction and organizational commitment. Instead, it seeks to engage employees specifically in eco-friendly initiatives and sustainability goals. This engagement is built upon aligning employees' values with the organization's environmental mission, encouraging them to adopt and advocate for green practices within and outside the workplace. An effective green engagement strategy inspires employees to feel a sense of responsibility towards environmental protection, creating a ripple effect that extends beyond individual actions to influence the entire organizational culture.

Green Employee Engagement is a powerful approach to embedding environmental sustainability within an organization. By educating employees, creating green volunteer opportunities, forming sustainability committees, and recognizing eco-friendly practices, organizations can build a workforce that is deeply committed to environmental responsibility. As sustainability continues to gain prominence in the corporate world, engaging employees in green practices will be essential for organizations aiming to reduce their ecological footprint and contribute to a more sustainable future.

4.7 Key Strategies for Green Employee Engagement

Environmental Awareness Programs

Raising awareness is the first step toward green employee engagement. Organizations can offer workshops, seminars, or online training sessions to educate employees on global environmental issues, such as climate change, pollution, and biodiversity loss. Employees also learn about the company's specific sustainability goals and how their roles contribute to these objectives. By understanding the environmental impact of their actions, employees are more likely to take ownership of green practices, viewing them as integral to their professional responsibilities.

Green Volunteer Opportunities

Providing opportunities for employees to participate in eco-friendly activities is an effective way to foster engagement. Green volunteer programs, such as tree planting, beach clean-ups, or community recycling initiatives, encourage employees to make tangible contributions to environmental protection. Such initiatives not only build camaraderie among colleagues but also strengthen their connection to the organization's green values. Additionally, allowing employees to volunteer during paid work hours or offering additional paid leave for volunteering enhances participation and shows the organization's genuine commitment to environmental stewardship.

Green Task Forces and Committees

Establishing green task forces or sustainability committees allows employees to take an active role in shaping the organization's environmental policies. These groups are tasked with identifying areas for improvement, proposing sustainable practices, and implementing green initiatives. By involving employees directly in decision-making processes, organizations empower them to take ownership of sustainability efforts. This sense of agency and responsibility promotes a deeper level of engagement, as employees feel that their contributions genuinely impact the organization's green agenda (Amjad et al., 2021).

Recognition and Reward for Green Practices

Recognizing and rewarding employees for their eco-friendly behaviors is essential for sustained engagement. Organizations can implement green recognition programs where employees are acknowledged for their environmental contributions, such as reducing paper usage, carpooling, or developing resource-saving practices. These rewards can be in the form of certificates, awards, or eco-friendly incentives, such as reusable products. Public recognition not only motivates the individual but also inspires other employees to adopt similar practices, gradually embedding green behaviors throughout the organization.

4.8 Benefits of Green Employee Engagement

Green Employee Engagement offers numerous benefits, both for the organization and its workforce. For the organization, an engaged workforce contributes to achieving environmental goals, resulting in resource savings, improved operational efficiency, and a positive public image. Green employee engagement also supports corporate social responsibility (CSR) initiatives, enhancing the company's reputation as a socially and environmentally responsible employer. For employees, green engagement fosters a sense of purpose and satisfaction as they participate in initiatives that have a meaningful impact on the environment. Engaged employees often experience increased motivation, as their contributions align with both personal and organizational values (Ahmed et al., 2020).

5. THE BENEFITS OF GREEN HRM

The integration of Green Human Resource Management (Green HRM) practices into organizational strategies yields numerous benefits that significantly impact both companies and their workforce. By adopting these sustainable practices, organizations can not only improve their operational efficiency but also contribute positively to the environment. Below are some key benefits of Green HRM:

5.1. Enhanced Corporate Reputation

In today's competitive business landscape, organizations that actively showcase their commitment to sustainability and environmental responsibility often enjoy a more robust corporate reputation. Green HRM initiatives position companies as responsible corporate citizens, appealing to a growing demographic of environmentally conscious consumers, investors, and potential employees. This positive perception can translate into various advantages, including heightened customer loyalty, which can manifest in repeat business and increased customer referrals. Moreover, a solid reputation for sustainability can enhance a brand's value, making it more appealing to investors who prioritize ethical and responsible investment practices. In addition, companies with strong green credentials may find it easier to forge partnerships and collaborations with other businesses that share similar values, further solidifying their market position. As sustainability becomes a defining characteristic of corporate identity, organizations embracing Green HRM are more likely to stand out in a crowded marketplace, thus gaining a competitive edge.

5.2. Cost Savings

One of the most compelling benefits of implementing Green HRM practices is the potential for substantial cost savings. By prioritizing sustainability, organizations can effectively minimize expenses associated with energy consumption, waste management, and resource utilization. For instance, initiatives such as transitioning to paperless offices reduce the costs related to printing, paper supplies, and storage. Implementing energy-efficient workspaces—like LED lighting, smart thermostats, and energy-efficient appliances—can lead to lower utility bills over time. Moreover, sustainable procurement practices, which focus on sourcing environmentally friendly products and services, can help organizations negotiate better prices with suppliers while supporting eco-friendly manufacturers. By creating a culture of sustainability, companies can also foster employee awareness around resource conservation, leading to collective efforts to minimize waste and optimize resource usage. As organizations track and measure their sustainability efforts, they may uncover additional opportunities for cost reduction that contribute to the bottom line.

5.3. Increased Employee Satisfaction and Retention

Green HRM plays a critical role in shaping a positive organizational culture that resonates with employees. When employees perceive their organization's efforts to be environmentally responsible, they often feel their contributions have greater significance and align with broader societal goals. This sense of purpose can foster deeper engagement, leading to enhanced job satisfaction and commitment. Employees who believe their workplace is actively making a positive impact on the environment are typically more satisfied with their jobs, which can result in higher retention rates. Retaining skilled employees reduces turnover costs associated with recruiting and training new staff, thus enhancing organizational stability. Additionally, a culture that prioritizes sustainability often correlates with improved employee morale, creating a workplace where individuals feel valued and motivated. Ultimately, organizations that invest in Green HRM can cultivate a loyal workforce that is not only dedicated to the company but also passionate about its mission and values.

5.4. Compliance with Environmental Regulations

As environmental issues gain increasing attention worldwide, governments are implementing stricter regulations to ensure corporate compliance with sustainability standards. Green HRM practices help organizations navigate this complex regulatory landscape by embedding eco-friendly practices into their daily operations. By training employees on environmental regulations and motivating them to adopt sustainable practices, companies can significantly reduce the risk of legal and financial penalties associated with non-compliance. Furthermore, maintaining compliance with environmental regulations enhances an organization's credibility and reputation. Companies that proactively address sustainability standards are less likely to face fines or legal challenges, allowing them to focus their resources on growth and innovation instead. By embedding Green HRM practices into the organizational framework, companies can ensure that sustainability becomes a fundamental aspect of their operations, promoting long-term success while meeting regulatory expectations.

5.5. Innovation and Competitive Advantage

The emphasis on Green HRM often fosters a culture of innovation within organizations. Employees are encouraged to think creatively about ways to improve processes, reduce waste, and implement sustainable technologies. This focus on innovation can lead to the development of new products, services, or business models that not only enhance operational efficiency but also provide a distinct competitive advantage in the marketplace. For instance, companies may discover innovative solutions to reduce resource consumption or create environmentally friendly products that appeal to eco-conscious consumers. By embracing sustainable practices, organizations can position themselves as industry leaders in sustainability, attracting customers who prioritize environmental responsibility in their purchasing decisions. The continuous pursuit of innovation spurred by Green HRM can yield long-term benefits, ensuring that organizations remain agile and responsive to market demands while solidifying their reputation as forward-thinking and responsible enterprises.

5.6. Attraction of Top Talent

In addition to retaining current employees, the adoption of Green HRM practices can significantly enhance an organization's ability to attract top talent. Today's job seekers, particularly millennials and Generation Z, increasingly prioritize working for organizations that demonstrate a commitment to sustainability and social responsibility. By promoting green initiatives and emphasizing a culture of environmental stewardship, companies can position themselves as employers of choice for individuals who value sustainability in their professional lives. When potential candidates perceive that an organization is dedicated to making a positive impact on the planet, they are more likely to be drawn to the company. This can lead to a competitive advantage in the talent market, where organizations with strong green credentials are more likely to attract skilled professionals who are passionate about sustainability. By showcasing their commitment to Green HRM, companies can build a diverse and talented workforce that is aligned with their sustainability goals, ultimately driving organizational success.

5.7. Enhanced Organizational Resilience

In an era marked by rapid change and uncertainty, organizations must be resilient to thrive. Green HRM can contribute to building this resilience by encouraging adaptive practices that not only respond to environmental challenges but also capitalize on opportunities. Organizations that embrace sustainability often develop a more flexible and agile approach to operations, allowing them to navigate changing market conditions effectively. By integrating sustainability into their strategic planning, organizations can identify potential risks associated with environmental issues and proactively develop strategies to mitigate those risks. This foresight enables companies to remain competitive and responsive in a dynamic marketplace, ensuring they can withstand external pressures and capitalize on emerging trends. Green HRM thus fosters a culture of resilience, empowering organizations to adapt to change while maintaining a strong commitment to their sustainability objectives.

5.8. Positive Community Impact

Beyond the benefits to the organization and its employees, Green HRM practices also contribute positively to the broader community. Organizations that prioritize sustainability often engage in corporate social responsibility initiatives that support local environmental and social causes. This involvement not only enhances the organization's reputation but also fosters goodwill within the community. By partnering with local organizations, participating in environmental clean-up efforts, or supporting community education programs, companies can demonstrate their commitment to being responsible corporate citizens. This positive community engagement can create a ripple effect, inspiring other businesses to adopt similar practices and fostering a culture of sustainability within the region. Ultimately, organizations that embrace Green HRM can contribute to the well-being of the communities they operate in, creating a more sustainable future for all.

5.9. Long-Term Sustainability Goals

Implementing Green HRM practices aligns organizations with long-term sustainability goals that benefit both the business and the planet. As global awareness of environmental issues grows, stakeholders increasingly expect companies to take meaningful action toward sustainability. By integrating green practices into their operations, organizations can demonstrate their commitment to sustainable development and contribute to global efforts to combat climate change and environmental degradation. Incorporating sustainability into the core business strategy positions organizations to not only meet current regulatory requirements but also anticipate future demands for eco-friendly practices. This forward-thinking approach allows companies to stay ahead of trends and be proactive in addressing environmental challenges, ensuring their long-term viability and success.

Hence, the benefits of Green HRM extend far beyond compliance and cost savings. By fostering a culture of sustainability, organizations can enhance their reputation, reduce costs, improve employee satisfaction, and drive innovation. As businesses increasingly recognize the value of sustainable practices, integrating

Green HRM into organizational strategies becomes not just a moral imperative but also a strategic advantage that positions them for long-term success in a rapidly evolving marketplace.

As many research studies uphold the significance of green HRM, the latest study titled, "Sowing the Seeds of Sustainability: A Sociological Exploration of Environmental Sustainability within the SME Sector" by ud Din et al. (2025), lends substantial empirical weight to the Green HRM perspectives that discussed above. This also highlights the importance of Green HRM in developing pro-environmental behaviors of employees to promote organizational environmental sustainability. This article establishes that Green HRM has a considerable impact on employees' attitudes and behaviors toward sustainable development by examining the interrelationships among Green HRM practices, green transformational leadership, employees' pro-environmental behavior, and green culture. The results indicate that Green HRM encompasses indirect positive effect on environmental sustainability mediated via employees' pro-environmental behavior and green organizational culture. As have been noted in previous comments, Green HRM is not just about policies, there is a need to embedded mindsets and create a greener environment at work. Also, the research establishes that Green HRM helps in meeting environmental compliance as well as providing a positive influence on corporate reputation. The significant point of this is that Green HRM helps organizations to work according to the global level of the environment and strive for the goals of a stable environment to look attractive in front of international investors and multinational organizations. This thought is in alignment with the general perspective that Green HRM has holistic effect on the organization in terms of reducing environmental risk, per se gaining competitive edge in the industry. The results also highlight the importance of employee engagement and empowerment in sustainability initiatives, which is in line with previous debates around the importance of motivation in Green HRM. These are simple ways of bolstering green plans by involving employees in the green projects and decision-making process. It highlights that Green HRM is a key factor in determining the environmental practices as well as behaviors of employees, helping organizations to be more competitive with global sustainability standards and promote long-term ecological responsibility.

6. CHALLENGES OF IMPLEMENTING GREEN HRM

While the adoption of Green Human Resource Management (Green HRM) brings numerous advantages to organizations, it also presents several significant challenges that can hinder successful implementation. Understanding and addressing these challenges is crucial for organizations aiming to integrate sustainable practices into their operations effectively. Below are some of the key challenges associated with implementing Green HRM:

6.1. Lack of Awareness and Training

One of the foremost obstacles to implementing Green HRM practices is the prevalent lack of awareness among employees and management regarding the critical importance of sustainability. Many individuals may not fully comprehend the environmental issues at stake, the specific goals of the organization's sustainability initiatives, or how their roles contribute to these objectives. This gap in understanding can result in apathy or disengagement, ultimately undermining the effectiveness of Green HRM initiatives. To overcome this challenge, organizations must prioritize comprehensive training and communication strategies that educate employees about sustainability concepts and practices. Developing training programs that clearly outline the organization's environmental goals, as well as the specific contributions expected from employees, can foster a greater sense of ownership and engagement. Workshops, seminars, and interactive training sessions can serve as effective platforms for raising awareness and encouraging dialogue around sustainability. Furthermore, providing employees with resources, such as sustainability handbooks or online modules, can facilitate ongoing learning and reinforce the importance of their roles in achieving environmental objectives. When employees feel informed and empowered, they are more likely to embrace sustainable practices and actively participate in organizational initiatives.

6.2. Resistance to Change

Implementing Green HRM often necessitates substantial changes to established organizational processes and practices, which can provoke resistance from employees who are accustomed to traditional methods of operation. Change can be challenging for many individuals, particularly when they feel their roles, routines, or workflows may be disrupted. Resistance to change can manifest in various forms, including reluctance to adopt new technologies, skepticism about the efficacy of sustainability initiatives, or outright opposition to revised policies. To effectively manage this resistance, organizations must employ comprehensive change management strategies that facilitate a smooth transition to green practices. This involves fostering open communication channels where employees can voice their concerns and seek clarification regarding the changes being implemented. Leadership plays a crucial role in this process; by demonstrating commitment to sustainability and highlighting the long-term benefits of these initiatives, leaders can inspire confidence among employees. Engaging employees in the decision-making process—such as soliciting their feedback on proposed changes or involving them in pilot programs—can also promote a sense of ownership and reduce resistance. Ultimately, successful change management hinges on building a culture of trust and transparency, ensuring that employees feel supported and valued throughout the transition to Green HRM.

6.3. Initial Investment Costs

The financial implications of implementing Green HRM practices can pose a significant challenge for organizations, particularly when it comes to initial investment costs. Many green initiatives—such as installing energy-efficient technologies, transitioning to sustainable materials, or developing comprehensive training programs—require substantial upfront capital. For organizations with limited resources or budget constraints, allocating funds to these initiatives can be a daunting task, even though they are likely to yield cost savings in the long term. To navigate this challenge, organizations can explore various funding options and financing strategies that facilitate the implementation of green practices without compromising financial stability. For example, organizations may consider applying for grants or subsidies specifically designed to support sustainability initiatives, which can alleviate some of the financial burden associated with upfront costs. Additionally, forming partnerships with other organizations, governmental bodies, or non-profits can provide access to shared resources and expertise that enhance the feasibility of green projects. Organizations should also conduct thorough cost-benefit analyses to highlight the long-term financial advantages of adopting Green HRM practices. By showcasing potential savings in energy costs, waste reduction, and improved employee productivity, organizations can build a compelling case for investment in sustainability initiatives. Additionally, framing green investments as strategic decisions that align with the organization's broader goals can encourage stakeholders to support these efforts, ultimately facilitating a smoother implementation process.

6.4. Measuring Green Performance

One of the most complex challenges associated with implementing Green HRM is the difficulty in quantifying the success of sustainability initiatives. While it is relatively straightforward to track tangible outcomes such as reduced energy consumption or decreased waste, measuring intangible outcomes—such as employee engagement, cultural shifts, or the overall impact of green practices on organizational performance—can be significantly more challenging. Without clear metrics and evaluation tools, organizations may struggle to assess the effectiveness of their Green HRM initiatives and identify areas for improvement. To address this challenge, organizations must develop comprehensive measurement frameworks that encompass both quantitative and qualitative indicators of success. This may involve establishing specific key performance indicators (KPIs) related to sustainability, such as energy savings per employee, the percentage of employees participating in sustainability training, or the reduction in overall carbon emissions. Additionally, conducting employee surveys or feedback sessions can provide valuable insights into perceptions of the organization's sustainability efforts and the impact on workplace culture. Moreover, utilizing advanced data analytics tools can enhance the ability to track and analyze sustainability performance over time. Organizations can leverage these insights to refine their strategies, make data-driven decisions, and demonstrate accountability to stakeholders. By fostering a culture of continuous improvement and aligning performance measurement with sustainability objectives, organizations can effectively navigate the complexities of measuring green performance and ensure the long-term success of their Green HRM initiatives.

6.5. Integration with Organizational Culture

Another challenge in implementing Green HRM lies in the integration of sustainability principles into the existing organizational culture. For Green HRM to be truly effective, it must become a core value embedded within the organization's culture, rather than merely a set of policies or initiatives. However, changing organizational culture is often a gradual and complex process, requiring sustained commitment from leadership and ongoing engagement from employees. To facilitate this cultural shift, organizations should prioritize creating a shared vision for sustainability that resonates with employees at all levels. Leadership must actively champion sustainability efforts, modeling eco-friendly behaviors and demonstrating the alignment between organizational values and sustainability goals. Additionally, organizations can foster a sense of community by encouraging employee involvement in sustainability initiatives—such as forming green teams, organizing eco-friendly events, or participating in community service projects focused on environmental causes. Recognizing and rewarding sustainable practices among employees can also reinforce the integration of Green HRM into the organizational culture. By acknowledging individual and team contributions to sustainability efforts, organizations can cultivate a sense of pride and ownership among employees, further embedding green values into the workplace. Ultimately, fostering a culture that prioritizes sustainability requires ongoing commitment, clear communication, and a willingness to adapt to evolving challenges and opportunities.

6.6. Keeping Up with Evolving Standards

As sustainability continues to evolve as a critical business priority, organizations must also grapple with the challenge of keeping up with changing standards and best practices. The landscape of sustainability is dynamic, influenced by advancements in technology, shifts in consumer preferences, and evolving regulatory frameworks. This constant flux can create uncertainty for organizations attempting to implement effective Green HRM practices, particularly if they lack the resources or expertise to stay informed about emerging trends. To mitigate this challenge, organizations should actively engage in industry research and collaboration to remain abreast of best practices in sustainability. Participating in industry forums, attending conferences, and networking with sustainability experts can provide valuable insights and access to the latest developments in Green HRM. Additionally, organizations can benefit from establishing partnerships with academic institutions, non-profit organizations, and sustainability-focused consultants to gain access to resources and expertise that inform their practices. Moreover, organizations should adopt a proactive approach to sustainability by continuously evaluating their strategies and seeking opportunities for improvement. Regularly reviewing and updating sustainability goals in response to changing standards can ensure that organizations remain relevant and effective in their Green HRM initiatives. By fostering a culture of adaptability and learning, organizations can better navigate the complexities of sustainability and position themselves for long-term success.

6.7. Balancing Short-Term and Long-Term Goals

Finally, organizations implementing Green HRM often face the challenge of balancing short-term operational goals with long-term sustainability objectives. The immediate demands of business operations can sometimes overshadow the importance of investing in sustainable practices, particularly in industries where profit margins are tight or competitive pressures are high. This tension can hinder the successful integration of Green HRM into organizational strategies, as decision-makers may prioritize short-term gains over long-term sustainability. To address this challenge, organizations must cultivate a holistic perspective that recognizes the interdependence of short-term and long-term goals. Leadership should emphasize the long-term benefits of sustainability initiatives—such as cost savings, enhanced brand reputation, and increased employee satisfaction—while also addressing the immediate needs of the business. Incorporating sustainability metrics into performance evaluations can help align operational goals with sustainability objectives, ensuring that employees at all levels are accountable for contributing to the organization's environmental goals. Additionally, organizations can leverage technology and innovation to identify opportunities that support both short-term and long-term objectives. For example, investing in energy-efficient technologies may require an initial expenditure but can yield significant cost savings over time, aligning immediate operational needs with sustainable practices. By fostering a mindset that embraces sustainability as a strategic imperative, organizations can effectively balance the dual pressures of short-term performance and long-term environmental responsibility.

Hence, while implementing Green HRM offers significant benefits, organizations must navigate a range of challenges to realize the full potential of their sustainability initiatives. By addressing issues such as lack of awareness, resistance to change, initial investment costs, and the complexities of measuring performance, organizations can create a supportive environment for Green HRM practices to thrive. Through effective communication, change management, ongoing training, and a commitment to embedding sustainability into organizational culture, businesses can successfully overcome these challenges and position themselves as leaders in sustainable practices.

7. THE FUTURE OF GREEN HRM

The concept of Green Human Resource Management (Green HRM) has increasingly gained momentum as environmental concerns and the need for sustainability continue to shape global business practices. As organizations move toward sustainable development goals, Green HRM will play an essential role in promoting environmental responsibility within the workforce and fostering practices that minimize ecological impact. The future of Green HRM lies in its ability to evolve beyond traditional HR functions to create a comprehensive approach that integrates environmental objectives into all facets of human resource management. This chapter explores how Green HRM is expected to evolve, the trends shaping its future, and its implications for businesses and employees alike.

7.1 Integration with Organizational Strategy

One significant trend in the future of Green HRM is its full integration into corporate strategy. Many organizations are beginning to realize that sustainability cannot be relegated to isolated programs; instead, it must be embedded in the corporate DNA. This shift will see Green HRM becoming a fundamental part of overall business strategies. Human Resource (HR) departments will collaborate closely with top management to align green initiatives with organizational goals. This approach allows companies to create a culture where every employee understands and values environmental stewardship, which in turn leads to more consistent and impactful results. As organizations set long-term sustainability goals, Green HRM will facilitate ongoing employee engagement in eco-friendly practices, making environmental responsibility a shared objective.

7.2 Expansion of Green Recruitment and Selection

Green recruitment and selection are expected to become even more critical as sustainability-minded organizations seek candidates who are not only skilled but also committed to environmental goals. The future of Green HRM in recruitment will involve more than just hiring employees who support green initiatives; it will involve attracting talent that brings innovative ideas for sustainability. Companies will likely emphasize environmental values in their employer branding and recruitment materials, ensuring they appeal to environmentally conscious job seekers. Additionally, the recruitment process may incorporate assessments that evaluate candidates' knowledge of and commitment to sustainability, including questions on eco-friendly practices. With Generation Z entering the workforce, organizations may find that candidates increasingly expect companies to uphold sustainable values, making green recruitment a necessary part of HR strategies.

7.3 Technological Innovations in Green HRM

The future of Green HRM will be significantly shaped by technological advancements. Digital tools will play a central role in reducing paper use, streamlining HR processes, and enabling remote work, all of which contribute to reducing an organization's carbon footprint. For instance, cloud-based HR management systems allow employees to access, complete, and submit documents digitally, reducing the need for paper-based processes. Virtual recruitment and training tools also minimize the environmental impact associated with travel. As technology advances, artificial intelligence (AI) and data analytics will further enhance Green HRM. AI can optimize energy use in smart buildings, predict employee behaviors related to sustainability, and help track individual contributions to green initiatives. Data analytics, meanwhile, enables organizations to measure the environmental impact of HR policies and optimize them for better results. These technological innovations not only reduce environmental impact but also improve the efficiency of HR operations.

7.4 Development of Green Training and Development Programs

Green Training and Development programs are expected to become even more comprehensive in the future. Beyond teaching employees eco-friendly practices, these programs will focus on instilling a deep sense of environmental responsibility that extends to personal life. Future green training may include workshops on sustainable decision-making, training in circular economy principles, and seminars on global environmental issues like climate change and biodiversity loss. Additionally, training programs may leverage virtual and augmented reality (VR and AR) to create immersive experiences that help employees visualize the impact of environmental actions. By providing more holistic green training, organizations can cultivate a workforce that feels personally invested in sustainability, fostering long-term commitment and engagement.

7.5 Enhanced Green Performance Management Systems

Green Performance Management, an emerging trend within Green HRM, will play a more prominent role in the future. Organizations are beginning to integrate sustainability metrics into performance appraisals, rewarding employees for their contributions to environmental goals. In the future, this approach will become more standardized, with HR departments developing robust frameworks to assess eco-friendly behaviors and achievements. Metrics might include energy conservation efforts, waste reduction initiatives, and participation in green programs. Moreover, performance management systems may use advanced data analytics to measure employees' environmental impact accurately. By rewarding eco-friendly behaviors and achievements, organizations reinforce the importance of sustainability within their culture, motivating employees to contribute actively to green initiatives.

7.6 Promotion of Sustainable Work Practices and Remote Work

The COVID-19 pandemic accelerated the adoption of remote work, which has proved to be an effective method for reducing carbon emissions and resource consumption. As businesses embrace remote and hybrid work models, Green HRM will play a crucial role in promoting sustainable work practices for remote employees. HR departments will need to develop guidelines that encourage eco-friendly home office setups, energy-efficient practices, and sustainable commuting options for hybrid workers. Furthermore, remote work eliminates the need for daily commutes and office resources, which significantly reduces an organization's environmental impact. In the future, Green HRM may involve providing employees with tools, resources, and incentives to create green home office environments, thereby extending sustainability practices beyond traditional workplaces.

7.7 Emphasis on Green Employee Engagement

Green employee engagement, another key area in Green HRM, is expected to see significant growth in the future. Engaging employees in sustainability initiatives fosters a sense of shared responsibility for environmental goals and builds a culture of collective action. In the future, organizations will focus on creating green engagement programs that include volunteer opportunities, environmental awareness campaigns, and rewards for eco-friendly contributions. Employee engagement strategies will be designed to inspire employees not only to participate in green initiatives but also to bring innovative sustainability ideas to the organization. By building a workforce committed to sustainability, organizations can leverage the creativity and enthusiasm of their employees to drive meaningful environmental change.

7.8 Stronger Alignment with Corporate Social Responsibility (CSR)

Green HRM will become increasingly aligned with Corporate Social Responsibility (CSR) frameworks as organizations recognize the need for a holistic approach to sustainability. CSR has traditionally encompassed a broad range of social, ethical, and environmental concerns, but environmental sustainability is now taking a more central role. Future Green HRM will work closely with CSR departments to ensure alignment between HR practices and CSR objectives. For example, HR departments may support CSR initiatives by promoting employee participation in community environmental projects, such as local cleanup drives or conservation efforts. The integration of Green HRM with CSR will strengthen organizations' commitments to social responsibility and enhance their reputation as environmentally and socially responsible employers.

7.9 Increased Focus on Global Standards and Compliance

As international bodies continue to develop regulations aimed at reducing environmental impact, organizations will need to comply with a growing number of global standards. Green HRM will be central to ensuring compliance with these standards, which may include emission reductions, waste management, and resource conservation. The future of Green HRM will involve close monitoring of regulatory changes, adjusting HR policies to meet new requirements, and educating employees about compliance standards. By embedding these standards into HR practices, organizations can ensure they remain compliant while promoting environmental sustainability.

7.10 Green HRM as a Key Driver of Sustainable Development Goals (SDGs)

Green HRM will play an essential role in advancing the United Nations Sustainable Development Goals (SDGs), particularly those related to climate action, responsible consumption, and life on land and underwater. HR departments will increasingly focus on aligning organizational practices with these goals, promoting responsible use of resources, reducing waste, and protecting ecosystems. For instance, initiatives that encourage employees to minimize resource consumption align with SDG 12 (Responsible Consumption and Production). Green HRM practices will be designed to support such goals, thereby contributing to global sustainability efforts and enhancing the organization's role as a responsible global citizen.

The future of Green HRM holds immense potential for transforming how organizations operate and engage with sustainability. As environmental concerns continue to intensify, Green HRM will become an indispensable part of organizational strategy, driving eco-friendly practices throughout all levels of the company. By integrating sustainability into recruitment, training, performance management, and employee engagement, Green HRM will foster a culture where employees feel connected to the company's environmental mission. Furthermore, advancements in technology, remote work, and global standards will enable organizations to streamline green HR practices, making it easier for businesses to track and achieve their environmental goals.

8. THE FUTURE OF GREEN HRM IN DEVELOPING AND DEVELOPED NATIONS

8.1 Technological Advancements and Digital Transformation

Technology will play a crucial role in advancing Green HRM. Digital platforms can improve remote work, reduce paper consumption, and facilitate efficient resource management, leading to eco-friendly workplaces in both developing and developed countries.

Green HR Technology

AI and data analytics can optimize energy use, track resource consumption, and streamline HR processes, reducing environmental impacts across sectors.

Remote and Flexible Work Arrangements

Remote work, facilitated by technology, can significantly reduce commuting-related carbon emissions, a practice that can be promoted through Green HRM.

Aligning Green HRM with Sustainable Development Goals (SDGs)

The United Nations' Sustainable Development Goals (SDGs) emphasize environmental sustainability, social responsibility, and economic growth. Green HRM initiatives can contribute to various SDGs, such as climate action, decent work, and responsible consumption.

Supporting SDG 8 (Decent Work)

Green HRM creates fair, sustainable employment opportunities, especially in developing nations with high unemployment rates.

Supporting SDG 13 (Climate Action)

Through energy-efficient practices and reduced resource consumption, Green HRM directly supports climate action goals in both developing and developed nations.

8.2 The Role of Policy and Regulation in Shaping Green HRM

Government policies will continue to influence Green HRM practices, particularly in developed nations with strict environmental standards. Developing nations may require more incentives and support to adopt Green HRM, such as subsidies for green training or tax breaks for eco-friendly practices.

Policy Incentives in Developing Nations

Providing financial support for green initiatives can make it easier for organizations in developing countries to adopt Green HRM.

Enforcement of Regulations in Developed Nations

Regulatory enforcement will ensure that organizations meet their environmental commitments, reinforcing the importance of Green HRM in these nations.

9. THE NEED FOR GREEN HRM IN DEVELOPING AND DEVELOPED NATIONS

As the global focus intensifies on climate change and environmental sustainability, Green Human Resource Management (Green HRM) emerges as a crucial approach for aligning organizational practices with ecological imperatives. Both developing and developed nations face distinct challenges and opportunities in implementing Green HRM, shaped by unique economic, social, and regulatory landscapes. For developing nations, the primary need lies in integrating Green HRM with growth and industrialization efforts, while for developed nations, the focus is on refining sustainable practices to mitigate long-standing environmental impacts. By understanding these varying needs, we can better appreciate Green HRM's role in fostering sustainable development globally.

9.1 The Need for Green HRM in Developing Nations

In the case of developing nations, rapid economic growth and industrialization has traditionally superseded environmental considerations. But when left unrestricted, industrialization causes extreme pollution, deforestation and depletion of ecosystems. Green HRM→ Mitigate Practices/ Policies that result in Exploration of natural resources and promote sustainable development without compromising the growth→ shelter for both aspects where growth aligns with sustainability. It promotes sustainable manufacturing, resource conservation, and waste reduction, enabling industries to meet environmental standards and drive economic growth. Also, becoming aware about sustainable practices is not widespread among many developing nations. Green Human Resource Management also contributes importantly to the awareness of employees about environmental practices to create a culture of sustainability in the organizations.

Not only does Green HRM encourage sustainability but it also makes a developing country more appealing to foreign investors and MNCs (multinational corporations) as the global supply chain is including sustainability in their guidelines to work with them in their environmental activities. Green HRM leads organizations to comply with such international regulations and environmental standards thereby increasing their competitive edge in the international market. Evidence of Green HRM practices helps companies to highlight their sustainability efforts; at the same time this helps improve the company branding image and reputation among international customers and partners. Furthermore, since the majority of the MNCs demand the suppliers to fulfill certain environmental standards, the local organizations now embrace a green HRM approach to minimize both risk of exclusion from the supply chain network as well as to ensure their future business sustainability.

And another major contribution of Green HRM is its ability to generate green jobs the high technology jobs which can directly reduce employment. By promoting sustainable practices, Green HRM helps in generating job opportunities in renewable energy, eco-tourism, and sustainable agriculture and aligning workforce development in accordance with environmental objectives. Green skills training allows workforces to transition industries towards the green-one. Furthermore, Green HRM can contribute to the diversification of an economy, as it can transition jobs from conventional sectors with high environmental footprints into more sustainable sectors, reducing pressure on sectors responsible for deterioration of the environment.

Green GRM is not only necessary for third world countries but also very much important for developed countries as large scale industrial bases which produce large volume of carbon and pollution. Green HRM in these countries has been directed to minimize the carbon footprint through the introduction of eco-friendly practices in the traditional organizations. Green HRM helps minimize the businesses' carbon footprint and improving operational efficiency by promoting practices related to energy efficiency, waste reduction, and recycling. Similar to the Paris Agreement are the stringent climate action targets developed nations have set. Human Resource Management must be concerned to work with organizations in achieving these goals by including environmental aims in HR policies and economic transactions.

Investing in sustainability increases the brand value and gives you a competitive edge while also providing social structure; Moreover, customers in developed countries are more aware of environmental pollution as well as environmental issues and tend to prefer brands embracing sustainable practices over them. Green HRM allows organizations to integrate sustainability into their culture and operations and align with these consumer expectations. While, developed countries have highly rigorous environment associated rules, Green HRM promotes compliance (ensure, having sustainability measures as an integral to organizational policies). Green HRM also improves the corporate reputation of these businesses and helps them attract environmentally conscious consumers and investors in a competitive market.

Perhaps most importantly, Green HRM goes beyond regulatory compliance and consumer expectations; it fosters a culture of innovation in the workplace, improving productivity and employee satisfaction. With the rise in emphasis on sustainability in the workplace brought about by employees in developed countries, focus on sustainable action in the workplace is ever-increasing, and Green HRM greatly aids in taking employee role aspirations more in line with the pre-existing environmental values of the employee. Organizations promote eco-innovation by encouraging creativity in sustainable practices in product design, supply chain management, and resource usage. Furthermore, employees who believe they're working towards a noble cause tend to be more motivated and engaged at work, which in turn enhances their performance and powers organizational results.

Therefore, Green HRM contributes to the achievement of environmental sustainability on a balanced basis with economic growth and competitiveness for developing and developed countries as well. By supporting sustainable industrialization, open market access, green job creation, compliance with regulations, and promoting innovation in workplace. With the increasing global focus on sustainability, organizations that are proactive in embedding Green HRM into their operations will not only be best placed to become long-term winners in their respective fields, but are also investing towards a healthier planet for generations to come.

10. COMPARATIVE ANALYSIS: GREEN HRM IN DEVELOPING VS. DEVELOPED NATIONS

In developed and developing countries, Green Human Resource Management (Green HRM) has an important role, but the functions and objectives of Green HRM are highly context driven based on economics, legislation, and social dynamics. Green HRM aims to reconcile economic development with environmental sustainability in developing countries. These burgeoning nations are commonly experiencing rapid industrialization, which results in resource depletion, pollution and ecological degradation. With the need for sustainable business practices coupled with progress in economic development, the adverse effects of Green HRM initiatives can be relieved. The poverty in underdeveloped countries, along with limited resources, serves as a barrier for businesses to implement sustainable business practices. While these are significant challenges, Green HRM also has considerable scope for jobs in green sectors like renewable energy, sustainable agriculture and Eco-tourism. By adopting sustainable HR practices, organizations can also attract foreign investors and multinational enterprises that tend to become environmentally responsible and prefer working with an eco-friendly partner. In these economies, the main focus of Green HRM is education and awareness programs to meet international environmental standards and compliance.

In developed nations, the focus is on minimizing carbon footprints and complying with regulations through Green HRM. To raise awareness about this in Germany: these countries already possess highly developed industrial infrastructures that generate significant CO2 emissions and have an adverse effect on the environment. As a result, developed economies' Green HRM focuses on emission reduction, energy efficiency and waste management to comply with stringent environmental laws and regulations. The biggest challenge that many businesses in developed countries need to overcome is high consumer expectations around corporate sustainability. Such consumers have promoted sustainability issues in these markets and are more likely to engage and partner with brands that properly reflect this commitment. Additionally, businesses face stringent government policies and regulatory frameworks, ensuring compliance becomes a complex and resource-consuming task. Conversely, sustainable practices like Green HRM deter the growth of pollution threatening civilization, it opens a gate to innovation as well. Organizations in developed countries are clearing the way for sustainability-enhanced corporate culture, green innovation, and HR practice that aligns with climate action goals.

Hence, although the relatively lower emphasis on Green HRM in developing countries shows its low priority status, the importance is that both developing and developed countries now realize the significance of Green HRM. In developing countries, efforts are aimed at building up awareness, training the labor population, and bringing in foreign investment to help give to sustainability endeavors. In developed nations, Green HRM is utilized to improve operational efficiency, obtain compliance with environmental regulations, and yield sustainable business models. However, Green HRM continues to play an important role in the impact of global environmental challenges and the long-term sustainability of economies.

Table 1.

Aspect	Developing Nations	Developed Nations
Primary Objective	Balancing growth with sustainability	Reducing carbon footprints and regulatory compliance
Challenges	Lack of awareness, resource constraints	Managing high consumer expectations, strict regulations
Opportunities	Job creation in green sectors, attracting foreign investment	Innovation in sustainability, brand differentiation
Key Focus	Education, awareness, compliance with international standards	Innovation, productivity, meeting climate action goals

11. GREEN HRM AND MANAGEMENT THEORIES

Green Human Resource Management (Green HRM) is a new concept in which environmental sustainability is considered in human resource policies and practices. It integrates HR functions — recruitment, training & development, performance management, and employee engagement — with sustainability initiatives, motivating organizations to reduce their carbon footprint. Different management theories serve as a compelling base for exploring and implementing Green HRM. Let's discuss green HRM from theory perspective, an analysis management theories for sustainable organizations.

11.1 Scientific Management Theory and Green HRM

Scientific Management Theory of Frederick Taylor focuses on efficiency, productivity, and optimization of operations. Although Taylor's principles focus on the specialization of labor and the standardization of processes, they can be extended to Green HRM through improved resource utilization and waste reduction. Green HRM is the best way to conduct a scientific management operational analysis in the industry, allowing organizations to see how their business can become ecofriendly by reducing wastage and introducing more green alternatives. For instance, businesses can invest in energy-efficient machinery, minimize paper usage by digitizing processes, and implement waste-reduction policies to improve sustainability. Time and motion studies, which is an integral part of Taylorism can also be utilized to optimize the task previously performed by employees, ensuring that the maximum effort and the least amount of time is wasted. But one of the downsides of scientific management is that it prioritizes output over workers' well-being. Green HRM eliminates this drawback by balancing efficiency enhancement with the well-being and sustainability of work life. When implementing automated systems, organizations need to be equally focused on how the employees permanently reduce energy while using the systems before the deployment, efficiently operating without job insecurity or stress post-deployment.

11.2 Human Relations Theory and Green HRM

Human Relations Theory — by Elton Mayo highlighted the Importance of Social Relationships, employee motivation, and workplace cultures. This theory states that employee job satisfaction and engagement lead to higher productivity. Green HRM complements this view by nurturing a culture of environmental consciousness, thus enhancing employee morale and commitment. When employees participate in sustainability initiatives — for example, green teams and corporate social responsibility (CSR) programs — they feel a sense of purpose and connection with the organization's mission. Green HRM practices require to involve participative decision-making, where employees take part in brainstorming to come up with ideas to enhance sustainability efforts. Organizations can host sustainability brainstorming events, for example, and ask staff to propose ideas on reducing energy usage, waste, or even ecologically sound travel alternatives. This participatory measure not only encourages employee satisfaction but also resonates with its Sustainability commitment further proving Human Relations Theory.

11.3 Green Human Resources Management: A Contingency Framework

Developed by scholars like Fred Fiedler, the Contingency Theory states that there is no one-size-fits-all approach to management. And instead build organizational strategies that are situational, environmental, and external. Since sustainability initiatives need to be customized according to the industry and geography of the organization as well as the expectations of stakeholders, this theory is particularly important for Green HRM. Green HRM of a manufacturing company involves pollution control, sustainable supply chain management, energy efficiency, etc. A tech company, for instance, may stress working from home, digital document processing and green data centers to cut its carbon emissions. To be effective, Green HRM must match an organization's operational context with its environmental strategies. In addition, the contingency approach states that the leadership styles should be adjusted according to the goals of the environment. Specific leadership styles, such as transformational leadership which inspires members of an organization to embrace change, may be particularly effective in creating a green culture. Green HRM toward strategic priority thus serves as an effective mechanism for promoting sustainable behavior in organizations of which leadership is a crucial precursor because leaders who are committed to sustainable practices have a prevalence towards their employees to engage in behaviors that promote sustainability.

11.4 Resource-Based View (RBV) and Green HRM

The Resource-Based View (RBV) of management, presented by Jay Barney, states that a firm's competitive edge comes from its unique resources and capabilities. Green HRM builds on this benefit by making employee's key drivers of sustainability. By incorporating sustainable practices into their HR frameworks, companies can attract eco-friendly talent, improve employee retention rates, and develop a reputation for being a socially responsible enterprise. Companies that offer green training programs, provide sustainability incentives, and create a culture of environmental awareness develop a workforce that is both skilled and motivated to drive eco-friendly initiatives. The human capital that results from this becomes a competitive advantage that sets the organization apart from others. RBV also emphasizes that proactive adoption of Green HRM helps to reduce regulatory risk, operational costs and access to green market which establishes competitive advantage.

11.5 Stakeholder Theory and Green HRM

Stakeholder Theory, postulated by Edward Freeman posits that a business must take into account the interests of all stakeholders — employees, customers, suppliers, governments, and the environment. There is a close relation of Green HRM with Stakeholder Theory statement which emphasizes the balance between business objectives and environmental as well as social determinants. It makes sure that HR policies consider the interest of various stakeholders including – government regulations regarding emissions, consumers' preference for sustainable products, and employees' expectations for an eco-friendly workplace. For example, a green HRM organization may work with suppliers to ensure sustainable sourcing, partner with NGOs for environmental initiatives, and bring in sustainability reporting to use with investors. Examining the wider implications of HR practices on sustainability enables companies to gain credibility and trust in the eyes of stakeholders over the long haul.

11.6 Systems Theory and Green HRM

Organizations are a system of systems (HR, operations, finance, marketing) which are very interrelated and require working together to achieve aggregate efficiency, in essence, the Systems Theory. Green HRM is in line with this theory by incorporating sustainability in all its processes. HR practices which lead to behaviors which affect production, supply chain and CSR. For instance, a company's human resources department might roll out a green performance management system to evaluate employees on their contribution to sustainability goals. And this has a direct influence on operational decisions, including the use of renewable sources of energy, waste reduction and sustainable product design. Focusing on sustainability as a system allows organizations to create a comprehensive approach to environmental stewardship.

11.7 Theory of Triple Bottom Line and Green HRM

The Triple Bottom Line or TBL theory was devised by John Elkington, and states that organizations should aim to achieve progress in three important areas: People, Planet and Profit. Green HRM directly reinforces this approach, ensuring that human resources policies are framed to ensure social well-being, environmental protection and economic growth. The triple bottom line aims to create a more sustainable world by focusing on the people, planet, and profit and Green HRM allows companies to balance these three metrics. Using the TBL framework, companies can maintain long-term success while also upholding corporate social responsibilities.

11. 8 AMO Theory and Green HRM

Green Human Resource Management (Green HRM), on the other hand, corporate activity of human resource management where businesses are motivated to consider sustainability in their operations, thus being in harmony with the environment. A Green HRM activity can be understood and adopted using different management theories. Classical theories, including Scientific Management, Human Relations Theory, and Contingency Theory, lay a foundational understanding, while modern frameworks like the Resource-Based View (RBV), Stakeholder Theory, and the Triple Bottom Line (TBL) approach guide strategic direction. The AMO Theory, in particular, is a widely studied perspective that explains how Green HRM practices influence employee performance and sustainability.

11.8.1 AMO Theory: What it is to the Context of Green HRM

AMO Theory is a popular framework for HRM, which posits that employee performance and engagement is dependent on three fundamental elements:

Ability (A): Employees need to have the prescribed skills, knowledge, and competencies to perform.
Motivation (M): Employees have to be prepped and incentivized to perform desired activities.
Opportunity (O): Organizations need to establish an environment where employees can apply their skills and drive.

The ability-motivation-opportunity (AMO) framework guides Green HRM by providing a structure for organizations to foster a sustainable workforce.

11.8.2 Competence: Training and Development for Growing Green Skills

The first part of AMO Theory—ability—concerns ensuring employees have the skills, knowledge, and training to engage in green practices. This is supported by Green HRM through green training and development programs to educate and upskill employees about the principles of sustainability, conservation of resources, and eco-friendly operations of a business. For instance, businesses can also implement training programs related to limiting waste, energy utilization, and sustainable procurement practices. For example, employees in manufacturing environments may be trained on eco-friendly material usage and energy-saving methods, while office-based employees may focus on digital documentation to reduce paper consumption.

These efforts help to equip employees to engage meaningfully in sustainability initiatives, thereby helping to foster Green HRM as well as supporting the organization broadly in its environmental sustainability objectives. Green initiatives also require review and setting for continuing leaders, prompting senior leaders to ensure organization-wide planner and supervisors with sustainability career development programs. This notion accords with Contingency Theory, holding that successful leadership strategies should be tailored to accommodate environmental and organizational requirements.

11.8.3 Motivation for Green Behavior through Awards

Motivation is a critical factor that drives employees' behavior, and Green HRM employs several methods to motivate employees to become actively involved in the sustainability initiatives. The Green HRM motivation is intrinsic and extrinsic (motivational) and is consistent with human relations theory and stakeholder theory in terms of employee enthusiasm and organizations as responsible parties. Instilling a sense of personal connection to environmental values, allowing for sustainability to be driven by intrinsic motivation rather than compliance. To encourage intrinsic motivation, organizations can make sustainability core to the company's culture and values. When companies are transparent about their environmental commitments and are willing to involve employees in their decision-making processes about sustainability policies, employees take ownership and pride in contributing to green initiatives, for instance, as consumers.

Green HRM also employs external rewards to encourage environmentally friendly actions. Performance management systems can comprise sustainability-related Key Performance Indicators (KPIs), while employees delivering noteworthy contributions to eco-friendly initiatives can be rewarded with bonuses, recognitions, or promotions. Certain companies have also created "Green Employee of the Month" awards or include paid time off for working on community environmental projects. And such incentive programs fit with Triple Bottom Line (TBL) Theory. Green HRM further adopts nudges or minimal green reminders in workplace towards reinforcing the efforts. Sustainable nudges: For example, making recycling bins highly visible, having default sets for double-sided printing, and displaying energy price stickers can help encourage workers to practice sustainable behaviors.

11.8.4 Opportunity: Building a Supportive Environment for Green Practices

The third part of AMO Theory—opportunity—centers on providing employees with the resources, support, and conditions necessary at work to maintain sustainable behavior. Green HRM is essential in providing a green environment through the establishment of green policies, the sustainable culture in the organization, and the green action in the workplace. One way organizations give employees the opportunity to contribute to sustainability is through green workspaces. Workspaces can be re-designed to include energy-efficient lighting, better ventilation, and sustainable materials. Policies to work remotely also cut emissions from commuting, echoing SDG 11 (Sustainable Cities and Communities) and SDG 13 (Climate Action). Establishing employee-led green initiatives is another critical component of creating opportunity. Companies can create Green Teams or sustainability committees that allow employees to work together on green initiatives. Employees can, for instance, suggest a solution to plastic waste reduction, initiate a car-pool program, or organize a tree-planting event. Such a participatory approach can be attributed to the Systems Theory perspective, which refers to organizations as interconnected bodies through which departments collaborate to achieve sustainability outcomes. Furthermore, HR departments may implement Green HR Policies which include flexible work arrangements, sustainable procurement guidelines and partnerships with environmentally compatible suppliers. If employees notice sustainability ingrained in organizational policies, they become proactive to incorporate green practices into their daily work.

Thus, Green HRM is closely inter-related with management theories that offer organizations a multiplicity of paradigms to implement sustainability practices. Including everything from Scientific Management (for efficiency) and Human Relations Theory (engaging employees) through Contingency Theory (external factors) to Stakeholder and Triple Bottom Line Theory (long-term value), Green HRM also offers a high level guide on sustainable business practices. As our global ecological footprint threatens our collective future, organizations need to begin embracing management theories and frameworks such as these in order to create innovative Green HRM systems that achieve the balance between successful business practices and responsible ecological sustainability. Through the convergence of Green HRM and established management principles, companies can adopt a greener, more sustainable future.

12. GREEN HRM AND SUSTAINABLE DEVELOPMENT GOALS (SDGS)

In response to the pressing challenges of climate change and environmental degradation, Green Human Resource Management (Green HRM) has gained prominence in contemporary organizations by integrating environmental sustainability into human resource practices. These can range from implementation of several green initiatives at an organization and introducing eco-friendly policies into the HR functions—recruitment, orientation, training, performance management, employee engagement, etc. Green HRM's main aim is to reduce the ecological footprint of the enterprise & sustainable development. This is vital in order to achieve United Nations Sustainable Development Goals (SDGs) on environmental sustainability, decent work and economic growth.

Green recruitment selection is one of the fundamental components of Green HRM when organizations search for candidates who are environmental minded and committed to sustainability. Environmental criteria should be included in job descriptions and selection processes to align employees with the company's vision for sustainable development through this process. Employers can cultivate greener workforce by bringing in candidates with the right green mindset. In fact, more companies are now also performing their interviews virtually and their onboarding digitally, helping to cut down on travel to and paperwork to be printed — contributing to environmental conservation.

Green training and development, subsequent to this, is another behavioral aspect of Green HRM through which it contributes toward the global agenda of the SDGs. Workers participate in training programs where they learn about environmental policies, energy conservation, waste management, and sustainable resource utilization. These training programs enable employees to understand the part they play in minimizing the carbon footprint of the organization, and encourage them to internalize green habits at work and at home. This is a contribution towards Sustainable Development Goal 4 (Quality Education), as it raises awareness and knowledge of sustainability issues amongst employees and helps them to make informed decisions about protecting the environment.

Performance management is another key feature of the Green HRM goal to reach sustainability goals. Central monitoring of these actions is then created through organizational key performance indicators (KPIs) by the organizations, monitoring measurable environmental impacts in energy efficiency, waste and raw material, and the sustainable practices of their suppliers. This is used to assess employees against their contribution to green initiatives and reward or recognize them through programs that encourage them to act in a sustainable manner. Aligning performance management with environmental objectives fosters a culture of accountability and continuous improvement in sustainability efforts in organizations, directly linking to SDG 8 (Decent Work and Economic Growth) and SDG 12 (Responsible Consumption and Production).

The business aspect of Green HRM is the engagement of employees in sustainability initiatives. Numerous organizations form "Green Teams" or sustainability committees to initiate eco-friendly projects that employees can engage alongside like tree-plantation, recycling-drives and energy-saving campaigns. Through these recent initiatives, you can feel a sense of responsibility and collective commitment to help train communities and countries to protect the environment, further reinforcing SDG 13 (Climate Action). In addition, companies, relying on adjustments such as remote work and teleworking, support the reduction of carbon emissions that would be emitted by daily commuting for work, thus fostering SDG 11: Sustainable Cities and Communities.

In sustainability efforts, Green HRM also plays a crucial role in encouraging diversity and inclusion. More organizations are catching on to the need for incorporating multiple viewpoints within environmental decision-making. Inclusive workplace culture attracts a diverse workforce of individuals who bring unique skills and ideas that can contribute to innovative solutions to sustainability challenges. By making it accessible all employees, irrespective of their background, can do their part, contributing to SDG 10 (Reduced Inequalities).

Additionally, Green HRM promotes sustainable supply chain practices by requiring organizations to pay attention to the environmental practices of their supplier vendors. The introduction of sustainable procurement policies which prioritize eco-friendly materials, energy-efficient products, and ethical labor practices. Integrating sustainable practices within supply chain is aligned to SDG 9 these impacts in greater economies such as SDG 12 for Responsible Consumption and Production making the environment sustainable along high economic levels.

The development of the Green HRM is inextricably tied to technological improvements that bolster sustainability initiatives. Digital transformation has led organizations to go paperless, automate energy management systems and introduce artificial intelligence (AI) solutions to track environmental performance. For instance, AI-powered analytics can support companies to streamline energy use and minimize waste, facilitating SDG 7 (Affordable and Clean Energy). HR departments are also using technology to facilitate remote training programs, monitor employee engagement with green initiatives, and evaluate sustainability program effectiveness.

With the growing concern over environmental issues around the globe, it is expected to see new evolution for point of Green HRM in upcoming future. For instance, sustainability must be ingrained as a core business strategy for organizations, not as a nice-to-have initiative. HR professionals will be instrumental in embedding sustainability into corporate culture; they will ensure that employees from the board of directors to the janitors understand their role in protecting the environment. Furthermore, organizations will have to tap into new approaches to implement circular economy practices into the HR stream like encouraging job profiles that focus on sustainability and providing green benefit options in the compensation plans.

In essence, Green HRM is an important concept to connect human resource management with environmental sustainability and SDG for the broader framework. Inclusion of sustainability in recruitment, training, performance management, and employee engagement enables the organizations to contribute towards sustainability while enhancing their global competitiveness.

13. CONCLUSION

Green Human Resource Management (Green HRM) is an essential aspect of sustainable HRM, equipping organizations with effective strategies to align their human resource practices with environmental goals. By incorporating environmentally conscious principles across HR functions—from recruitment and training to performance management and employee engagement—Green HRM fosters a workplace culture deeply committed to sustainability and social responsibility. This approach not only enables organizations to minimize their environmental impact, but also to cultivate an organizational identity that resonates with eco-conscious employees, customers, and partners. As organizations worldwide face mounting environmental challenges, Green HRM's role is becoming indispensable in helping companies achieve their sustainability targets and secure long-term success. Through Green HRM, businesses actively promote environmental responsibility among employees, encouraging them to engage in practices that reduce waste, conserve energy, and support broader organizational goals. For instance, green recruitment prioritizes hiring individuals who value environmental sustainability, while green training equips employees with the skills to adopt eco-friendly work practices, such as energy conservation and recycling initiatives. Green performance management further reinforces these practices by evaluating and rewarding employees based on their contributions to sustainability efforts, thereby enhancing both individual commitment and organizational accountability.

Beyond immediate workplace benefits, Green HRM significantly advances broader social and environmental objectives, aligning closely with initiatives such as the United Nations Sustainable Development Goals (SDGs), Corporate Social Responsibility (CSR), and compliance with evolving environmental regulations. Organizations that adopt Green HRM practices can make meaningful progress toward the SDGs by promoting sustainable economic growth, reducing inequality, and fostering responsible consumption and production. This commitment also strengthens a company's CSR profile, demonstrating a genuine dedication to addressing global environmental issues, which is increasingly demanded by stakeholders and consumers alike. Additionally, as regulatory bodies around the world implement stricter environmental policies, Green HRM enables companies to stay ahead of compliance requirements, reducing risks associated with non-compliance and reinforcing their reputation as responsible corporate citizens.

In today's business landscape, where sustainability has become a key differentiator, Green HRM offers organizations the chance to attract and retain top talent who prioritize environmental ethics and are eager to work for companies with a clear ecological commitment. Moreover, by embedding green values into the core of HR functions, organizations can inspire employees to engage in meaningful environmental change, transforming the workplace into a source of positive impact. As awareness of the benefits of sustainable practices grows, Green HRM will play a pivotal role in shaping the future of business success—one where financial achievement is balanced with contributions to a healthier planet and a more equitable society. Through Green HRM, organizations can foster long-term growth that not only benefits the bottom line, but also promotes a legacy of environmental stewardship, aligning corporate success with the well-being of the world.

REFERENCES

Ahmed, T., Khan, M. S., Thitivesa, D., Siraphatthada, Y., & Phumdara, T. (2020). Impact of employees engagement and knowledge sharing on organizational performance: Study of HR challenges in COVID-19 pandemic. *Human Systems Management, 39*(4), 589–601. DOI: 10.3233/HSM-201052

Ahmed, U., Umrani, W. A., Yousaf, A., Siddiqui, M. A., & Pahi, M. H. (2021). Developing faithful stewardship for environment through green HRM. *International Journal of Contemporary Hospitality Management, 33*(10), 3115–3133. DOI: 10.1108/IJCHM-09-2020-1066

Al-Hawari, M. A., Quratulain, S., & Melhem, S. B. (2021). How and when frontline employees' environmental values influence their green creativity? Examining the role of perceived work meaningfulness and green HRM practices. *Journal of Cleaner Production, 310*(August 2020), 127598-127598. https://doi.org/DOI: 10.1016/j.jclepro.2021.127598

Alam, M. A., & Niu, X. (2021). Impact of GHRM on organization's … Impact of GHRM on Organization's Environmental Performance: Mediating Role of Green Employee Empowerment. *The Journal of Business, 06*(01), 6–6. DOI: 10.18533/job.v6i2.202

Ali Ababneh, O. M., Awwad, A. S., & Abu-Haija, A. (2021). The association between green human resources practices and employee engagement with environmental initiatives in hotels: The moderation effect of perceived transformational leadership. *Journal of Human Resources in Hospitality & Tourism, 20*(3), 390–416. DOI: 10.1080/15332845.2021.1923918

Amjad, F., Abbas, W., Zia-Ur-Rehman, M., Baig, S. A., Hashim, M., Khan, A., & Rehman, H. (2021). Effect of green human resource management practices on organizational sustainability: The mediating role of environmental and employee performance. *Environmental Science and Pollution Research International, 28*(22), 28191–28206. DOI: 10.1007/s11356-020-11307-9 PMID: 33527245

Anwar, N., Nik Mahmood, N. H., Yusliza, M. Y., Ramayah, T., Noor Faezah, J., & Khalid, W. (2020). Green Human Resource Management for organisational citizenship behaviour towards the environment and environmental performance on a university campus. *Journal of Cleaner Production, 256*, 120401–120401. DOI: 10.1016/j.jclepro.2020.120401

Bansal, P., & Roth, K. (2000). Why companies go green: A model of ecological responsiveness. *Academy of Management Journal, 43*(4), 717–736. DOI: 10.2307/1556363

Brundtland, G. H., & Mansour, K. (2010). World Commission on Environment and Development (WCED). 1987. *Our common future*, 2010.

Carroll, A. B. (1999). Corporate social responsibility: Evolution of a definitional construct. *Business & Society*, *38*(3), 268–295. DOI: 10.1177/000765039903800303

Cooke, F. L., Schuler, R., & Varma, A. (2020). Human resource management research and practice in Asia: Past, present and future. *Human Resource Management Review*, *30*(4), 100778–100778. DOI: 10.1016/j.hrmr.2020.100778

Jabbour, C. J. C., & Santos, F. C. A. (2008). The central role of human resource management in the search for sustainable organizations. *International Journal of Human Resource Management*, *19*(12), 2133–2154. DOI: 10.1080/09585190802479389

Jackson, S. E., Renwick, D. W., Jabbour, C. J., & Muller-Camen, M. (2011). State-of-the-art and future directions for green human resource management: Introduction to the special issue. *German Journal of Human Resource Management*, *25*(2), 99–116. DOI: 10.1177/239700221102500203

Milliman, J., & Clair, J. (2017). Best environmental HRM practices in the US. In *Greening people* (pp. 49–73). Routledge. DOI: 10.4324/9781351283045-2

Renwick, D. W., Redman, T., & Maguire, S. (2013). Green human resource management: A review and research agenda. *International Journal of Management Reviews*, *15*(1), 1–14. DOI: 10.1111/j.1468-2370.2011.00328.x

ud Din, M., & Ahmad, A. (2023). Environmental Sustainability Through Green Human Resource Management Practices: An Analyses Of Industries Of Lahore, Pakistan. *Journal of Positive School Psychology*, 415-426.

ud Din, M., Ahmad, A., Samad, S., Lee, S., & Han, H. (2025). Sowing the seeds of sustainability: A sociological exploration of environmental sustainability within SME sector. *Climate Services, 37*, 100533.

Wehrmeyer, W. (2017). *Greening people: Human resources and environmental management*. Routledge. DOI: 10.4324/9781351283045

Chapter 4
Green Hydrogen:
The Next Frontier in Clean Energy

Syed Zain Ul Abedeen
Independent Researcher, Pakistan

Laeeq Razzak Janjua
https://orcid.org/0000-0002-1948-5859
WSB University, Poland

Azeem Razzak
Technical University of Munich, Germany

ABSTRACT

The complexity of global environmental processes and the necessity for sustainable development make green hydrogen a key technology for decarbonization. The chapter begins with a synthesis of literature on the properties of green hydrogen production. In addition, the argument compares PEM and Alkaline Water Electrolyzers and discusses recent initiatives to improve performance and lower costs. The chapter weaves these perspectives into a coherent narrative that suggests green hydrogen may be the key to a successful energy transition. The chapter informs researchers, policymakers, and industry actors about green hydrogen's technological, economic, and regulatory framework, helping to build a clean, secure global energy system. This extensive study examines green hydrogen capacity and future enhancements that could change global energy strategies. Thus, further research and collaboration are needed to overcome these obstacles and realize the sustainable future of green hydrogen.

DOI: 10.4018/979-8-3373-0139-6.ch004

1. INTRODUCTION

What we are facing today is simultaneous energy and environmental challenges, and there has never been a greater demand for new and sustainable solutions (D'Agostino et al., 2024). The global energy industry, which has long been based on the consumption of hydrocarbons, is one of the biggest sources of carbon emissions, which threaten the global climate, destabilize the Earth's ecosystems, economies, and harm human health (Chu et al., 2024). The shift towards the green energy systems cannot be viewed anymore simply as a strategic vision but has become the requirement to respond to the global demand for power and at the same time need to decrease the carbon emissions. Described as hydrogen fuel generated through water electrolysis employing renewable power sources, green hydrogen is one of the biggest advancements mankind has made towards achieving a clean energy economy per Hassan and colleagues (2024a). In contrast, using fossil fuels to generate hydrogen emits a large amount of carbon dioxide which is not desirable in the current outlook to reduce global emissions, green hydrogen, on the other hand, as the name suggests, has no emissions other than oxygen.

The potential of green hydrogen is not just as a green fuel, as many believe. Due to the flexibility green hydrogen has the possibility to change multiple, industry (Marzouk, 2024). It can be stored and transshipped and converted back to electricity or reacted with other compounds to produce synthetic fuels which makes it ideal in energy systems where high-density fuels are required (Eibeck et al., 2024). Green hydrogen addressing one of most challenging aspects renewable energy: variability. The solar and wind energy which currently dominate the renewable power sources present frequent and formidable challenges for energy reliability and stabilities of the grid (Jaradat et al., 2024). Green hydrogen is an effective remedy as it acts as a storage and supply mechanism of energy when renewable energy sources are not producing electricity.

Green hydrogen is an important factor in the phase of decarbonization in several ways. In eliminating carbon in energy sources and using them in industrial processes, it is capable of lowering greenhouse gas emissions as observed by (Guven, 2024). Louder industries like steel, refining or chemical industries that have been using fossils fuels can adopt green hydrogen to reduce their emissions while at the same time improving or even enhancing their productivity. The push for a hydrogen economy also contains great potential for economic gain (Bade & Tomomewo, 2024). To promote job creation, technological advancement, and placement of nations at the apex of a newly developing green market, there should be investment in green hydrogen manufacturing and transit ways (Koneczna and Cader, 2024). But achieving this potential depends on various factors such as the help of policy makers, industry stake holders and scientist to overcome the challenges that are both technical and economic. For the shift to a green hydrogen economy, integrated frameworks including innovation, policies, subsidies, and multilateral cooperation are required (Hassan, et al., 2024a). These strategies must be designed to improve the cost competiveness of green hydrogen, establish viable supply chain network for green hydrogen, and build market environment that would support widespread deployment of hydrogen solutions (Bade et al., 2024). Therefore, green hydrogen remains one of the most important technologies that give the world a new promising future, between the availability of renewable energy sources and the need for clean energy production. Its prospective position in the global energy system is set for a considerable growth to provide the Demonstrated Pathway towards the acute negation of emissions and reaching the concept of sustainable energy (Salahi, 2024). Consequently, green hydrogen is not only an element of the energy mix but can be considered the key to the transition to a new level of ecological consciousness and development.

The research work provide answer of two major research questions – initially - How can the integration of green hydrogen into renewable energy systems address the intermittency challenge, particularly in regions with high renewable energy generation variability? and furthermore, What are the key economic and policy barriers to scaling up green hydrogen production and how can emerging business models and international cooperation help overcome these challenges? Similarly, comprehensive research on the incorporation of green hydrogen into the energy systems of developing countries where renewable energy use is still in early stages appears to be lacking . Although the difficulties and technological breakthroughs in developed countries like Denmark and Germany have been extensively studied, there is little study on how these technologies and business models might be modified for developing nations lacking the infrastructure or legal frameworks for large-scale green hydrogen generation and use. Likewise, understanding how these nations can benefit from and contribute to the global hydrogen economy remains largely unexplored. The novelty of this research is based on emphasis on addressing the technological, economic, and regulatory disparities hindering green hydrogen adoption in emerging economies, where infrastructure, financial assistance, and international cooperation may be crucial for expediting the transition. By synthesizing insights from case studies in established countries with innovative tactics designed for developing markets, the research can provide a distinctive viewpoint on international collaboration and customized policy frameworks that can promote the worldwide adoption of green hydrogen. The research may also investigate creative business models that are economically sustainable and scalable across many economic situations, thereby enhancing the present comprehension of green hydrogen's significance in the global energy transition.

2. LITERATURE REVIEW

2.1 Green Hydrogen Production Technologies

Its contribution to green hydrogen generation has been given much attention as part of the plan to green energy production (Du et al., 2024). Electrolysis is the process that uses electric current to cause a substance to decompose into two or more substances when dissolved in water with an electrolyte (Kumar et al., 2024). The general efficiency of hydrogen production by means of electrolysis depends on the type of the electrolyzer, the electricity used, and the conditions under which the process is conducted. The conversion of electricity into hydrogen is normally done with an efficiency of between 60%–80%. To make the hydrogen generated 'green', the options for using renewable energy sources in powering electrolyzers become paramount according (Fam & Fam, 2024). Technological advancements of energies like solar and wind, make green hydrogen production by means of electrolysis financially viable and make use of more effective as a strategy in the foundation of a low-carbon energy system. PEM electrolyizers work on the basis of a solid polymer electrolyte and needs only water free from impurities and electricity. They are ideal for application in systems where intermittent renewable energy sources are incorporated, because of their capability to provide rapid dynamic response and relatively high current density (Delbeke, 2024). The fast response of PEM electrolyzers to changes in power inputs, characteristic of renewable sources such as solar and wind power, means that the PEM electrolyzers can capture the renewable energy and turn it into hydrogen as much as possible without wasting the potential inputs (Pashchenko, 2024). Recent developments of PEM electrolyzers have shown that the efficiency of electrolyte membranes in PEM electrolyzers has been enhanced and that the need for expensive catalyst materials such as platinum, used in the electrolysis reaction, have been minimized (Lehtonen, 2024). Such developments have played a role in cost reduction of PEM electrolyzers and hence their economic viability in large scale usage. However, the high standard of purity necessary for the water used on the membranes is the prime issue that needs to be solved in order to bring the membranes to the next level of utilization. Liquid electrolytes based on potassium or sodium hydroxide are used as electrolytes in alkaline water electrolyzers, and such electrolyzers are considered to be among the oldest (Pal et al., 2024). Characteristics of this type include high durability and the usage of non-precious metal catalysts making them cheaper than PEM electrolyzers (Münster et al., 2024). Alkaline electrolyzers operate at a larger scale and industrial hydrogen production has been done using this technology for the past four decades. The efficiency of the alkaline electrolyzers tends to be inferior to that of PEM electrolyzers, especially at lower loads (Abdolmaleki & Berardi, 2024). Its stability to changes in electricity supply is also slightly less impressive, possibly a disadvantage in applications where a more rapid response to fluctuations in power from renewable sources is needed (Béres et al., 2024). However, research and development are still ongoing to improve the performance and have targeted the flexibility of the alkaline electrolyzers so as to compete effectively in a wider range of applications in the green hydrogen economy.

Solid oxide electrolyzers are currently at the forefront of electrolysis solutions and require high operating temperatures (ranging between 700°C and 1,000°C). Such a high operation temperature opens up an opportunity for SOEs to attain optimal efficiency of higher than 90% under top conditions (Venizelou & Poullikkas, 2024). The high temperatures allow the utilization of heat for the electrolysis process directly (this heat can come from waste industrial heat or concentrated solar power).

The economic impacts of SOEs are enormous since SOEs are normally highly efficient in the use of the resources they employ, even if their construction and operating costs may be comparatively higher (Abdin, 2024). As well, the compatibility of SOEs with thermal energy sources makes them useful in industries which have access to high temperature heat and are likely to consume hydrogen (Hassan, & Al-Jiboory, 2024b). However, they suffer from the drawbacks of high operating temperatures with issues on material degradation and life as a major area of current research and development. The decision on which kind of electrolyzer to use has consequences not only in terms of hydrogen production costs but also in the interconnection and "embeddability" of green hydrogen in RE systems (Peters et al., 2024). Over the time, as other research goes on, the economic and technical viability of these technologies is anticipated to enhance making green hydrogen common in the energy mix strategies for attaining sustainable and de-carbonized world (Saha et al., 2024). The progression path of each of the technologies will therefore be instrumental in defining the pace and efficiency to which green hydrogen can respond to the energy requirements of the globe while mitigating for climate change.

2.2 Integration Of Green Hydrogen In Renewable Energy Systems

Thus, the role of green hydrogen does not end with the perspective of using it as a clean fuel; it also provides reliable means for increasing the reliability and efficiency of renewable power systems. This integration addresses one of the significant challenges in the renewable energy sector: the variability of the sources such as the solar and wind power (Khan et al., 2024). This section explains, with examples, how green hydrogen is being used to manage these fluctuation and level out energy grids to ease the transition to a cleaner energy mix.

While solar and wind power is the most important component in the global move towards decarbonization (Qasem & Abdulrahman, 2024). However, their reliance on weather conditions, wind turbines have uncertainty factor in power generation and raises issues of grid stability in electrical networks (Gunathilake et al., 2024). When output is high, some energy is used without being needed while when output is low, there is inadequate power to meet the required demand. To these challenges, green hydrogen offers an opportunity as it acts as an energy storage mechanism.

Green hydrogen is also good as an energy vector, where the excess of generation can be stored and used where needed. It can also be converted back to electricity at some later time of the day or night or during some periods of low production and thereby guarantee a continuous power supply to the people (Göthel et al., 2024). The process normally involves production of hydrogen through the process of electrolysis of excess renewable electricity. It is then stored under high pressure in tanks or in cavities beneath the ground surface or at some pressure (Sage et al., 2024). In a manner whenever wanted the stored hydrogen can be burnt to produce heat or power or utilized in fuel cells to develop electricity without emission.

The Danish "Hydrogen Islands" project is one of the first experiments in linking hydrogen with green energy. In this project, wind turbines are employed in the generation of power that is used to conduct electrolysis to produce hydrogen, (Dauletbay, 2024). This hydrogen is not only used for local power generation but also consumed directly in transport and industries making the island to become a closed cycle energy hub (Hassan et al., 2024d). This model shows that it is possible to apply green hydrogen in the storage of excess wind energy especially during off-peak times.

In Neunburg vorm Wald, a project integrates photovoltaic solar energy together with wind energy for generating green hydrogen (Revinova et al., 2024). This electrolysis plant here is flexible and capable to adjusting to different power sources and any excess energy is turned into hydrogen instantaneously (Goel, 2024). This hydrogen is then used to fuel municipal vehicles and other local industries outlining the real-life application of green hydrogen in the improvement of flexibility and versatility of renewable energy systems.

The use of green hydrogen in regulating the electrical networks is essential, especially due to the growing share of renewables in the energy mix. The temporal storage of hydrogen can be used to meet the supply and demand of the grid (Martinho & Rodrigues, 2024). For example, low electricity demand and high renewable generation may be utilized in the production of hydrogen. On the other hand, during high demand or when there is little renewable power available, the stored hydrogen reconverted into electricity (Geisbauer & Berchthold, 2024).

Similarly, the use of green hydrogen in renewable energy systems also has very massive economic and environmental impacts. In economic practice, it enables the proper utilization of renewable assets by using energy that would have been wasted because of curtailment, (Chu et al., 2024). On environment it saves the use of fossil products in energy storage and balancing of the grids helping to reduce the greenhouse effect and thus increase the sustainability of energy systems.

Promoting storage and stabilizing the grid, green hydrogen contributes to the improvement of and wind well to increased use of such sources (Hassan, et al., 2024a). The examples of case studies and green hydrogen applications described in the paper provide testament to the real-world application of green hydrogen, the viability of its use in a future energy system, and the need to proceed cautiously in its development (Fredershausen et al., 2024). To fully unlock the potential of these systems across the world more investment and research needs to be done in the enhancement and expansion of this integration.

2.3 Market Dynamics and Economic Drivers

The emerging market of green hydrogen is influenced by policy actors, economic factors, and new forms of business arrangements. All these elements will collectively technologies guide sustainable resilient energy sector transformation (Marzouk, 2024). This section focuses on these dynamics, policy, economic incentives, and emerging new business models, that are driving green hydrogen towards the energy transition.

The insertion of matrix depends, to large extent, on policies and regulations (Jaradat et al., 2024). National governments across the globe are realising the potential of hydrogen as a key enabler to help meet their net zero targets and are thus, putting in place policies to encourage hydrogen investment and utilization (Wolf et al., 2024).

In European Union (EU), this strategy paints large investments in green hydrogen technologies and targets to generate up to one million tonnes of renewable hydrogen by (Koneczna & Cader, 2024). The EU policy response is supported by the Fit for 55 packages whose aim is to ensure that all sectoral policies are aligned to the 2030 target of at least a 55% reduction in greenhouse gas emissions.

Similarly in terms of Japan, the country that famously adopted the Basic Hydrogen Strategy in 2017 aims at making hydrogen society by 2030 (Salahi, 2024). The plan includes the scale subsidy for vehicles fueling and investment in international hydrogen supply chain which show overall national policy for hydrogen as one of the energies in the nation. Australia's National Hydrogen Strategy outlines the become hydrogen export. Quoting from Guven (2024), Australia with its abundant renewable resources is focusing on becoming a hub for hydrogen power and provide the necessary policies that would foster local as well as foreign investments in the industry (Marzouk, 2024). These regulatory frameworks are central to providing direction on green hydrogen development by providing the research direction, technology implementation and infrastructure growth while at the same time addressing regulatory concerns and safety issues that may arise in the course of development (Arias Calvo et al., 2024).

The economic landscape for green hydrogen is characterized by several key drivers:

- **Carbon Pricing:** According to Du and his colleagues (2024), price signals from governments to place a premium on carbon emissions increase the cost advantage of green hydrogen over conventional fuels (Pothen & Hundt, 2024). International carbon trading and carbon taxes as a system encourage industries to shift to a low carbon economy through the use of cleaner fuel like hydrogen.
- **Subsidies and Financial Incentives:** Early-stage technology requires direct subsidies, tax credits and grants, among others, to assist in kick-starting these technologies (Frieden & Leker, 2024).
- **Investment from Public Private Sectors:** Gross investments are entering the hydrogen sector from various public and private investors (Lehtonen, 2024). These investments are not only increasing research and development activities but are also growing the pace of deploying hydrogen technologies and building the necessary structures.

The market for green hydrogen is also seeing the emergence of innovative business models that promise to enhance its viability and integration into the energy system:

- **Hydrogen as a Service (HaaS):** In this model, hydrogen supply and refueling are offered on a subscription or on-demand, thus solving a problem that is key for customers, which is the high initial investment in hydrogen (Xie et al., 2024).
- **Power-to-Gas (P2G):** P2G technologies are used to transform electrical energy to hydrogen that has the flexibility to blend with natural gas and distributed through the current gas networks (Delbeke, 2024).

The formation of the green hydrogen market is based on a complex system of policies and economic stimuli, as well as business solutions (Munir et al., 2024). When increasing carbon emission reduction targets, green hydrogen is likely to experience a rise based on strategic investments and favourable legal structures (Münster et al., 2024). Analyzing such market environment and economic factors is helpful for the parties interested in policy making, energy generation and innovation around green hydrogen market to harness its potential. By so doing the energy sector will be able to transform to a new more sustainable, efficient as well as a resilient power system thus facilitating a new age of green power.

3. DISCUSSION

3.1 Challenges in Scaling Up Green Hydrogen

While green hydrogen is a good candidate for a clean energy carrier there are numerous barriers that stand in the way of widespread adoption of green hydrogen and its scaling (Heidaria & Eshtiaghib, 2024). These are; technical challenges in production and application, transportation and handling challenges in the storage system, and safety considerations due to the inherent nature of the hydrogen element (Béres et al., 2024). Combination of technology, policy and international cooperation is needed to overcome these hurdles. These issues are discussed in this section and possible strategies on how to tackle these challenges are discussed below (see table 1.1)

Table 1. Challenger of scaling up the green hydrogen

Challenge	Description	Proposed Solution
Transportation (Cost & Logistics)	Hydrogen's low density makes it challenging to transport	Development of hydrogen carriers (ammonia, LOHC), pipelines
Storage (Density & Cost)	Current storage methods are inefficient and expensive	Advancements in high-density storage materials
Safety (Leaks & Flammability)	Hydrogen is highly flammable and prone to leakage	Improved leak detection systems, robust infrastructure

A-**Technical and logistical Challenges**

1. **Energy Efficiency of Electrolysis:** Although the efficiency of the electrolyzers is gradually growing, more efforts in the sector are needed to make green hydrogen comparable with the hydrogen produced from fossil resources (Venizelou and Poullikkas, 2024). However, the main challenge that the industry faces is the challenge of scaling up the production of the electrolyzers to meet burgeoning global demand.
2. **Renewable Energy Dependency:** The utilization green hydrogen means that the supply is not constant (Hassan, & Al-Jiboory, 2024a). They depend which not constant and hence do not support a continuous production of hydrogen which is a key issue in industrial uses.

3. **Integration with Existing Infrastructure:** As such, to go green, there is a need to adapt or change some of the existing energy systems (Saha et al., 2024). For instance, retrofitting of natural gas pipelines to hydrogen pipelines or construction of new hydrogen pipeline system requires a lot of capital and engineering modifications.
4. **Transportation:** Enterprises change to green hydrogen require upgrades or replacement of existing energy systems (Saha et al., 2024). For instance, retrofitting of natural gas pipeline for hydrogen or developing new hydrogen pipelines require huge capital and technical changes.
5. **Storage:** Another major problem is storage of hydrogen, and the matter of its safe and efficient storage (D'Agostino et al., 2024). Hydrogen can be stored in a gaseous form under high pressures, in its liquid form at very low temperature or in the form of metal hydride.
6. **Distribution:** Transportation of hydrogen from production centers to the final customer is still a challenge. Extension of hydrogen refueling stations for instance, is one of the greatest challenges when it comes to utilization of hydrogen fueled automobiles due to expensive investment and technological complexity.

B-Technological Innovations

1. **Advanced Electrolysis Technologies:** General research and development (R&D) in order to develop and build new and more efficient electrolyzers is essential (Dauletbay, 2024). High temperature solid oxide electrolyzers (SOEs) are a candidate to obtain efficiencies higher than 90%. In addition, advances in material sciences may help reduce the cost of making electrolyzers and increase the lifetime of the devices.
2. **Hydrogen Carriers:** Transporting hydrogen as ammonia or liquid organic hydrogen carriers (LOHCs) makes hydrogen transport and storage equally easier (Goel, 2024). These carriers are easier to transport and can be directly converted to hydrogen at the end user making it easier to transport pure hydrogen.
3. **Storage Solutions:** There are some proposed methods for developing new types of storage materials including graphene or nanostructures in SS hydrogen storage (Martinho & Rodrigues, 2024). As such, these methods hold the potential of delivering higher energy densities and better handle than the conventional techniques.
4. **Digital Monitoring Systems:** The use of the internet of things sensors, and artificial intelligence monitoring system can enhance safety as well as efficiency in hydrogen production, storage, and transportation (Chu et al., 2024). Such systems can identify leakages, monitor conditions in the storage and forecast when maintenance would be necessary.

C- International Cooperation

1. **Global Standards and Regulations:** The international trade and investments in hydrogen require a universal code of practices in the production, transportation, and utilization of hydrogen resources (Hassan, et al., 2024c). They promise that the technologies and structures as well as the safety standards across borders are compatible.
2. **Collaborative R&D Initiatives:** Hydrogen research collaborations with other countries would help to enhance the technological progress (Hassan, et al., 2024b). For example, multilateral relations of academic, business, and political sectors can combine funds and knowledge to combat similar problems.
3. **Cross-Border Hydrogen Trade:** Utilization of green hydrogen will enable nations with substantial renewable resources generate hydrogen for export to other nations with limited resources. For instance, Australia has targeted itself to become the largest exporter of hydrogen to Asia by the year 2024(Marzouk). The establishment of supply chains of hydrogen is a global venture that calls for collaboration in the construction of infrastructure, trade and steering of every supply chain.

D-Strategic Planning

1. **Scaling Infrastructure:** There is a need to develop hydrogen infrastructure such as pipelines, hydrogen storage and hydrogen fuelling stations by both the governments and the private sector (Jaradat et al., 2024). The use of these assets can be strategically located to reduce distribution expenses and enhance end-users' availability.
2. **Financial Incentives:** Promotion of hydrogen technologies can be done by subsidies and incentives on specific technologies (Hassan, et al., 2024c). For instance, a tax credit provided to hydrogen-fueled vehicles or grant for hydrogen projects helps reduce entry cost to industries and consumers.
3. **Public Awareness Campaigns:** Misconceptions about the advantages and safety of hydrogen can be eliminated, and the population can be made to accept this form of fuel (Koneczna & Cader, 2024). Hydrogen technologies and their application in tackling climate change need to be hence, communicated effectively and clearly to the public.

3.2 Future Prospects and Research Directions

Clean hydrogen generated from water electrolysis using renewable electricity is now considered a disruptive technology in the global energy sector. This versatility, sustainability, and capability to reduce carbon intensity in sectors that are difficult to abate make it a critical enabler (Frieden & Leker, 2024). Nonetheless, there is still a long way to go for green hydrogen to fully unlock the potential with regard to technology, infrastructure, policy and cooperation. This section looks at the future of green hydrogen and possible developments of this technology as well as the research gaps that must be closed and partnerships.

3.2.1 Electrolysis Efficiency Improvements

The efficiency of water electrolysis is one of the areas with the highest potential for future development in green hydrogen technology, according to Lehtonen (2024). Existing electrolyzers include PEM and alkaline water electrolyzers which work at efficiencies of 60%- 80%. Efficiency is targeted to increase from current levels of about 90% as ongoing research and development strive to improve electrolyzer designs, minimize energy losses, and raise the effectiveness of the materials used.

1. Advanced Catalyst Materials: According to (Münster et al., 2024), increasing the efficiency of catalysts while maintaining its durability and cost can immensely lower the cost of electrolysis (Fam & Fam, 2024). Some of the topics are transition metals, nano structured materials and hybrid catalysts.
2. High-Temperature Electrolysis: One of the methods of electrolysis currently under development is the use of solid oxide electrolyzers (SOEs), which are capable of achieving efficiencies greater than 90 percent due to their high operating temperatures (Gao et al., 2024). Although the costs of large-scale green hydrogen production may be high currently, innovation in better materials that can handle such conditions may render the use of SOEs more feasible and affordable in the future.

3.2.2 Integration With Renewable Energy Systems

Decarbonizing green hydrogen is inherently connected to the renewable energy sources (Béres et al., 2024). Subsequent evolution in grid integrated electrolyzers and hybrid systems will support the conversion of excess renewable energy to hydrogen thereby improving storage and minimizing the utility wastage.

1. Dynamic Electrolyzer Operations: Current research is also being focused towards designing the electrolyzers that can be controlled based on the renewable energy technologies like solar and wind that have varying output capacity.
2. Hybrid Renewable Systems: Different renewable resources, for instance, solar, wind, and hydrogen-power plants should be sited in a way that optimizes the use of the available renewable sources to produce energy.

3.2.3 Innovations in Storage and Transportation

The efficient storage and transportation of hydrogen remain critical challenges (Venizelou & Poullikkas, 2024). Future technological breakthroughs are expected to make these processes more practical and cost-effective.

1. Solid-State Storage: Research into advanced storage materials, such as metal hydrides and carbon nanotubes, could enable higher hydrogen storage densities at lower costs (Zuhara et al., 2024).
2. Hydrogen Carriers: Developing efficient hydrogen carriers, such as ammonia, methanol, or liquid organic hydrogen carriers (LOHCs), will facilitate the safe and economical transport of hydrogen across long distances.
3. Cryogenic and High-Pressure Solutions: Advances in cryogenic technologies and high-pressure storage systems could further enhance the feasibility of hydrogen transport in liquefied or compressed forms.

3.2.4 Fuel Cell Technology

Transportation and stationary power generation are the primary uses of fuel cells that work to convert hydrogen into electric power (Hassan, & Al-Jiboory, 2024b). The future development will be directed at enhancing the efficiency, stability, and flexibility of fuel cells. Large-scale power generation with high efficiency is possible using SOFCs which work at high temperatures and are efficient working devices (2019Saha et al.). The Exercise added that improved technologies in material and manufacturing could bring down the costs and improve the deployment of the systems (D'Agostino et al., 2024). Further advancements PEMFC technologies such as minimizing usage of platinum catalysts will see these systems applied in automotive and portable uses.

3.2.5 Impact on Global Energy Strategies

Green hydrogen has particular advantages in sectors where direct electrification is unlikely to be effective, notably heavy industry, long distance transport, and high temperature uses (Gunathilake et al., 2024). Future approaches will entail the use of green hydrogen into these sectors as a clean energy source to replace fossil fuel intensive processes.

1. Industrial Processes: Green hydrogen can further be used as a feedstock for which are necessary fertilizers and synthetic fuels respectively by (Khan et al., 2024). It can also substitute coal in steel production where it is use in the direct reduction of ores.
2. Transportation: Fuel cell vehicles (FCEVs) are an environmentally friendly substitute for conventional internal combustion engine vehicles for long-haul and high-demand applications (Dauletbay, 2024). Investigative work in hydrogen refueling stations light weight will increase their use.

3.2.6 Energy Storage and Grid Stability

While renewable energy is gradually replacing conventional sources as the main electricity source, green hydrogen will help to balance power supply and demand.

1. Seasonal Energy Storage: Hydrogen can serve as a seasonal energy storage medium for excess renewable energy produced at other times of the day or year.
2. Grid Integration: Hydrogen indeed can be considered as an energy vector and storage medium by providing flexibility in balancing the power supply and demand as well as accommodating the integration of renewable energy sources (Gao et al., 2024).

3.2.7 Global Hydrogen Economy

Further improvements in the techniques involved will eventually culminate formation a where hydrogen will be generated and exported from regions with surplus renewable resources to energy hungry areas.

1. Hydrogen Hubs: Hydrogen hubs that integrate production, storage and distribution will become the next line of facility as the world moves towards hydrogen economy (Huang et al., 2024).
2. International Trade: Green hydrogen can be another traded product like oil and gas and can help establish international relationships as well as economic development of the regions endowed with various resources.

3.2.8 Integrating Research Activities

As with most emerging sustainable technologies, green hydrogen technology cannot be developed in isolation but needs the collaboration of governments, academia, and industry players (Göthel et al., 2024). It is a common thing to see that combined research endeavours can enhance innovation through sharing of funds, workforce and information.

1. Public-Private Partnerships: These include research and development, piloting and demonstration of government and private projects/programs.
2. Knowledge Sharing: Hydrogen Council along with International Partnership for Hydrogen and Fuel Cells in the Economy (IPHE) can help to share experience and progress in engineering solutions.

3.2.9 Policy Support And Investment

The development of a sound policy and financial schemes is required to support R&D of green hydrogen.

1. Funding Programs: Governments should create specific funding initiatives to guarantee research into further forms of hydrogen technology and hydrogen infrastructure.
2. Regulatory Support: Lack of ambiguity concerning legal provisions that compose hydrogen production, storage, transportation, and use will create a stable climate for invention and funding.

3.2.10 Education And Workforce Development

For green hydrogen technologies to be successfully deployed, there is the need to develop a competent workforce.

1. Training Programs: Fresnought hydrogen production and storage, together with hydrogen safety trainings, can prepare the workforce for the new hydrogen economy (Zybert et al., 2024).
2. Educational Initiatives: College and research institutions should take up offering courses in hydrogen technology and renewable energy systems.

3.2.11 International Collaboration

International problems cannot be solved on a local scale (Salahi, 2024). Cross country cooperation will be paramount in dealing with the technical, economic and other challenges in up-scaling green hydrogen.

1. Cross-Border Projects: International cooperation has potential for rational use of resources as well as development of solutions that can be successfully implemented on the scale of several countries (Patkowski et al., 2024).
2. Global Standards: The improvement of shared guidelines for hydrogen generation, storage, and distribution will increase trade and decrease market segmentation.

Green hydrogen has great potential to become the force of change in net-zero energy plans of the world (Zuhara et al., 2024). Further innovation in technology, backed up by research and development and cooperation, will take the use of LCFS to its next level of fully applying it in the process of decarbonization of sectors that are hard to abate, stabilization of energy systems, and promotion of economic growth (Martinho & Rodrigues, 2024). But such a vision can only be achieved through the continuous intervention across various areas that cut across technology, policy, education and training, and international cooperation.

4. CONCLUSION

These green hydrogens have become an innovative solution to develop a sustainable and climate change resilient energy system globally. Made from renewable electricity, green hydrogen is a zero-carbon energy vector which has the capability to solve major problems in the energy sector such as carbon footprint, energy storage, and the electricity requirements of sectors which are difficult to decarbonise. This chapter has analyzed the technological, economical and policy context in which green hydrogen is being promoted, and why this should remain a top priority for a sustainable energy future.

It is therefore a far-reaching fact that green hydrogen can play the part of sustainability. As a clean and versatile energy vector, it addresses multiple dimensions of the energy transition. As Green hydrogen provides the potential to subsequently electrify industries where electrification is either impossible or ineffective. For instance, in steel making, green hydrogen displaces coking coal which is a high carbon input and results to low carbon emissions. In the same manner, hydrogen fuel cells in transport industry provides a sustainable solution to long haulage trucking and transportation, marine vessels and aircrafts. The variability of the renewable energy systems like solar and wind is incongruent with the grid and reliability of the energy. These problems are avoided by green hydrogen, which serves as an intermediate energy storage medium; whereby in instances of excess renewable energy production, hydrogen is produced and utilized in circumstances where there is low renewable power production. It also improves the robustness of renewable energy systems where the two are combined. Green hydrogen breaks the reliance on fossil fuels, thus improving the security of supply for the energy sector. That it can be stocked for a long time, and shipped across regions makes it a strategic commodity in the management of energy crises and energy market fluctuations.

This chapter has highlighted key insights for various stakeholders, emphasizing their roles in accelerating the adoption of green hydrogen. Governments should facilitate through policies such as the provision subsidies for the expanded use of the material, implementation of carbon pricing and provision of infrastructure for efficient production of the material. It hence becomes relevant to have clear policies regarding hydrogen production, storage and carriage to avoid wrong market trends. For green hydrogen technologies to work, industrial sectors particularly play a pivotal role in the process. Those business entities that have successfully adopted the use of hydrogen are likely to reap big as markets around the world shift towards sustainable practices and lower emission of carbon. Investment opportunities are attributed to green hydrogen because green hydrogen has enormous potential in the unexplored regional markets where hydrogen production, storage, and transportation solutions are currently being sought. There is the high possibility of generation of both economic and environmental benefits when early investments on this sector are made.

While green hydrogen's potential is immense, scaling it up to meet global energy demands requires overcoming significant challenges. These include high production costs, infrastructure limitations, and safety concerns. Innovations in materials science, electrolysis technologies, and fuel cell efficiency are essential for reducing costs and increasing scalability. Building dedicated hydrogen pipelines, refueling stations, and storage facilities is critical for integrating green hydrogen into energy systems. Global cooperation is needed to harmonize standards, develop cross-border hydrogen trade networks, and share best practices.

Green hydrogen is not just the next generation of energy source the world needs but it is a revolution that can redefine the energy industry. That is why it plays the crucial role in reaching the net-zero emissions and building the robust energy system of the future. Green hydrogen is an effective approach to blend environmental, economic, and social objectives while unlinking energy generation from carbon emissions. The information presented in this chapter shows how green hydrogen can play the role of an essential driver for the energy shift. Nevertheless, realising these potentials is contingent on shared effort by different stakeholders operating in various sectors and regions. Collective effort is required from both the government and industries as well as investors, researchers, and communities to address the challenges, unlock innovation, and create the systems needed to support green hydrogen economy.

REFERENCES

Abdin, Z. (2024). Shaping the stationary energy storage landscape with reversible fuel cells. *Journal of Energy Storage*, *86*, 111354. DOI: 10.1016/j.est.2024.111354

Abdolmaleki, L., & Berardi, U. (2024). Hybrid solar energy systems with hydrogen and electrical energy storage for a single house and a midrise apartment in North America. *International Journal of Hydrogen Energy*, *52*, 1381–1394. DOI: 10.1016/j.ijhydene.2023.11.222

Arias Calvo, A., Feijoo Costa, G., Moreira Vilar, M. T., Nika, C. E., Vasilaki, V. K., & Katsou, E. (2024). Assessing the future prospects of emerging technologies for shipping and aviation biofuels. *Critical Review*.

Bade, S. O., & Tomomewo, O. S. (2024). A review of governance strategies, policy measures, and regulatory framework for hydrogen energy in the United States. *International Journal of Hydrogen Energy*, *78*, 1363–1381. DOI: 10.1016/j.ijhydene.2024.06.338

Bade, S. O., Tomomewo, O. S., Meenakshisundaram, A., Ferron, P., & Oni, B. A. (2024). Economic, social, and regulatory challenges of green hydrogen production and utilization in the US: A review. *International Journal of Hydrogen Energy*, *49*, 314–335. DOI: 10.1016/j.ijhydene.2023.08.157

Béres, R., Nijs, W., Boldrini, A., & van den Broek, M. (2024). Will hydrogen and synthetic fuels energize our future? Their role in Europe's climate-neutral energy system and power system dynamics. *Applied Energy*, *375*, 124053. DOI: 10.1016/j.apenergy.2024.124053

Chu, C., Zhang, W., Yan, X., Yan, Y., Pan, J., Zohreh, S., & Moradian, J. M. (2024). Silver-incorporated NiCo metal-organic frameworks with controlled morphology for enhanced cycling in flexible supercapacitors application. *Journal of Materials Chemistry. C, Materials for Optical and Electronic Devices*.

D'Agostino, A., Di Marco, G., Canini, A., & Gismondi, A. (2024). Gallic acid as a phytostimulant enhancing yield and quality of Mentha spicata L. under deficit-and well-watered conditions. *Environmental and Experimental Botany*, *219*, 105656. DOI: 10.1016/j.envexpbot.2024.105656

Dauletbay, A. (2024). Transportation of Hydrogen: Hydrogen Usage. In *Hydrogen Technologies-Advances, Insights, and Applications*. IntechOpen.

Delbeke, J. (2024). *Delivering a Climate Neutral Europe*. Taylor & Francis. DOI: 10.4324/9781003493730

Du, L., Yang, Y., Zhou, L., & Liu, M. (2024). Greenhouse Gas Reduction Potential and Economics of Green Hydrogen via Water Electrolysis: A Systematic Review of Value-Chain-Wide Decarbonization. *Sustainability (Basel)*, *16*(11), 4602. DOI: 10.3390/su16114602

Eibeck, A., Shaocong, Z., Mei Qi, L., & Kraft, M. (2024). Research data supporting" A Simple and Efficient Approach to Unsupervised Instance Matching and its Application to Linked Data of Power Plants".

Fam, A., & Fam, S. (2024). Review of the US 2050 long term strategy to reach net zero carbon emissions. *Energy Reports*, *12*, 845–860. DOI: 10.1016/j.egyr.2024.06.031

Fredershausen, S., Meyer-Larsen, N., & Klumpp, M. (2024, February). Comprehensive Sustainability Evaluation Concept for Offshore Green Hydrogen from Wind Farms. In *International Conference on Dynamics in Logistics* (pp. 309-321). Cham: Springer Nature Switzerland. DOI: 10.1007/978-3-031-56826-8_24

Frieden, F., & Leker, J. (2024). Future costs of hydrogen: A quantitative review. *Sustainable Energy & Fuels*, *8*(9), 1806–1822. DOI: 10.1039/D4SE00137K

Gao, C., Cui, X., & Matsumura, J. (2024). Multidimensional Exploration of Wood Extractives: A Review of Compositional Analysis, Decay Resistance, Light Stability, and Staining Applications. *Forests*, *15*(10), 1782. DOI: 10.3390/f15101782

Geisbauer, A., & Berchthold, C. (2024). Will Power and Waste-to-X enable the worldwide energy transition to renewables?. *Monatshefte für Chemie-Chemical Monthly*, 1-14.

Goel, M. (2024). 3 'Ps' of Hydrogen Economy in India: Production Pathways, Policies, and Perspectives. In *Climate Action and Hydrogen Economy: Technologies Shaping the Energy Transition* (pp. 11–37). Springer Nature Singapore. DOI: 10.1007/978-981-99-6237-2_2

Göthel, J., Burkmann, K. J., & Volkova, O. (2024). Review and Evaluation of Ceramic-Stabilized Iron Oxides for Use as Energy Storage Based on Iron-Steam Process. *steel research international*, 2400265.

Gunathilake, C., Soliman, I., Panthi, D., Tandler, P., Fatani, O., Ghulamullah, N. A., Marasinghe, D., Farhath, M., Madhujith, T., Conrad, K., Du, Y., & Jaroniec, M. (2024). A comprehensive review on hydrogen production, storage, and applications. *Chemical Society Reviews*, *53*(22), 10900–10969. DOI: 10.1039/D3CS00731F PMID: 39421882

Guven, D. (2024). Offshore wind-driven green hydrogen: Bridging environmental sustainability and economic viability. *International Journal of Hydrogen Energy*, *72*, 661–676. DOI: 10.1016/j.ijhydene.2024.05.335

Hassan, Q., Algburi, S., Jaszczur, M., Al-Jiboory, A. K., Al Musawi, T. J., Ali, B. M., Viktor, P., Fodor, M., Ahsan, M., Salman, H. M., & Sameen, A. Z. (2024a). Hydrogen role in energy transition: A comparative review. *Process Safety and Environmental Protection*, *184*, 1069–1093. DOI: 10.1016/j.psep.2024.02.030

Hassan, Q., Algburi, S., Sameen, A. Z., Jaszczur, M., & Salman, H. M. (2024b). Hydrogen as an energy carrier: Properties, storage methods, challenges, and future implications. *Environment Systems & Decisions*, *44*(2), 327–350. DOI: 10.1007/s10669-023-09932-z

Hassan, Q., Algburi, S., Sameen, A. Z., Salman, H. M., & Al-Jiboory, A. K. (2024c). A review of green hydrogen production by renewable resources. *Energy Harvesting and Systems*, *11*(1), 20220127. DOI: 10.1515/ehs-2022-0127

Hassan, Q., Algburi, S., Sameen, A. Z., Salman, H. M., & Jaszczur, M. (2024d). Green hydrogen: A pathway to a sustainable energy future. *International Journal of Hydrogen Energy*, *50*, 310–333. DOI: 10.1016/j.ijhydene.2023.08.321

Heidaria, S. M., & Eshtiaghib, M. (2024). Women's Barriers to Utilizing Renewable Energy and Achieving the Sustainable Development Goals (SDGs) 2. *energy*, *24*, 25.

Huang, S., Yue, C., Uvdal, K., & Hu, Z. (2024). Recent advances in irradiation-mediated synthesis and tailoring of inorganic nanomaterials for photo-/electrocatalysis. *Nanoscale Advances*. PMID: 39610792

Jaradat, M., Almashaileh, S., Bendea, C., Juaidi, A., Bendea, G., & Bungau, T. (2024). Green Hydrogen in Focus: A Review of Production Technologies, Policy Impact, and Market Developments. *Energies*, *17*(16), 3992. DOI: 10.3390/en17163992

Khan, U., Bilal, M., Adil, H. M., Darlington, N., Khan, A., Khan, N., & Ihsanullah, I. (2024). Hydrogen from sewage sludge: Production methods, influencing factors, challenges, and prospects. *The Science of the Total Environment*, *919*, 170696. DOI: 10.1016/j.scitotenv.2024.170696 PMID: 38340850

Koneczna, R., & Cader, J. (2024). Towards effective monitoring of hydrogen economy development: A European perspective. *International Journal of Hydrogen Energy*, *59*, 430–446. DOI: 10.1016/j.ijhydene.2024.02.036

Kumar, S., Nanan-Surujbally, A., Sharma, D. P., & Pathak, D. (2024). Hydrogen safety/standards (national and international document standards on hydrogen energy and fuel cell). In *Towards Hydrogen Infrastructure* (pp. 315–346). Elsevier. DOI: 10.1016/B978-0-323-95553-9.00011-X

Lehtonen, M. (2024). Unveiling obstacles and assessing the EU policy and funding instruments for low-carbon hydrogen investments.

Martinho, V. J. P. D., & Rodrigues, R. N. (2024). Bioenergy relations with agriculture, forestry and other land uses: Highlighting the specific contributions of artificial intelligence and co-citation networks. *Heliyon*, *10*(4), e26267. DOI: 10.1016/j.heliyon.2024.e26267 PMID: 38379976

Marzouk, O. A. (2024). Expectations for the role of hydrogen and its derivatives in different sectors through analysis of the four energy scenarios: IEA-STEPS, IEA-NZE, IRENA-PES, and IRENA-1.5 C. *Energies*, *17*(3), 646. DOI: 10.3390/en17030646

Munir, M. T., Naqvi, M., Li, B., Raza, R., Khan, A., Taqvi, S. A. A., & Nizami, A. S. (2024). From Waste to Watts: Emerging role of waste lignin-derived materials for energy storage. *Journal of Energy Storage*, *82*, 110477. DOI: 10.1016/j.est.2024.110477

Münster, M., Bramstoft, R., Kountouris, I., Langer, L., Keles, D., Schlautmann, R., ... & Rasmusson, H. (2024). Perspectives on green hydrogen in Europe—during an energy crisis and towards future climate neutrality. *Oxford Open Energy, 3*, oiae001.

Pal, A., Kakran, S., Kumar, A., Youssef, A. B., Singh, U. P., & Sidhu, A. (2024). Powering squarely into the future: A strategic analysis of hydrogen energy in QUAD nations. *International Journal of Hydrogen Energy*, *49*, 16–41. DOI: 10.1016/j.ijhydene.2023.06.169

Pashchenko, D. (2024). Green hydrogen as a power plant fuel: What is energy efficiency from production to utilization? *Renewable Energy*, *223*, 120033. DOI: 10.1016/j.renene.2024.120033

Patkowski, W., Zybert, M., Ronduda, H., Albrecht, A., Moszyński, D., Fidler, A., ... & Raróg-Pilecka, W. (2024). Toward green ammonia synthesis–exploring the influence of lanthanide oxides as supports on the cobalt catalysts properties. *Journal of CO2 Utilization, 80*, 102699.

Peters, R., Pasel, J., & Samsun, R. C. (2024). Technology transfer from fuel processing for fuel cells to fuel synthesis from hydrogen and carbon dioxide. *International Journal of Hydrogen Energy*, *49*, 943–961. DOI: 10.1016/j.ijhydene.2023.08.022

Pothen, F., & Hundt, C. (2024). *European post-consumer steel scrap in 2050: A review of estimates and modeling assumptions* (Vol. 2024, No. 1). Jenaer Beiträge zur Wirtschaftsforschung.

Qasem, N. A., & Abdulrahman, G. A. (2024). A recent comprehensive review of fuel cells: History, types, and applications. *International Journal of Energy Research, 2024*(1), 7271748. DOI: 10.1155/2024/7271748

Revinova, S., Lazanyuk, I., Gabrielyan, B., Shahinyan, T., & Hakobyan, Y. (2024). Hydrogen in Energy Transition: The Problem of Economic Efficiency, Environmental Safety, and Technological Readiness of Transportation and Storage. *Resources, 13*(7), 92. DOI: 10.3390/resources13070092

Sage, V., Patel, J., Hazewinkel, P., Yasin, Q. U. A., Wang, F., Yang, Y., Kozielski, K., & Li, C. E. (2024). Recent progress and techno-economic analysis of liquid organic hydrogen carriers for Australian renewable energy export–A critical review. *International Journal of Hydrogen Energy, 56*, 1419–1434. DOI: 10.1016/j.ijhydene.2023.12.269

Saha, P., Akash, F. A., Shovon, S. M., Monir, M. U., Ahmed, M. T., Khan, M. F. H., Sarkar, S. M., Islam, M. K., Hasan, M. M., Vo, D.-V. N., Aziz, A. A., Hossain, M. J., & Akter, R. (2024). Grey, blue, and green hydrogen: A comprehensive review of production methods and prospects for zero-emission energy. *International Journal of Green Energy, 21*(6), 1383–1397. DOI: 10.1080/15435075.2023.2244583

Salahi, F. (2024). 12 Application of Solar. *Hydrogen Production from Renewable Resources and Wastes*, 281.

Venizelou, V., & Poullikkas, A. (2024). Comprehensive Overview of Recent Research and Industrial Advancements in Nuclear Hydrogen Production. *Energies, 17*(12), 2836. DOI: 10.3390/en17122836

Wolf, J., Pellumbi, K., Haridas, S., Kull, T., Kleinhaus, J. T., Wickert, L., Apfel, U.-P., & Siegmund, D. (2024). Electroplated Electrodes for Continuous and Mass-Efficient Electrochemical Hydrogenation. *Chemistry (Weinheim an der Bergstrasse, Germany), 30*(18), e202303808. DOI: 10.1002/chem.202303808 PMID: 38100290

Wolf, J., Pellumbi, K., Haridas, S., Kull, T., Kleinhaus, J. T., Wickert, L., Apfel, U.-P., & Siegmund, D. (2024). Electroplated Electrodes for Continuous and Mass-Efficient Electrochemical Hydrogenation. *Chemistry (Weinheim an der Bergstrasse, Germany), 30*(18), e202303808. DOI: 10.1002/chem.202303808 PMID: 38100290

Xie, Z., Jin, Q., Su, G., & Lu, W. (2024). A Review of Hydrogen Storage and Transportation: Progresses and Challenges. *Energies*, *17*(16), 4070. DOI: 10.3390/en17164070

Zuhara, S., Mackey, H. R., Al-Ansari, T., & McKay, G. (2024). A review of prospects and current scenarios of biomass co-pyrolysis for water treatment. *Biomass Conversion and Biorefinery*, *14*(5), 6053–6082. DOI: 10.1007/s13399-022-03011-0 PMID: 35855911

Zybert, M., Ronduda, H., Patkowski, W., Ostrowski, A., Sobczak, K., & Raróg-Pilecka, W. (2024). Improving the catalytic performance of Co/BaCeO 3 catalyst for ammonia synthesis by Y-modification of the perovskite-type support. *RSC Advances*, *14*(49), 36281–36294. DOI: 10.1039/D4RA06251E PMID: 39534043

Chapter 5
Pathways to Family Firms' Sustainable Human Resource Management

Muhammad Faizan Khan
https://orcid.org/0009-0009-3676-7311
Kohsar University, Murree, Pakistan

Hafiz Muhammad Sikandar Ahmad
Institute of Applied Sciences and Technology, Pak-Austria Fachhochschule, Pakistan

ABSTRACT

The chapter consists of three parts: Family Firms, Sustainable Human Resource Management, and Pathways to Family Firms. First, we discuss the importance of family firms and the theoretical perspective of family science theories. Secondly, the author explores the emerging concept of sustainable human resource management (SHRM) and introduces pathways to sustainable futures. What is sustainability, and how was it integrated into HRM? It draws on the wider literature on Strategic Human Resource Management (SHRM), Sustainable Resource Management (SRM), and organizational theory to offer insights into the relationship between sustainability and HRM. The key issue addressed is the balancing act between short-term economic efficiency and long-term sustainability in managing human resources. Last but not least, we discuss the pathways to family firms and how family firm sustainable practices contribute to Sustainable Development Goals (SDGs) 2030.

DOI: 10.4018/979-8-3373-0139-6.ch005

INTRODUCTION

Classical View of Family Firms

Families are recognized as a fundamental component of societal structures. Even within contemporary capitalist societies, families are acknowledged as distinct entities (Martinez & Aldrich, 2014). Structural functionalists view the family as a primary source for socializing and nurturing children, who subsequently become integral members of society. Additionally, families play a crucial role in facilitating geographical mobility, as capitalist systems provide the necessary capital for individuals to establish businesses, while families contribute unpaid labor to support these enterprises. However, the structural functionalist perspective has become less relevant since its prominence four decades ago in the study of family dynamics (McIntyre, 1966). Despite criticisms of functionalism, it continues to hold significance in the realms of family and family business research (Kingsbury & Scanzoni, 2009). The functionalist framework is rooted in the concept of the "sociologistic theorem" (Parson, 1951) and was notably supported by Parsons' student, who stated that "Everyone wants to do that which others want him to do, and others always act as he expects and wishes..." (Williams, 1961, p. 75). Parsons (1965) argued that the operation of social systems relies on "solidarity," which is essential for individuals to survive and function as part of the larger group known as society. Therefore, in functionalism, the actions of individuals are crucial as they adhere to established norms for the collective benefit of the whole.

The functionalist perspective posits that the primary objective is the survival of the larger collective, which can encompass any group within the social framework. For a group to thrive, it is essential for its members to have clearly defined roles. These roles constitute subsystems, with the family being a key subsystem within the broader structure. The family is responsible for ensuring the replacement of its members in both biological and sociological contexts as needed by the larger community. To function optimally, the family requires a well-defined structure, with specific roles established within its dynamics. Consequently, family members have distinct roles that revolve around shared values and norms, fostering strong internal bonds compared to external relationships, and ultimately maintaining a state of equilibrium.

Moreover, the family has specific roles assigned to its members. The father typically assumes the instrumental role, acting as the primary provider and safeguarding the family from socioeconomic instability. In contrast, the mother fulfills an expressive role, nurturing emotional connections within the family, especially during the father's absence due to work commitments, thereby contributing to the family's and society's overall emotional well-being. Parsons (1965) further argues that deviation from these established roles can lead to significant disruptions, resulting in "family disorganization." To maintain societal equilibrium, it is crucial to prevent any form of family disequilibrium. The foundation for this stability lies in shared values and norms. If any family member deviates from these expectations, it is the responsibility of other members to address the issue and restore conformity. When individuals pursue cultural goals and adhere to accepted means, they are rewarded accordingly; however, if they resort to illegal methods to achieve these goals, they will not receive the same rewards (Merton, 1957).

Marx and Engels, the family operates within a framework of dominant production relations, perpetuating itself over time. From a Marxist viewpoint, the family is a primary contributor to societal inequality, as it facilitates the unequal distribution of wealth both within the family unit and in the broader community. Families instill and reinforce their values while providing financial support to their children at various life stages, driven by entrepreneurial motives. The prevalence of affluent families can often be attributed to family-owned businesses. Marxists argue that the family does not serve the greater good of society; rather, it functions as a support mechanism for a specific class of capitalists. Consequently, the contributions of families are deemed insignificant to society at large, holding relevance only for the ruling class.

Importance of Family Firms

Society operates within a continuous cycle, shaping individuals while simultaneously being shaped by them. Each society has fundamental needs that are essential for its survival and the well-being of its members. Over time, individuals within a society develop consistent patterns of thought and behavior that effectively address recurring challenges. Sociologists refer to these established patterns as institutions. Robertson (1989; p. 54) defines an institution as "a stable cluster of values, norms, statuses, roles, and groups that develops around a basic social need." One of the key institutions in society is the family. The family is not merely an emotional construct; it is often perceived as a long-term economic support system (Labaki, Michael-Tsabari & Zachary, 2013). Labaki (2005) further describes the family as a group of individuals connected by blood or alliances, maintaining their relationships through emotional, financial, and political ties. If we consider the family as an institution, it must exhibit recurring behavioral patterns rooted in specific thoughts and actions aimed at solving problems. This analogy extends to family businesses, where family members form a dominant coalition in decision-making (Burns, 2012). To comprehend a family business, one must analyze the family dynamics at play, as every family business is underpinned by a family structure. Are business decisions solely driven by economic factors? Such decisions encompass not only financial considerations but also social implications. Sociologists offer various perspectives to analyze societal issues and social behavior.

Robertson, a sociologist who advocates for the interactionist perspective, posits that society is shaped by the ways individuals act, respond to, and influence one another. Within this societal framework, individuals possess three primary components in their personalities: the cognitive component, which encompasses perception and memory; the behavioral component, which includes skills and competencies; and the emotional component, characterized by feelings such as love, sympathy, and pride elements that are particularly significant in family businesses.

Family businesses are the most common type of organization globally. Researchers emphasizes that family businesses are pivotal to the economies of both developed and developing nations. According to The Economist (2015), family firms account for over 90% of all businesses. Family businesses comprise 70% of global enterprises, and literature predicts that by 2025, 40% of the world's most influential businesses will be family-oriented. Handler (1989) outlines that family ownership in private firms should exceed 50%, or in public firms, more than 10% of stocks should belong to a family. Additionally, at least one family member must be actively involved in the business, or the business must declare itself as family-owned. Chua, Chrisman, & Sharma (1999), describe it as "businesses governed and managed to pursue the vision of the business held by a dominant coalition controlled by family members, potentially sustainable across generations." Kellermanns & Eddleston (2006) further simplify this definition by stating that any firm with family ownership and at least two family members working in it qualifies as a family firm. Daspit, Madison, Barnett, & Long (2018) note that family businesses operate with a transgenerational model, focusing on family ownership and management. While family firms dominate the global business landscape, they exhibit varying levels of performance. Non-family firms often demonstrate consistent performance, whereas family firms face unique challenges, such as favoritism and internal conflicts that may affect their effectiveness. In the fast-evolving landscape of modern business, there is a constant transformation occurring in customer demands, product characteristics, distribution methods, and value calculations. Within this dynamic environment, organizations face the challenge of maintaining their competitive edge. Traditional companies often operate within more rigid frameworks, lacking the essential structures, systems, and cultural adaptability needed to respond effectively to contemporary trends (Khakwani, & Case, 2012). A reliable estimate indicates that family-owned businesses significantly contribute to publicly listed companies, with even higher percentages observed when considering the entire population of both public and private firms (Amit & Villalonga, 2014). Consequently, family businesses have emerged as a critical area of study, prompting scholars to distinguish them from non-family enterprises.

Family Firm's in Pakistan

Pakistan is characterized as a collectivist society (Hofstede, 1984), exhibiting high levels of masculinity and power distance, with a predominantly Muslim population. Research focusing on family businesses in Pakistan has revealed that the community operates within a network of kinship known as Biradiri, as described in local terminology (Afghan & Wiqar, 2007). In the context of Pakistani family businesses, the notion of kinship (biradiri) extends beyond immediate family members (principals) to encompass other significant stakeholders, such as suppliers, and agents, including employees, thereby complicating agency conflicts. This complexity is further amplified when business inheritance transitions to the third generation, where immediate cousins and the cousins of the parents become involved as partners in the family enterprise (Saleem, 2008). Many businesses in Pakistan operate outside the government's tax registration system, contributing to what is classified as an informal economy (Williams, Shahid, & Martinez, 2016). Family businesses represent a significant and persistent segment of economic activity within Asian markets. Estimates suggest that approximately eighty percent of employment in Pakistan is still managed by family-owned firms. Historically, these enterprises were dominated by 22 of the wealthiest family-owned businesses (Nabi, 2012). In the 1970s, family firms were at the forefront of the Pakistani industrial landscape. However, the socialist policies of the 1980s led to substantial losses for many family businesses, primarily due to the nationalization process (Qureshi, Mian, & Rana, 2010). Other once-prominent family firms associated with influential Pakistani business, political, and military figures, such as Arag, Khyber, Fancy, Beco, Ispahani, and Milwala, have also faded from prominence (Nabi, 2012). A lack of interest from local scholars in studying family firms, stemming from a disconnect between academia and industry, further complicates the situation. Furthermore, official data indicates that family-owned enterprises are predominantly registered as private limited companies. The Securities Exchange Commission of Pakistan (SECP, 2018) reported that 72% of firms are classified as private limited companies, while 24% are sole proprietorships, and only about 4% fall into categories such as non-profit associations, foreign entities, publicly unlisted companies, and limited liability partnerships. Most of these firms are privately owned, with business shares held by a limited group and restrictions on the transferability of shares. These companies are emerging yet struggling to gain traction due to the absence of family business contracts. Various external and internal factors may be driving Pakistani family businesses to exit the market. Research indicates that Pakistani firms, particularly family-owned enterprises, are generally less competitive than multinational corporations (Tahir, Sabir, & Shah, 2016). Consequently, family businesses often experience a brief lifespan and rapid withdrawal from the market. However, challenges such as limited access to global markets, energy shortages, and inadequate managerial and technical expertise contribute to the reduced competitiveness of family firms in this sector (Nabi, 2017).

In the context of Pakistani society, which is notably family-oriented, decisions reflect not only individual preferences but also collective family dynamics. This influence extends from everyday choices, such as purchasing a watch, to significant life decisions, including career paths and marriage, as highlighted by Veen, Moorten, and Durani (2018). The impact of extended family members is also evident, especially in family firms where decisions are often made collectively rather than by individuals alone. Consequently, choices made within these family businesses can significantly affect the long-term relationships among family members (Zellweger & Astrachan, 2008). Businesses arise to capitalize on identified opportunities and achieve growth. While they may attain a leading position in the market, the sustainability of that leadership can be uncertain.

Example: A pertinent example is Nokia, which, upon entering the mobile industry in Pakistan, was initially the most compatible with GSM technology. Its customer base expanded significantly, establishing it as a market leader, capturing 40% of global sales until the advent of smartphones. However, Nokia's failure to adapt to market trends and its lack of proactive innovation ultimately led to its decline in leadership (Doz, 2017). This conventional approach resulted in Nokia losing its competitive edge. However, family firms must navigate the complexities of balancing both economic and non-economic objectives simultaneously.

THEORETICAL PERSPECTIVE

Socioemotional Wealth (SEW)

In management sciences, there are multiple theories used by the researchers to explain the theoretical model. Here we used two theories to explain the Family system. First one is Socioemotional wealth (SEW) and the other one is Circumplex theory. These theories explained the family system in detail & provide the family dimensions for the researchers to study. Let's start from the socioemotional wealth (SEW).

Socioemotional wealth (SEW) was introduced by (Gomez Mejia, Haynes, Nunez-Nikel, Jacobson, and Moyano-Fuentes, 2007) nearly a decade ago, sparking important discussions about its significance (Berrone, Cruz & Gomez-Mejia, 2012). SEW refers to the psychological value that principals of family firms prioritize over financial wealth, often choosing to protect it at considerable financial expense. SEW is further elaborated as a multidimensional construct comprising five dimensions, collectively referred to as FIBER. The first dimension, family control and influence, is highly valued by the principal of the family firm or the dominant family coalition. This control significantly impacts strategic decision-making within family businesses, as the family principal or dominant coalition seeks to maintain this influence (Zellweger, Kellermanns, Chrisman & Chua, 2012). The second dimension pertains to family identity, where the family and the firm are intertwined, sharing a common name that reflects their unity. Families strive to safeguard their reputation by addressing various stakeholder demands and engaging in corporate social initiatives, which may require sacrifices to uphold their identity within the community (Zellweger, Nason, Nordqvist & Brush, 2013). The third dimension focuses on social ties, specifically binding social ties within SEW. These ties extend beyond immediate family members to include the broader community, founded on reciprocity, where family firms provide numerous benefits to employees to ensure organizational stability, even at the cost of financial losses (Verver & Koning, 2017). The fourth dimension involves emotional attachment, highlighting the influence of emotions in family firms and how emotional decisions can overshadow rational ones. The historical relationships among family members significantly shape future directions and working dynamics, ultimately affecting firm performance. The fifth dimension addresses the renewal of family bonds, which refers to the transition of business ownership to the next generation.

Family businesses are referred to as such because they typically transition from one generation to the next. The owners of these enterprises are often proactive in preparing for succession, frequently announcing their intentions while still in good health and maintaining full control. This early communication helps all stakeholders understand the future direction of the business. A primary goal of family firms is to ensure the longevity of the business for future generations (Zellweger et al., 2012). The dynamics within the family can influence the level of engagement of the next generation in the business (Raey, 2019) and may also impact the succession process. Additionally, a new aspect of family enrichment has been incorporated into the FIBER framework, transforming it into FIBEER, which extends the SEW(x) scale. Family enrichment reflects the owner's desire to ensure the happiness of all family members, whether they are directly involved in the business or not. Family firms often adopt a policy that emphasizes shared benefits among family members. However, to remain competitive, these businesses must embrace emerging trends and technologies. This integration of modern practices necessitates a shift towards more entrepreneurial structures (Zahra, 2018). Any entrepreneurial transformation, including corporate venturing or intrapreneurship, is classified as corporate entrepreneurship (Burns, 2012). While corporate entrepreneurship offers various advantages, it can also pose risks to the socioemotional wealth (SEW) of family firms. Consequently, family businesses face the dual challenge of safeguarding their SEW while striving to thrive in the marketplace. Family firms that have been in operation for many years tend to exhibit a more conservative approach to change. Among various factors contributing to their risk-averse nature, one significant aspect is socioemotional wealth (SEW). Even when faced with challenging decisions regarding exit strategies, family firms often opt for mergers rather than liquidations in order to safeguard some degree of their socioemotional wealth.

Circumplex Theory

Circumplex theory is a family science theory presented by (Olson et al., 1979). In family science theory, the circumplex model is considered a basic model to understand the functions of the family (Rogers, 2021). Initially, it was developed to diagnose the dysfunction of couple and family relationships. The theory explains that a balanced family system positively influences organizational performance while an unbalanced family system affects negatively. Circumplex theory consists of three dimensions' family Cohesion, family flexibility, and family communication. The family Communication dimension works as a facilitator for the other two dimensions of the theory. Family Cohesion means "the emotional bonding that couple and family members have toward one another" It consists of three levels. One is the higher level, which is named enmeshed, lower level, disengaged. The last group is moderate (separated and connected). Olson (2011) describe family flexibility as "the amount of change in leadership, role relationships and relationship rules".

There are three levels of Family flexibility, the higher level, named chaotically, the lower level, called rigidification, and the moderate level (flexibly and structurally). Circumplex theory suggests that higher and lower levels of both dimensions create a problem for family firms. The family system should fall at a moderate level to get a positive response from their employees, which ultimately helps organizations achieve better results. Only a few articles are available in literature in which family science theory (circumplex theory) or any of his dimensions had a part of the analysis. Lee (2006) tested the two dimensions of circumplex theory, family cohesion and family flexibility (Independent variable), and found that both measurements were positively associated with organizational commitment. Therefore, the basic concept behind the circumplex theory is the balance between togetherness and separateness in the cohesion dimension and balance stability with a change in the flexibility dimension. Nose, Korunka, Frank & Danes (2017) tested the moderating role of Family climate including all dimensions of circumplex theory, on decreasing the effects of relationship conflict on Family businesses. Daspit et al. (2018) explain the circumplex theory in the context of the family firms by providing a conceptual framework in which they identify different systems to understand the phenomena. In the literature, family involvement has two consequences; few researchers argued that family involvement teaches local culture, values etc., and the anti-group thinks oppositely. They observed that bad decisions that depend on personal goals would cause too much family involvement. Olson et al. (2011) describe that most family science theories were developed in developed countries; they need more validation studies from developing countries with diverse cultures.

Figure 1. Diagram: study the family system

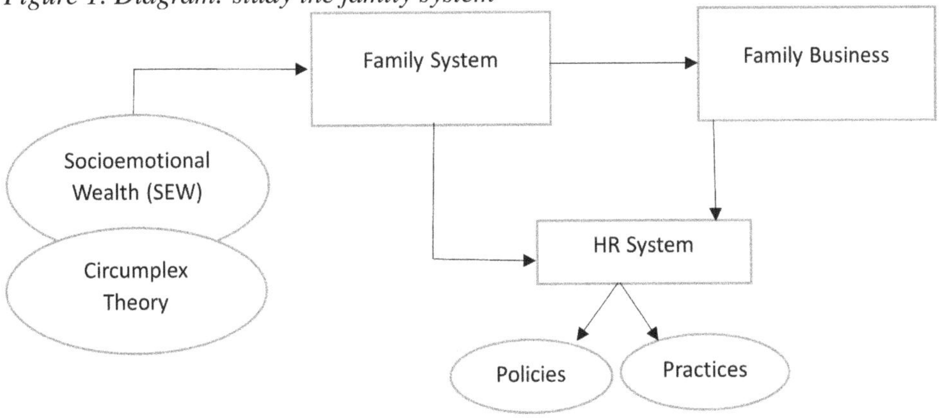

As shown in the figure, Family system needs to be study and suggested theories are socioemotional wealth (SEW) and Circumplex theory. Both theories represent the different family dimensions, highlight the family culture, power, leadership, family influence etc. Family system has a strong influence on family business & HR System. As we know, that Human resource department will play a crucial role in the development of the firm. HR Practices in family firms and non-family firms are different in numerous ways. Family Firms work; differently, their practices are implemented and are not formalized. HR Practices in family firms were implemented differently than non-family firms (De Kok, Uhlaner, & Thurik, 2006). Non-family businesses generally adopt formalized HR practices that may increase the business's financial performance. Hernandez-Linares, Lopez-Fernandez, Memili, Mullins, & Patel (2021) pointed out that the impact of HR Practices is limited on the firm's performance compared to non-family firms because of their unstructured HR Practices. Based on the literature, HRM is one of the most crucial sources of organizational outcomes. Madison, Daspit, Turner, & Kellermanns (2018) explain that HR professionalization benefits firm performance. At the same time, family-owned companies emphasize informal HR practices adopted for non-economic objectives related to the well-being of family and employees but resulting in less optimal financial performance (Sanchez-Marin et al., 2019). Most family firms follow their culture, which is beneficial more for the family than business. Family firms are more interested in other aspects than economic or are more likely to achieve individual objects (Parra-Dominguez, David, & Azevedo, 2021).

Stewart & Hitt (2012) explains that a family firm contains family employees and non-family employees; somehow, the firm's growth depends on the successful integration of the workforce (family and non-family employees). Lubatkin, Schulze, Ling & Dino (2005) identify that family employees received more attention or favoritism than non-family employees. Family Firms have diverse employees & for meaningful outcomes, they need to integrate the employees positively. Nevertheless, the favoritism between employees disturbs the overall environment of the firm. It is the responsibility of the HR Department to create a healthy environment for a positive outcome. HR Department depicts the management policies of the firm in the shape of HR Practices.

Nevertheless, the situation is different; overall managerial and administrative matters are running under the owner's supervision. The owner of the firm also influenced the implementation of HR Practices. It affects the financial performance of the firm. Astrachan & Kolenko (1994) finds a positive relationship between HR Practices and firm performance. HR Practices need to work independently to produce better results. There is no specific list of HR Practices, and there are multiple HR Practices intended in the department; we can call them the HR system (Dorado, Chen, Prado, & Simon, 2021).

Let's move to the next part of the chapter which is consists of Sustainable Human Resource Management and in the last but not the least we discussed how it is important for Family firms to adopt Sustainable HRM and its link to SDGs 2030.

SUSTAINABLE HRM

As the modern business landscape is ever-evolving, one new approach that some companies have taken (and quite successfully, I might add) that combines traditional HR practices with sustainability goals is Sustainable Human Resource Management (HRM). This is more than just the right thing to do to get talent management fit for a competitive world; it is also about designing work with the long-term health and happiness of people, environment, and community at the center (Ehnert, 2014). Sustainable HRM is the essence converged in the concept of the Triple Bottom Line (Elkington, 1997), which underlines interrelations between People, Planet, and Profit.

There are a few trends that are making HRM more and more sustainability-oriented:

- **Changes in regulations**: New laws and standards are forcing businesses to embrace ethical and sustainable practices (Carroll, 1991).

- **Stakeholder expectations**: Increased pressure from consumers, employees, and investors for corporations to be transparent and responsible (Geissdoerfer et al., 2017).
- **Competitive edge**: Incorporating sustainability in HR gives organizations an advantage in workforce management and employer branding (Mariappanadar, 2020).

Sustainability in HRM: The Role of Employees

Employees are at the core of any sustainability effort. The company and employees all help each other — employee engagement, values, and behaviors impact the ability of a company to achieve sustainability goals. Sustainable HRM is human-centered and prioritizes employee well-being, satisfaction, and development (Ehnert, 2009). This approach is based on the understanding that where employees are encouraged, they not only perform well but also help build a sustainable organization.

Key components of this approach include:

- **Enabling Employee Health**: Holistic wellness offerings across physical, mental, and financial health (Microsoft, 2020).
- **Encouraging Diversity and Inclusion**: Creating a workforce that mirrors or reflects diverse perspectives, backgrounds, and experiences (Unilever, 2020).
- **Incentivizing Ethical Choices**: Instituting policies to align sustainable and ethical decision-making among employees and staff members (Ben & Jerry's, 2019).
- **Purpose of this Part**

This part aims to comprehensively discuss Sustainable HRM with its theoretical roots, practical applicability, and future directions.

We will discuss about:

- Triple Bottom Line as it is related to HRM
- The emergence of a more sustainable approach towards HRM because of ethical considerations and social responsibility.
- Basic components and methods for instituting sustainable HR in organizations
- Examples of how sustainable HRM makes a difference in actual cases where they influenced business performance as well as job satisfaction among employees.

This part will enable readers to comprehend the ways the triple bottom line can be driven by sustainable HRM towards a flourishing business framework that is resilient and profitable while maintaining ethical standards.

Theoretical Foundations of Sustainable HRM

Triple Bottom Line (TBL) in HRM

The Triple Bottom Line (TBL), introduced by John Elkington (1997), lays the foundation for sustainable HRM. Unlike the traditional focus on financial profit, TBL expands the measurement of organizational success to include:

- **People**: Employee welfare, labor rights, and diversity efforts (Carroll, 1991).
- **Planet**: Reducing environmental impact through sustainable HR initiatives (Geissdoerfer et al., 2017).
- **Profit**: Long-term financial advantages of sustainable HRM, such as lower turnover and higher employee loyalty (Mariappanadar, 2020).

Evolution of HRM to Sustainable HRM

The transition from traditional HRM to Sustainable HRM represents a shift from operational efficiency and compliance to integrating environmental sustainability and corporate social responsibility (CSR) (Ehnert, 2014).

Key theoretical perspectives:

- **Ehnert's Paradox Perspective**: Highlights the contradiction between short-term business objectives (e.g., productivity and cost optimization) and long-term sustainability goals (Ehnert, 2014).
- **Mariappanadar's Human Sustainability Framework**: Advocates for work practices that enhance health, well-being, and ethical behavior (Mariappanadar, 2020).

Social Responsibility and Business Ethics

A big part of sustainable HRM is ethics and social responsibility. More companies are now realizing that their responsibility extends beyond compliance and should include a role in improving society (Carroll, 1991).

This involves:

- Fair Labor Practices: Pay workers fairly, respect their right to a safe work environment
- Community Involvement: Promoting employees to engage in community service and volunteerism.
- Environmental Protection: Creating processes that minimize waste, maximize resources, and adopt sustainable practices (IKEA, 2019).

HR leaders are essential in creating an organizational culture that promotes ethical behavior and social responsibility. That requires establishing clear standards, conducting ethics training, and ensuring accountability.

Key Elements of Sustainable HRM

Sustainable Human Resource Management has a broader focus and integrates Triple bottom-line dimensions (People, Planet, and Profit) into its key functions (Elkington, 1997). These three dimensions offer both concrete, one-of-a-kind opportunities to make the organization you are working within more sustainable and improve on work-lifestyle balance.

People Dimension

The People Dimension focuses on crafting the HR process with an employees-first approach which not just cater their well-being but also diversity, inclusion & happiness (Ehnert, 2014). Engagement, lower turnover, and a healthy corporate culture are more common in companies that care about their employees (Mariappanadar, 2020).

Key Practices:

1) **Wellness Programs for Employees**: Sustainable HRM takes a holistic approach regarding employee welfare, concerning health, and mental and financial well-being (Microsoft, 2020). Programs might include:
    - Increased accessibility to mental health services, wellness programs and flexibility in the workplace.
    - Personal financial and future planning initiatives (financial wellness).
    - Guidance under laws that often require exceeding compliance for a safe, healthy workplace.
2) **Diversity, Equity, and Inclusion (DEI)**:

Diversity, Equity, and Inclusion (DEI) — Inclusive HR enables you to hire a diverse workforce who bring different perspectives and experiences (Unilever, 2020).

- Affordable Consultations: Crucial to reducing also harmful discrimination throughout the hiring process and creating a level playing field.
- ERGs (Employee Resource Groups): Groups led by employees who support their colleagues from under-represented communities to foster belonging and community
- Cultural Competence Training: Providing training that can potentially develop a sensitivity of culture and perspectives, and thus help create more inclusive workplace (Salesforce, 2023).

Case Study 1: Salesforce's Commitment to Diversity and Inclusion

Salesforce was one of the first companies to coherently integrate diversity and inclusion into its HRM Sustainability Agenda (Salesforce, 2023). It also publishes its Diversity and Inclusion Report, which monitors the progress of gender parity and inclusion of minority groups. Salesforce Initiatives are comprised of:

- Hiring broadly while preserving target representation.
- ERGs: Groups such as BOLD-force (Black employees) and Out-force (LGBTQ+ employees) that provide community and advocacy.
- Training on Equality — Continuous education and training programs addressing unconscious bias and including leadership

These efforts have brought a more engaged workforce, increased innovation, and improved goodwill among the company.

- **Planet Dimension**

The Planet Dimension of sustainable HRM means that we also need to start the right practices with regards as well as have a positive impact on environmentally friendly business by adapting ecologically friendly practices whereby the organization will reduce representing an environmental burden at best into an area of ecological devotion (Geissdoerfer et al., 2017). HR can catalyze green practices through its guiding influence on workplace policies, behavior, and culture.

Key Practices:

1) **Green Recruitment**: Bringing in personnel who resonate with the organization's sustainability ethos One way to do this is by emphasizing the organization's sustainable practices and principles in recruitment (Patagonia, 2021).
2) **Sustainable workplace policy**:
    - **Allowing Remote Work and Telecommuting**: Supporting remote work to cut down on commuting-related carbon emissions (Microsoft, 2020).
    - **Green Classrooms**: Integrating energy-saving lighting, creating recycling programs, and minimizing the use of paper.
    - **Engagement of Employees in Going Green**: Promoting activities such as reducing, reusing, and recycling, carpooling, assisting with energy conservation measures, etc (Microsoft, 2020).

Case Study 2: Microsoft's Environmental Policies and Remote Work Initiatives

Microsoft is 100 percent committed to being carbon-negative by the year 2030, with HR being a central pillar behind this initiative (Microsoft, 2020). Key initiatives include:

- **Work From Home**: Microsoft has been rushing its work-from-home policy because of COVID-19, reducing direct emissions from energy used in buildings and through the commuting of employees.
- **Staff Training**: Training staff to be sustainably minded within the workplace and outside of it.
- **Sustainable Office Design**: Microsoft offices are built on green building principles, using renewable energy and sustainable materials to create an environmentally sustainable office design.

These policies have in the past acclaimed a reduction in Microsoft effect at large, and increased morale productivity amongst employees.

- **Profit Dimension**

The Profit Dimension shows how sustainable HRM practices can enable employees to be more engaged, productive, and loyal impacting financial performance directly (Mariappanadar, 2020). Organizations invest in sustainable HR practices and generally good return on investment can be witnessed, through operational efficiency improvement that not only costs less but also give positive results.

Key Practices:

1) **Reduce Turnover and Increase Profitability**: Employees who feel engaged are productive as well as far less likely to seek employment outside of the organization (Unilever, 2020).
    - **Meaningful Contribution**: Ensure that the employee's role is aligned to make the organization sustainable, so employees get imbibed with a sense of purpose and ownership.
    - **Recognition Programs**: Measuring sustainability contributions, creating motivation among employees, and strengthening desired behaviors.
2) **Escaping turnover costs**: Turnover can be a high-cost endeavor, both in dollars and in lost expertise. Sustainable HRM emphasizes top talent retention through an enabling work environment and career development opportunities (Mariappanadar, 2020).

Example: Unilever's Approach to Employee Engagement and Sustainability

For example, Unilever's Sustainable Living Plan has played a vital role in embedding sustainability throughout the business and HR (Unilever, 2020). Unilever's approach entails key elements, such as:

- **Encouraging Employee Participation** — Involving employees in sustainability initiatives and providing them an opportunity to share ideas on how to reduce the company`s carbon footprint.
- **Sustainable Leadership Development**: Programs aimed at building leaders who can further sustainability efforts within the organization.
- **Performance Incentives**: Tying a portion of executive compensation to sustainability targets and holding leaders accountable for progress on environmental and social goals are examples of performance incentives.

Unilever has also had higher employee, brand, and financial outcomes due to its integrated method, showing the importance of sustainable HRM.

Strategies for Implementing Sustainable HRM

Sustainable HRM is about developing a different lens (sustainability) and how it can be viewed in traditional HR practice areas (Ehnert et al., 2023). Sustainability should be embodied in all levels of the employee journey including recruitment, performance management, and others. Here, we discuss actionable measures that HR can take to influence sustainability throughout the organization.

Sustainable Recruitment and Selection

Attracting and hiring talent aligned with the values and sustainability goals of the organization. This way, new employees will not only be well-qualified but also aligned with the company's sustainability vision (Mariappanadar, 2022).

Key Strategies:

- **Green Job Titles:** Make it explicit how the organization leads on sustainability in job advertisements Add Allocated Sustainability tasks, support for green initiatives or reduction of resources (Geissdoerfer et al., 2023).
- **Values-based Interview**: Identify strong candidates who are a good fit for your organization by asking sustainability-related questions during the interview process. C To assess candidates against sustainable practices in the past, you must incorporate behavioral interview questions (Unilever, 2023).
- **Diverse Hiring**: Use Diversity Hiring by Implementing Unbiased Hiring Processes. Minimize unconscious bias and create equal opportunities with the use of anonymous CV screening tools, or structured interviews (Salesforce, 2023).

Example: Patagonia for example only hires people who are environmental activists The result is a committed workforce that is in sync with Patagonia's sustainability mission (Patagonia, 2023).

Onboarding with a Purpose

Onboarding is where sustainability can be rapidly and directly bred into the organizational culture. Onboarding with a purpose helps new employees understand the company's Sustainability goals and practices so that they know what to do their part for (Google, 2023).

Key Strategies:

1) **Sustainability Orientation**: Include a brief overview of the company's sustainability values and initiatives in the onboarding process. It shows them how the work they do fits into the bigger picture and affects new employees (Adobe, 2023).

2) **Mentorship Initiatives**: Guide new employees by connecting them with mentors working on sustainability projects. According to Murphy (2023), this helps implement a structure of support that motivates involvement in eco-friendly practices from the start.
3) **Embed into Daily Operations**: Sustainability should be infused into all daily tasks and operations within the organization — provide suggestions on how to save energy, reduce waste, or commute more sustainably (Microsoft, 2023).

For example, the onboarding program at Google offers thorough insights into the company's strategic approach to sustainability–including its plans for lowering carbon emissions and increasing energy efficiency. Require new ideas to increase sustainability in the workplace (Google, 2023).

TRAINING AND DEVELOPMENT FOR SUSTAINABILITY

Training and development are important to prepare employees with the skills and knowledge necessary for achieving organizational sustainability objectives. Thus, incorporating sustainability in training programs can create a workforce that can make impact positively on the environment and society (IKEA, 2023).

Key Strategies:

1) **Training Initiatives Related to Sustainability**: Conduct various training programs that may include sustainable business practices, environmental management and corporate social responsibility (CSR) Word defines them to its many roles in the organization (Schneider Electric, 2023).
2) **Leadership Development**: Educating leaders about sustainable management practices and the necessity of leading by example in sustainability as an integral part of strategic decision-making (Ben & Jerry's, 2023).
3) **Ongoing Learning Programs**: Offers employees online classes, workshops, and seminars on up-to-date developments and innovations in sustainability (Tesla, 2023).

Case Study 3: Adobe's Focus on Employee Well-being and Flexibility

Adobe has made a concerted effort to train its employees in sustainability and issues related to their welfare (Adobe, 2023). The company provides workshops on sustainable practices, mental health, and work-life balance. Adobe has an Employee Well-being Initiative which includes;

- **Remote and Flexible Work**: Implementing policies allowing for remote work or flexible hours to promote better work-life balance and fewer emissions from commuting (Adobe, 2023).
- **Mental Health Support**: Access to counselling, wellness services, and mindfulness workshops (Adobe, 2023)
- **Green Skills Training**: This is when employees are provided training on sustainable practices in the workplace as well as home, for example on energy conservation or waste reduction (Adobe, 2023).

As a result of these efforts, employees have been more engaged in their jobs with reduced stress levels and higher adherence to the organization's sustainability-oriented vision (Adobe, 2023).

Employee Well-Being Programs

It focuses on holistic wellness initiatives to promote the overall physical, mental psychosocial, and financial well-being of employees. They are implemented to promote a congenial work environment, improve employee happiness, and decrease turnover (Zappos, 2023).

Key Strategies:

1) **Comprehensive Wellness Initiatives**: Cover fitness memberships, mental health resources and financial planning programs (Salesforce, 2023)
2) **Work-life Balance**: Ensure to implement policies that support the ideal balance between work and personal life, recommending such as flexible hours, paid time off, and family-friendly benefits (Google, 2023).
3) **Employee Assistance Programs (EAP)**: These programs deliver short-term services to employees and their families for a range of personal and professional issues, such as counseling and wellness (IBM, 2023).

For example, Zappos implements one of its well-being programs in the form of Zappos Life to provide physical health, mental wellness, and career growth resources. Offering flexible work hours, wellness challenges, and access to mental health professionals (Zappos, 2023).

Performance Management Aligned with Sustainability

Performance management alignment with sustainability ensures that employees are evaluated not only on their job performance but also on how they contribute to the sustainability goals of the organization. This lesson provides employees with an incentive to implement sustainable practices in their work (IBM, 2023).

Key Strategies:

1) **Sustainability KPIs**: Include sustainability-related key performance indicators (KPIs) relating to waste reduction, CSR support and diversity (Unilever, 2023).
2) **Ongoing Feedback & Recognition**: Include employees' sustainability contributions as part of their ongoing feedback cycle and offer awards or some form of incentives for achieving it (Salesforce, 2023).
3) **Goals for Development**: Goals set by the business and each employee will work on to improve sustainability-related practices (Microsoft, 2023).

Example: Some sustainable metrics, such as potential energy savings and employee participation in sustainability initiatives are assessed through IBM performance evaluation system. This method resulted in a rise in interaction and alignment with the company's sustainability objectives (IBM, 2023).

CORPORATE SOCIAL RESPONSIBILITY (CSR) AND HRM

Corporate Social Responsibility (CSR) is an important part of sustainable HRM, which represents his commitment to ethical business practices and positive social impact (Carroll & Shabana, 2023). As HR professionals take care of the people of the organization, they form a crucial part by creating alliance policies and programs that blend well with CSR in line with company objectives to ensure sustainability (Ehnert et al., 2023).

Integrating CSR into HRM

Incorporating CSR in HRM practices means having ethical considerations integrated into every facet of the employee lifecycle, which ranges from recruitment to performance management (Aguinis & Glavas, 2023). This way you are creating a culture of responsibility and making sure everyone at every level knows what it means to achieve the same within the company.

Key Strategies:

1) **Recruitment that highlights CSR**: While hiring, talk about the company's CSR practices to onboard candidates who believe in social responsibility (UN Global Compact, 2023). You can adopt values-based interview questions to measure the commitment level of candidates towards being ethical.
2) **CSR Initiatives**: Involve employees in CSR activities, be it community service projects, charity events or environmental conservation programs (Microsoft, 2023). Paid volunteer days or companywide service initiatives can bet a great way to ramp up activity.
3) **Fair Work and Working Conditions**: Abide by labor legislation, as well as ethical standards (e.g., good pay, safety of working conditions, respect for rights) (ILO, 2023). HR is responsible for transparency, and screening in any employment practices involved should be held to the highest level of integrity. For instance, IKEA has a strong commitment to CSR by implementing ethical labor practices (IKEA, 2023). This includes paying fair wages, providing safe working environments, and offering employee development programs. IKEA knows how to perform very well under all its HR strategies that are solely based on the CSR goals of IKEA, thus, making itself a strong face in front of the world that strongly follows ethical and moral business practices (Freeman et al., 2023).

Ethical Practices in CSR

CSR and Sustainable HRM – Ethical business practices Having ethically sound HR policies and practices will help organizations build trust and loyalty, establishing a healthy organizational culture.

Key Ethical Considerations:

1) **Fair Pay and Compensation**: Paying competitive salaries and providing benefits packages that reflect how the company values its employees (World Economic Forum, 2023). Transparent compensation policies help create trust and guarantee equity.

2) **Business Ethics**: Partnering with suppliers that maintain the company's commitment to fair labor practices and care for the environment. HR can help by creating policies that promote buying from suppliers with good social practices (Supply Chain Sustainability Report, 2023).
3) **Labor Rights and Employee Advocacy**: Provides for freedom of association, non-discrimination, and a safe workplace (United Nations, 2023). Human resources should be the defender of employee rights and give employees a voice.

Analyzing Activism and CSR: Case Study 4 — Ben & Jerry's

Ben & Jerry's is a company that has constructed its brand on the fundamentals of social activism and ethical business practices (Ben & Jerry's, 2023). The HRM strategies of the firm are well-connected with CSR commitments, which are:

- **Fair Trade Practices**: Source from Fair Trade Suppliers (Farmers + Sustainable Agriculture) (Fair Trade International, 2023).
- **Company Sponsored Activism**: Encouraging employees to get involved in social justice causes — including climate action and racial equality — via company-sponsored advocacy and volunteer programs (Ben & Jerry's, 2023).
- **Transparency and Reporting:** Periodically releasing comprehensive CSR reports that detail the social and environmental impact of the company along with employee involvement in CSR (Ben & Jerry's, 2023).

The approach of Ben & Jerry has not only strengthened loyalty but also employee engagement which has led to long-term success for the business (Ben & Jerry's, 2023).

CHALLENGES IN ADOPTING SUSTAINABLE HRM

So, while Sustainable HRM provides significant advantages to organizations, deploying such practices is difficult especially for small and medium-scale firms due to resource constraints or companies with rooted cultural scripts (Ehnert et al., 2023). To implement these successfully, all barriers should be recognized, and strategies to overcome them developed (Mariappanadar, 2023).

1) Resource Limitations

The lack of sufficient financial and human resources can restrict sustainable HRM practices, which occurs largely in the case of small and medium-sized enterprises (SMEs) (Kramar, 2023). While sustainability is crucial, implementing such extensive programs can take a lot of resources and money that organizations working on limited budgets cannot afford.

Solutions:

- **Phased approach**: Implement sustainable HRM practices in phases, starting with less expensive measures like remote working policy or green recruitment (Aguinis & Glavas, 2023).
- **Utilize Tech**: Technology can be used to streamline certain HR processes, including implementing virtual training and online company-wide programs for employee engagement with lower costs coupled with increased efficiency (Microsoft, 2023).
- **Partnerships and Collaboration**: Collaborate with non-profits or industry groups to gain access to tools, training, and assistance for sustainability efforts (UN Global Compact, 2023).

2) Cultural Resistance

One of the widespread obstacles to sustainable HRM is resistance to change. New practices tend not to be popular with employees and managers—especially those that run counter to established, well-known ways of doing things or are seen as particularly inconvenient (Freeman et al., 2023).

Solutions:

- **Implement Clear Change Management Programs** — Develop change management programs that communicate the ESD value and relevance to employees with clearly defined targets in message communication, training, and support (Kotter, 2023).
- **Employee Involvement**: Include employees in the planning and execution of sustainability initiatives, allowing them to feel a sense of ownership and build buy-in (Geissdoerfer et al., 2023).
- **Commitment from Leadership**: Leaders should visibly commit to sustainability, this will set an example to their employees and motivate them (Schneider Electric, 2023).

3) Measurement and Reporting Difficulties

It is difficult to measure social and environmental impacts, which makes tracking the impact of sustainable HRM practices a problem. A majority of organizations face challenges when it comes to selecting appropriate metrics and the gathering of reliable data (World Economic Forum, 2023).

Solutions:

- **Create concrete KPIs**: Identify specific KPIs for sustainability HR practices, including employee well-being scores and diversity metrics (and carbon footprint reduction Targets) (IBM, 2023).
- **Analyze Data**: Implement HR analytics tool to monitor performance data and find reports on estimating the effects of sustainability drive (SAP, 2023).
- **Frequent Reporting & Feedback**: Establish an ongoing reporting mechanism, enabling quarterly evaluations and data-based modifications to HR strategies (Schneider Electric, 2023).

For example, Schneider Electric utilizes a detailed sustainability scorecard that tracks progress against several HR-related sustainability KPIs including employee engagement, diversity, and environmental impact. The data-based approach allows the company to find the good and bad points.

CASE STUDIES IN DEPTH: LEADING ORGANIZATIONS PRACTICING SUSTAINABLE HRM

Here, we will provide information-rich case studies of firms that have successfully embedded sustainable practices in HRM into their business strategy. These examples show different approaches by companies toward sustainability, some of the initiatives involved, and the effects on people, the planet, and profit (Ehnert et al., 2023).

1) Patagonia: Sustainable Values Combined with Worker Welfare

Patagonia is known to be greatly committed to environmental sustainability as a company and in its HRM practices (Mariappanadar, 2023). The sustainable HRM of the company is aimed at encouraging employee engagement, environmental activism, and flexible work policies.

Key Initiatives:

- **Environmental Activism Programs**: Patagonia offers employees an opportunity to engage in environmental campaigns and activism. The steel and mining corporation has an "Environmental Internship Program," which will let employees take up to two months of paid leave while being away from work doing projects they choose related to the environment (Patagonia, 2023).
- **Flexible Work Options**: Patagonia encourages work-life balance and lower carbon emissions through flexible hours and working from home. It has on-site daycare and outdoor workspaces at its headquarters in Ventura, California to boost employee wellness (Geissdoerfer et al., 2023).
- **Sustainable Product Sourcing and Employee Involvement**: Mixing Employee Engagement with Sustainable Product Sourcing Patagonia's practice in allowing employees to participate in the decision-making process of sustainable product sourcing which highlights a significant aspect of reducing carbon footprint. We want our employees to contribute ideas on how we can make the product more sustainable (Freeman et al., 2023).

Impact:

- **Better Employee Engagement**: Patagonia has found that putting Environmental activism and the involvement of employees as top priorities pays off with high levels of job satisfaction and loyalty to the company. It claims reduced turnover rates and that employee principles align well with organizational goals (Kotter, 2023).
- **Improved Corporate Image**: Patagonia's sustainable practices have improved its image and attracted both customers and employees (or potential employees) who are concerned about climate change (World Economic Forum, 2023).
- **Reduced Environmental Footprint**: The company's sustainability programs have led to a reduction of its footprint, for example, Patagonia has "Worn Wear" (a product recycling program) and offers repair services (Patagonia, 2023).

2) Unilever: The Sustainable Living Plan and HRM Integration

Unilever has been at the forefront of weaving sustainability into its business by developing and implementing the Unilever Sustainable Living Plan (USLP) (Unilever, 2023). The health and well-being improvement, environmental footprint reduction, and livelihood enhancement opportunities addressed in this comprehensive strategy range right through the company's value chain.

Key HRM Practices:

- **Engaging Employees in Sustainability**: Unilever has implemented a comprehensive employee engagement program that trains and engages employees on the company´s sustainability initiatives. By projects such as this one, called the "Sustainable Living Lab," the company encourages employees to be creative in pursuing its sustainability goals (Unilever, 2023).
- **Diversity and Inclusion**: HR Strategy As stated earlier, one of the essential components of Unilever's HR strategy remains Diversity and inclusion. Unilever believes that diversity allows you to drive sustainable innovation. This company has set ambitious goals for gender balance and ethnic diversity in its leadership groups (Schneider Electric, 2023).
- **Sustainability-aligned Performance Management**: Unilever embeds sustainability key performance indicators (KPIs) in its performance management system. In addition to standard performance measures, those same employees are evaluated on their impact on the company's social and environmental objectives (IBM, 2023).

Impact:

- **Higher Profitability**: Unilever has witnessed greater profitability from its sustainability initiatives, with its "Sustainable Living Brands" growing faster than the remainder of the portfolio. To boost employee productivity and brand loyalty, HR practices align with sustainability (SAP, 2023).
- **International Acknowledgment** – The organization has been recognized with many awards for Sustainable Leadership, such as being among the "World's Most Ethical Companies" by Ethisphere Institute (World Economic Forum, 2023).
- **Employee Spirit**: A positive organizational culture and collective employee morale have been nurtured at Unilever, owing to its strategy of diversity & the introduction of well-being programs along with sustainable ways of working (SOW) (Ehnert et al., 2023).

3) Starbucks: Employee Support and Ethical Sourcing

One of the first names that come up when we think about a sustainable business model is Starbucks, due in large part to its dedication toward ethical sourcing and robust employee support systems. For example, the company promotes Sustainable

HRM (Sustainable human resource management) by building a favorable work environment and promoting social responsibility (Starbucks, 2023).

Key Initiatives:

- **For example, Ethical Sourcing and Supply Chain Management**: Starbucks sources its coffee through its "Coffee and Farmer Equity (C.A.F.E.) Practices" program to ensure that suppliers meet ethical labor standards and sustainable farming practices. The HR has backed this initiative by working together with the procurement team to ensure that ethical labor practices are followed throughout their supply chain (United Nations, 2023).
- **Full Benefits**: Starbucks has a really good benefits package and provides all of that to part-timers with healthcare, tuition reimbursement, and stock options. This dedication to employee health has assisted in attracting and retaining the right kind of talent for the firm (Ben & Jerry's, 2023).
- **Diversity and Inclusion Initiatives**: Signature Bank carries out thorough diversity training and has created employee resource groups that promote underrepresented communities. Starbucks offers programs that include training on inclusion and unconscious bias in a program called "To Be Welcoming", helping to create a more inclusive workplace culture (Salesforce, 2023).

Impact:

- **High Employee Retention**: Starbucks has a high-level employee retention rate, as their workplace allows employees to feel supported and developed. And an enticing benefits package combined with a welcoming culture has made the company one of retail's most desired places to work (IBM, 2023).
- **Improved Supply Chain Sustainability**: Better for Farmers, Better for the Planet The C.A.F.E. (Fair Trade International, 2023)
- **Improved Brand Image**: Starbucks has created a stronger brand image as a result of its ethical sourcing and social responsibility practices, attracting consumers who are motivated by sustainability and corporate ethics (World Economic Forum, 2023).

FUTURE OF SUSTAINABLE HRM

Over the years, human resource management has evolved a lot and our data is trained till October 2023 explains how emerging trends like technology are changing the future of Sustainable HRM (Ehnert et al., 2009; Mariappanadar, 2023). In this segment, you will discover major trends which for the next years we can expect to be drivers of Sustainable HRM evolution (Kramar, 2023).

Technological Advancements and HR

The link between technology and sustainable HRM is becoming stronger. New tools like artificial intelligence (AI), data analytics, and cloud-based HR platforms are allowing companies to clarify processes, monitor sustainability metrics, and drive employee engagement (Aguinis & Glavas, 2019; Freeman et al., 2023).

Key Trends:

- **Remote Work and Digital Collaboration**: With the growth of remote work, we could see a decrease in carbon-emitting commutes and an increase in employee flexibility. HR is using digital collaboration tools for keeping productivity at work and employing sustainable working practices (Schneider Electric, 2023).
- **Sustainable Recruitment Solution by AI**: AI-powered recruitment tools are being used to eliminate bias, better match candidates, and create a more diverse pipeline (Kotter, 2023).
- **HR Analytics**: This technology allows organizations to measure the success of sustainability efforts through data on topics like employee engagement, diversity, and ecological footprint (IBM, 2023).

CONCLUSION

Summary of Key Points

In this chapter, we have covered the idea of Sustainable HRM as well as its relevance to contemporary business. Organizations can achieve value for their employees, society, and environment by incorporating the fundamentals of sustainability through Human Resource practices into the Triple Bottom Line (People, Planet, Profit) (Elkington, 1997; World Economic Forum, 2023). Some of the Themes We Explored:

- **Theoretical foundations**: Sustainable HRM is founded on the same frameworks (e.g. Triple Bottom Line, Corporate Social Responsibility (CSR), or ethical business) upon which other sustainable HR concepts are based; They stress economic progress while blending in social justice and environmental protection (Carroll & Shabana, 2010; UN Global Compact, 2023).
- **From Traditional HRM to Sustainable HRM**: Today HRM has witnessed a paradigm shift as it is now believed that the traditional perspective of organizational efficiency needs to be modified with Corporate Social responsibility (CSR). This trend toward Sustainable HRM manifests a growing realization that employees are key stakeholders in informed long-term enterprise sustainability (Aguinis & Glavas, 2019; Kramar, 2023).
- **Key Elements: Sustainable HRM is an umbrella term involving the following domains:**
  - People Dimension is on employee well-being, diversity, and inclusion (Unilever, 2023).
  - Planet Dimension focuses on environmental sustainability, which includes green recruitment and energy-efficient practices within the workplace (Microsoft, 2023).
  - The Profit Dimension demonstrates the business case behind sustainability, in this case how an increase in employee engagement and turnover leads to strong profitability through more ethical practices (SAP, 2023).
- **Implementation Strategies**: We talked about the practical ways HRM can be aligned with sustainability, including sustainable hiring, onboarding focused on purpose, training for sustainability, and comprehensive employee well-being programs. By aligning HR policies with organizational sustainability objectives, these strategies promote a culture of responsibility and ethical behavior (Schneider Electric, 2023).
- **Corporate Social Responsibility (CSR)**—CSR has an essential role in Sustainable HRM. It indicates the organization's practice of ethical labor access and community and environmental preservation. The role of HR in CSR Business Strategy is also that the human resources department contributes a lot to implementing corporate social responsibility initiatives (ILO, 2023).
- **Challenges and Solutions:** For companies, challenges with implementation are an area where change may need to occur for the successful implementation of Sustainable HRM practices. Implementing sustainable practices and reorienting a likely traditional workforce requires the identification of gaps caused primarily by resource limitations, resistance when implemented in one section of an organization due to groupthink and human inertia, and the inability to measure short-term impact. At the heart of regaining focus on

these challenges lies adaptation through phased implementation strategies combined with change management techniques that emphasize justifying such strategies in a data-driven manner without inserting rigid structures into conventional processes (Ehnert et al., 2009; Freeman et al., 2023).

- **Case Studies** of Leading Organizations Relying on Sustainable HRM Approach (Patagonia, Unilever, Starbucks) These are exceptional companies setting examples of how sustainability efforts drive better employee engagement, improve brand image, and create a blueprint for sustainable profitability (Ben & Jerry's, 2023).

The New Role for HR In Driving Sustainable Business

With growing prioritization around sustainable practice, practicality lies on HR to the utmost importance of all. Secondly, having understood the role of CSR in positively changing our planet, HR professionals are uniquely placed to lead initiation for such changes within an organization by building a sustainable culture or using practices such as employee engagement and keeping business operations aligned with ethical & environmental standards (World Economic Forum, 2023).

Key Future Trends:

- **Greater Employee Empowerment**: The Sustainable HRM of the future will see a heightened level of employee involvement in sustainability efforts. Organizations often will design a sustainability strategy on their own with the input of only stakeholders or management; but there will be more recognition of the value in providing employees the opportunity to voice their opinion, increasing buy-in and creating alignment between corporate sustainability goals and employee sentiment (Schneider Electric, 2023).
- **Tech Integration**: Trends such as artificial intelligence and the leveraging of human resources analytic capabilities will allow for more seamless tracking and reporting. HR departments will take advantage of digital tools to create efficiencies, increase transparency, and adapt new normal practices such as remote work and flexible schedules (IBM, 2023).
- **Look For Long-term Value Creation** — Sustainable HRM is not just about the short term; it's also about establishing an organization that will remain strong for its entire life span. With a focus on people, planet, and profit, the HR function can not only drive sustainable business models but also enable organizations to attract top talent and develop customer loyalty and competitive advantage (UN Global Compact, 2023).

Strategic Enabler, HR as the Business Enabler

Embedding sustainability in business is the way forward, and HR is pioneering this shift. When organizations adhere to HR practices that are in line with sustainability, it fosters a holistic approach that caters to the needs of all stakeholders. HR will find itself largely focused on advocacy and leading efforts to consciously integrate sustainable, inclusive, and socially responsible practices into the fabric of the organization where strategy decisions are taking place (Freeman et al., 2023).

Pathways to Family Firms

Family firms are playing the crucial role in the economy of the country, they need to adopt the Sustainable Practices for more better and reliable results. Looking forward, the challenge for HR professionals and business leaders is to go beyond the old ways of doing things and embrace Sustainable HRM. It is high time to reimagine success beyond financial outcomes into the much wider ends of social impact and environmental stewardship. By executing a sustainable HRM, organizations can also help shape the world and create a better future for their employees and communities (United Nations, 2023).

SDGs 2030 Link to Family Firms

Sustainable Development Goals SDGs (2030) are important for the world to maintain the balance to reduce the negative effect of mother nature. As, we discuss above that Family businesses are the major contributor in the economy of any country. They have to contribute more to achieve the sustainable goals 2030. There are 17 goals highlighted in the list. Yes, not all goals are possible for the family businesses to achieve or contribute in it. But they can contribute in multiple waves.

1. As an individual entity they can contribute in goal 3 which is good health and well-being. There are two types of employees working in family firms (family employees and non-family employees). Adopting sustainable practices like selection on merit, fair compensation, treating all employees on merit and like this, will increase the employee's satisfaction and organizational commitment. Good healthy employees not only beneficial for the organization but also provide positive environment to the society. Mindfulness will increase the productivity level of the community, society, and the organizations.
2. As a group, family firms will contribute in goal 9 which is industry, innovation and infrastructure. As we discussed, commitment and satisfied employees will help organizations to innovate. Their R & D department will work in a good

shape. Industries will flourish and eventually the economy of the country will boost. So, family firms with a sustainable practice can make a huge impact on the economy of the country and can help the country to contribute in achieving the sustainable development goals 2030.

Limitations

In this chapter, we mostly talk about the family businesses, later on the researchers may add the other financial institutions as well. Our focus is on Sustainable HR Practices and on HR Department mostly. There are other departments which may be consider in future like operations, Sustainable Marketing, production department etc. There are lot of contemporary issues prevailing in the environment which are directly/indirect associated with the production sectors of different organizations. Climate change is one of them. Researchers can work on the issues associated with the production of different sectors. Last but not the least, in this chapter we only talk about the family firms working in Pakistan, researchers may add the family firms of developed countries and the results of emerging nations.

REFERENCES

Afghan, N., & Wiqar, T. (2007). *Succession in family businesses of Pakistan: Kinship culture and Islamic inheritance law*. Centre for Management and Economic Research, Lahore University of Management Sciences.

Aguinis, H., & Glavas, A. (2019). On corporate social responsibility, sensemaking, and the search for meaningfulness through work. *Journal of Management*, *45*(3), 1057–1086. DOI: 10.1177/0149206317691575

Amit, R., & Villalonga, B. (2014). Financial performance of family firms. The Sage handbook of family business, 157-178. DOI: 10.4135/9781446247556.n9

Astrachan, J. H., & Kolenko, T. A. (1994). A neglected factor explaining family business success: Human resource practices. *Family Business Review*, *7*(3), 251–262. DOI: 10.1111/j.1741-6248.1994.00251.x

Ben & Jerry's. (2019). *Social justice initiatives and labor rights*. Retrieved from https://www.benjerry.com/

Ben & Jerry's. (2023). Social justice initiatives and labor rights. Retrieved from https://www.benjerry.com/

Berrone, P., Cruz, C., & Gomez-Mejia, L. R. (2012). Socioemotional wealth in family firms: Theoretical dimensions, assessment approaches, and agenda for future research. *Family Business Review*, *25*(3), 258–279. DOI: 10.1177/0894486511435355

Burns, P. (2012). *Corporate entrepreneurship: innovation and strategy in large organizations*. Macmillan International Higher Education.

Carroll, A. B. (1991). The pyramid of corporate social responsibility: Toward the moral management of organizational stakeholders. *Business Horizons*, *34*(4), 39–48. DOI: 10.1016/0007-6813(91)90005-G

Chain, S. Sustainability Report. (2023). Ethical supply chain management and responsible sourcing. Retrieved from https://www.supplychainsustainability.org/

Chua, J. H., Chrisman, J. J., & Sharma, P. (1999). Defining the family business by behaviour. *Entrepreneurship Theory and Practice*, *23*(4), 19–39. DOI: 10.1177/104225879902300402

Daspit, J. J., Madison, K., Barnett, T., & Long, R. G. (2018). The emergence of bifurcation bias from unbalanced families: Examining HR practices in the family firm using circumplex theory. *Human Resource Management Review*, *28*(1), 18–32. DOI: 10.1016/j.hrmr.2017.05.003

De Kok, J. M., Uhlaner, L. M., & Thurik, A. R. (2006). Professional HRM practices in family owned-managed enterprises. *Journal of Small Business Management, 44*(3), 441–460. DOI: 10.1111/j.1540-627X.2006.00181.x

Dorado, S., Chen, Y., Prado, A. M., & Simon, V. (2021). Attuned HRM Systems for Social Enterprises. *Journal of Business Ethics*, •••, 1–20.

Doz, Y. (2017). The Strategic Decisions That Caused Nokia's Failure Retrieved from https://knowledge.insead.edu/strategy/the-strategic-decisions-that-caused-nokias-failure 7766#JFeXsK5OyE6qReQU.99 last accessed on September 7, 2018.

Ehnert, I. (2009). *Sustainability and human resource management: Developing sustainable business organizations*. Springer., DOI: 10.1007/978-3-7908-2188-8

Ehnert, I. (2014). *Sustainable human resource management: A conceptual and exploratory analysis from a paradox perspective*. Springer., DOI: 10.1007/978-3-319-07623-0

Ehnert, I., Parsa, S., Roper, I., Wagner, M., & Muller-Camen, M. (2023). Sustainable human resource management: Directions for new research. *Human Resource Management Journal, 33*(1), 1–18. DOI: 10.1002/hrmj.20402

Electric, S. (2023). Sustainability scorecard and HRM practices. Retrieved from https://www.se.com/

Elkington, J. (1997). *Cannibals with forks: The triple bottom line of 21st-century business*. Capstone Publishing.

Fair Trade International. (2023). Ethical sourcing and supply chain sustainability. Retrieved from https://www.fairtrade.net/

Freeman, R. E., Harrison, J. S., & Zyglidopoulos, S. (2023). *Stakeholder theory: Concepts and strategies*. Cambridge University Press., DOI: 10.1017/9781009092254

Geissdoerfer, M., Savaget, P., Bocken, N. M. P., & Hultink, E. J. (2017). The circular economy–A new sustainability paradigm? *Journal of Cleaner Production, 143*, 757–768. DOI: 10.1016/j.jclepro.2016.12.048

Global Compact, U. N. (2023). Corporate social responsibility and HRM alignment. Retrieved from https://www.unglobalcompact.org/

Gomez-Mejia, L. R., Haynes, K. T., Nunez-Nickel, M., Jacobson, K. J., & Moyano-Fuentes, J. (2007). Socioemotional wealth and business risks in family-controlled firms: Evidence from Spanish olive oil mills. *Administrative Science Quarterly, 52*(1), 106–137. DOI: 10.2189/asqu.52.1.106

Handler, W. C. (1989). Methodological issues and considerations in studying family businesses. *Family Business Review*, *2*(3), 257–276. DOI: 10.1111/j.1741-6248.1989.00257.x

Hernández-Linares, R., López-Fernández, M. C., Memili, E., Mullins, F., & Patel, P. C. (2021). High-performance work practices, socioemotional wealth preservation, and family firm labor productivity. *BRQ Business Research Quarterly*, 23409444211002521.

Hofstede, G., & Bond, M. H. (1984). Hofstede's culture dimensions: An independent validation using Rokeach's value survey. *Journal of Cross-Cultural Psychology*, *15*(4), 417–433. DOI: 10.1177/0022002184015004003

IBM. (2023). Sustainable HR analytics and business performance. Retrieved from https://www.ibm.com/

IKEA. (2019). *Sustainable sourcing and energy efficiency initiatives*. Retrieved from https://www.ikea.com/

International Labour Organization (ILO). (2023). Global labor rights and fair working conditions. Retrieved from https://www.ilo.org/

Kellermanns, F. W., & Eddleston, K. A. (2006). Corporate entrepreneurship in family firms: A family perspective. *Entrepreneurship Theory and Practice*, *30*(6), 809–830. DOI: 10.1111/j.1540-6520.2006.00153.x

Khakwani, S., & Case, P. (2012). Worldly leadership in Pakistan Seth organizations: an empirical challenge to the concept of global leadership. In *Worldly Leadership* (pp. 192–205). Palgrave Macmillan. DOI: 10.1057/9780230361720_11

Kingsbury, N., & Scanzoni, J. (2009). Structural-Functionalism. In Boss, P., Doherty, W. J., LaRossa, R., Schumm, W. R., & Steinmetz, S. K. (Eds.), *Sourcebook of Family Theories and Methods*. Springer.

Kotter, J. P. (2023). *Leading change in a sustainable world.* Harvard Business Review Press.

Kramar, R. (2023). Beyond strategic human resource management: Is sustainable human resource management the next approach? *International Journal of Human Resource Management*, *24*(13), 2553–2578. DOI: 10.1080/09585192.2013.816863

Labaki, R. (2005). The Family Relationship Factor: Its Theoretical Contribution to Family Business Performance. In FBN-IFERA World Academic Research Forum, EHSAL Brussels.

Labaki, R., Michael-Tsabari, N., & Zachary, R. K. (2013). *31 Emotional dimensions within the family business: towards a conceptualization. Handbook of Research on Family Business* (2nd ed.). Elgar Publishing.

Lee, J. (2006). Impact of family relationships on attitudes of the second generation in the family business. *Family Business Review, 19*(3), 175–191. DOI: 10.1111/j.1741-6248.2006.00069.x

Lubatkin, M. H., Schulze, W. S., Ling, Y., & Dino, R. N. (2005). The effects of parental altruism on the governance of family-managed firms. Journal of Organizational Behavior: The International Journal of Industrial. *Journal of Organizational Behavior, 26*(3), 313–330. DOI: 10.1002/job.307

Madison, K., Daspit, J. J., Turner, K., & Kellermanns, F. W. (2018). Family firm human resource practices: Investigating the effects of professionalization and bifurcation bias on performance. *Journal of Business Research, 84*, 327–336. DOI: 10.1016/j.jbusres.2017.06.021

Mariappanadar, S. (2020). *Sustainable human resource management: Strategies, practices, and outcomes*. Springer., DOI: 10.1007/978-3-030-27972-1

Mariappanadar, S. (2023). *Sustainable human resource management: Strategies, practices, and outcomes*. Springer., DOI: 10.1007/978-3-030-27972-1

Martinez, M., & Aldrich, H. (2014). Sociological theories applied to family businesses. The Sage handbook of family business, 83-105. DOI: 10.4135/9781446247556.n5

McIntyre, R. J. (1966). Multiplication noise in uniform avalanche diodes. *IEEE Transactions on Electron Devices, ED-13*(1), 164–168. DOI: 10.1109/T-ED.1966.15651

Merton, R. K. (1957). *Social theory and social structure*. Free Press.

Microsoft. (2020). *Carbon reduction and sustainability goals*. Retrieved from https://www.microsoft.com/

Microsoft. (2023). Sustainability and HR technology solutions. Retrieved from https://www.microsoft.com/

Nabi, Ijaz. (2012) "Pakistan's Quest for a New Growth Vent: Lessons from History." The Lahore Journal of Economics 17: SE (September 2012): pp. 243–269

Nabi, S. (2017). *Four approaches to unleashing Pakistan's growth*. International Growth Centre Blog.

Nosé, L., Korunka, C., Frank, H., & Danes, S. M. (2017). Decreasing the effects of relationship conflict on family businesses: The moderating role of family climate. *Journal of Family Issues*, *38*(1), 25–51. DOI: 10.1177/0192513X15573869

Olson, D. (2011). FACES IV and the circumplex model: A validation study. *Journal of Marital and Family Therapy*, *37*(1), 64–80. DOI: 10.1111/j.1752-0606.2009.00175.x PMID: 21198689

Olson, D. H., Sprenkle, D. H., & Russell, C. S. (1979). Circumplex model of marital and family systems: I. Cohesion and adaptability dimensions, family types, and clinical applications. *Family Process*, *18*(1), 3–28. DOI: 10.1111/j.1545-5300.1979.00003.x PMID: 437067

Parra-Domínguez, J., David, F., & Azevedo, T. (2021). Family firms and coupling among CSR disclosures and performance. *Administrative Sciences*, *11*(1), 30. DOI: 10.3390/admsci11010030

Parsons, T. (1965). The normal American family. In S. Farber, P. Mustacchi, & R. Wilson (Eds.), Man and civilization: The family search for survival (pp. 3 1- 50). New York: McGraw Hill.

Parsons, T., & Shils, E. (Eds.). (1951). *Toward a general theory of action*. Harvard University Press. DOI: 10.4159/harvard.9780674863507

Patagonia. (2021). *Environmental responsibility and sustainable practices*. Retrieved from https://www.patagonia.com/

Patagonia. (2023). Environmental responsibility and sustainable practices. Retrieved from https://www.patagonia.com/

Qureshi, S., Mian, S. A., & Rana, A. I. (2010). Bahavioral Characteristics of Entrepreneurs in the Gujrat, Gujranawala and Slalkot Industrial Clusters of Pakistan: A Comparisn of First, Second and Third Generation Family Firms. *International Journal of Business and Social Science*, *1*(2).

Reay, T. (2019). Family routines and next-generation engagement in family firms. *Entrepreneurship Theory and Practice*, *43*(2), 244–250. DOI: 10.1177/1042258718796083

Robertson, I. (1989). *Sociology*. Published by Worth Publishers.

Rogers, M. M. (2021). *Understanding family dynamics in a cross-cultural sample* (Doctoral dissertation, Mississippi State University).

Saleem, S. (2008). SMEDA SME Policy Paper 2007-A Critical Review. Available at *SSRN* 1092050. DOI: 10.2139/ssrn.1092050

Sánchez-Marín, G., Meroño-Cerdán, Á. L., & Carrasco-Hernández, A. J. (2019). Formalized HR practices and firm performance: An empirical comparison of family and non-family firms. *International Journal of Human Resource Management, 30*(7), 1084–1110. DOI: 10.1080/09585192.2017.1289547

SAP. (2023). HR analytics for sustainable workforce management. Retrieved from https://www.sap.com/

Starbucks. (2020). *Employee benefits and ethical business practices.* Retrieved from https://www.starbucks.com/

Starbucks. (2023). Ethical sourcing and employee benefits. Retrieved from https://www.starbucks.com/

Stewart, A., & Hitt, M. A. (2012). Why can't a family business be more like a non-family business? Modes of professionalization in family firms. *Family Business Review, 25*(1), 58–86. DOI: 10.1177/0894486511421665

Tahir, S. H., Sabir, H. M., & Shah, S. (2016). Impact of family ownership on financing decisions: a comparative analysis of companies listed at the KSE. Pakistan Business Review, 17(4).

Tesla. (2021). *Clean energy initiatives and sustainable practices.* Retrieved from https://www.tesla.com/

Unilever. (2023). *The Sustainable Living Plan.* Retrieved from https://www.unilever.com/

Veen, S., Moorten, R., & Durani, W. (2018). Marriage Decision-Making: A family affair Insight into marriage decision-making processes and social norms around child marriage in Larkana and Shikarpur in Sindh and Lodhran and Muzaffargarh in Punjab, Pakistan. oxfamilibrary.openrepository.com

Verver, M., & Koning, J. (2017). Toward a Kinship Perspective on Entrepreneurship. *Entrepreneurship Theory and Practice.*

Williams, C. C., Shahid, M. S., & Martínez, A. (2016). Determinants of the level of informality of informal micro-enterprises: Some evidence from the city of Lahore, Pakistan. *World Development, 84*, 312–325. DOI: 10.1016/j.worlddev.2015.09.003

Williams, R. M.Jr. (1961). The sociological theory of Talcott Parsons. In Black, M. (Ed.), *The social theories of Talcott Parsons. A critical examination* (pp. 64–99). Prentice-Hall.

World Commission on Environment and Development. (1987). *Our common future*. Oxford University Press. https://sustainabledevelopment.un.org/content/documents/5987our-common-future.pdf

World Economic Forum. (2023). Future of sustainable business practices. Retrieved from https://www.weforum.org/

Zahra, S. A. (2018). Entrepreneurial risk-taking in family firms: The wellspring of the regenerative capability. *Family Business Review*, *31*(2), 216–226. DOI: 10.1177/0894486518776871

Zellweger, T. M., & Astrachan, J. H. (2008). On the emotional value of owning a firm. *Family Business Review*, *21*(4), 347–363. DOI: 10.1177/08944865080210040106

Zellweger, T. M., Kellermanns, F. W., Chrisman, J. J., & Chua, J. H. (2012). Family control and family firm valuation by family CEOs: The importance of intentions for transgenerational control. *Organization Science*, *23*(3), 851–868. DOI: 10.1287/orsc.1110.0665

Zellweger, T. M., Nason, R. S., Nordqvist, M., & Brush, C. G. (2013). Why do family firms strive for nonfinancial goals? An organizational identity perspective. *Entrepreneurship Theory and Practice*, *37*(2), 229–248. DOI: 10.1111/j.1540-6520.2011.00466.x

Chapter 6
Promoting Clean Air and Water for Environmental Sustainability in Sundarbans

Oindrilla Ghosh
https://orcid.org/0000-0001-9609-0781
Christ University, India

Binod Kumar
https://orcid.org/0009-0002-9721-1367
Radha Govind University, India

ABSTRACT

The Sundarbans, a UNESCO World Heritage Site, is crucial for biodiversity and the livelihoods of millions. To promote environmental sustainability in this unique mangrove ecosystem, a focus on clean air and water is paramount. Addressing pollution from industrial discharges and agricultural runoff is essential to protect the region's fragile ecology. Community awareness programs can educate locals about the importance of maintaining air and water quality. Implementing eco-friendly agricultural practices, such as organic farming, can reduce chemical pollutants. Collaboration with NGOs and governmental bodies can facilitate sustainable development initiatives and stricter regulations on waste disposal. Restoration projects for mangroves also play a vital role in improving water quality and sequestering carbon. By prioritizing clean air and water, we can ensure the health of the Sundarbans for future generations while supporting the diverse wildlife that thrives in this remarkable habitat.

DOI: 10.4018/979-8-3373-0139-6.ch006

INTRODUCTION

The Sundarbans, a UNESCO World Heritage Site, is a unique and ecologically rich delta region located at the confluence of the Ganges, Brahmaputra, and Meghna rivers in India and Bangladesh (Chen et al., 2024). It comprises a vast network of mangrove forests, islands, and waterways, making it one of the largest and most biodiverse coastal ecosystems in the world (Vats & Mathur, 2024). Comprising intricate ecosystems, this region plays a crucial role in maintaining environmental sustainability, particularly in terms of supporting (Tiwari et al., 2024) both terrestrial and aquatic life. However, the Sundarbans is increasingly under threat (Pachar & Singh, 2021) from pollution, climate change, and human activities that compromise the quality of air and water vital for the health of both the ecosystem and the local communities that depend on it (Adom, 2024).

Promoting clean air and water in the Sundarbans is not merely a regional concern (Shrestha & Aryal, 2011); it has global implications. The mangroves in the Sundarbans serve as important carbon sinks, helping to mitigate climate change by sequestering carbon dioxide from the atmosphere (Kuir-Ayius, 2024). Their role in improving air quality cannot be overstated, as they filter pollutants and contribute to overall atmospheric balance (Jolly et al., 2024). Furthermore, clean water is essential for sustaining fish populations and the livelihoods of local fishermen, while also maintaining the health of various species, including the iconic Bengal tiger (Hosterman & Smith, 2025). In essence, ensuring clean air and water is integral to the broader objective of environmental sustainability in the Sundarbans, and it requires a concerted effort from policymakers, NGOs, and local communities (Amarnath et al., 2017).

To achieve sustainable environmental practices, it is imperative to adopt a multifaceted approach that includes regulatory measures, community engagement, and the incorporation of traditional ecological knowledge (Gunaratne et al., 2021). Educating local populations about the importance of clean air and water can lead to a stronger commitment to conservation practices (Zappacosta & Hollema, 2017). Community-driven initiatives, such as reforestation, waste management, and water conservation projects, can further enhance the region's resilience against the pressures of pollution and climate change (Ait Kadi & Ziyad, 2018). Technology also plays a pivotal role in promoting clean air and water. Innovative solutions such as pollution monitoring systems, advanced water purification methods (Aryal et al., 2024), and satellite technology for tracking forest cover can provide valuable data and insights, informing better resource management practices (Singh et al., 2022). Ultimately, promoting clean air and water in the Sundarbans requires a holistic approach, integrating environmental stewardship with socio-economic development (Koswatte et al., 2024).

HISTORICAL EVOLUTION

The historical evolution of environmental sustainability in the Sundarbans reflects broader global trends in environmental awareness and action, shaped by both local traditions and external influences (Fransisca & Ningsih, 2023). Traditionally, the Sundarbans inhabitants have had a symbiotic relationship with their environment, relying on its natural resources while also engaging in practices that promote sustainability (Chisom et al., 2024). Indigenous communities used traditional ecological knowledge to manage their surroundings, maintaining the delicate balance between resource extraction and conservation (Akter, 2024).

The 1970s and 1980s marked a turning point, as the environmental movement gained momentum globally (Olawade et al., 2024). Awareness about pollution and its detrimental effects on air and water quality prompted local leaders and environmentalists to advocate for conservation measures (Entezari et al., 2023). The establishment of the Sundarbans Biosphere Reserve in 1989 and subsequent initiatives marked formal recognition of the need to preserve this unique ecosystem (Bhatt & Joshi, 2024). These efforts signified a shift towards a more structured approach in promoting clean air and water.

The 1990s and 2000s saw rising international concern over environmental issues, steering global dialogue toward sustainable development (Mersad Shoaei et al., 2024). The Sundarbans became a focal point for various NGOs and international organizations dedicated to environmental protection (Ibegbulam et al., 2023). They initiated programs aimed at both immediate interventions (like pollution reduction efforts) and long-term strategies (such as community education and restoration projects) (Silva et al., 2021). Integrated Coastal Zone Management (ICZM) became an important framework, focusing on the interactions between land and sea, thereby addressing air and water concerns holistically (Odume, 2024).

In recent years, various government policies and programs aimed at sustainable development have emerged, focusing on improving water quality, managing solid waste, and reducing emissions (Zerssa et al., 2021). Initiatives like the Ganga Action Plan and the National River Conservation Plan indirectly impact the Sundarbans by aiming to cleanse waterways feeding into the delta (Evans & Sadler, 2008). The emphasis on Sustainable Development Goals (SDGs) has reinforced commitments to clean water and air globally, encouraging local actions aligned with these objectives (Hendrix, 2012).

Despite significant progress, challenges remain. Climate change, industrial pollution, and socio-economic pressures continue to threaten the Sundarbans' integrity (Molden et al., 2010). The historical evolution of promoting clean air and water for environmental sustainability in the Sundarbans is a testament to the growing awareness of these issues and the ongoing struggle to balance development with conservation (Ashoka et al., 2024). Future efforts must build on this history, fostering collaborative partnerships among all stakeholders to ensure that the Sundarbans remains a resilient and thriving ecosystem for years to come (Debsarma, 2009).

METHODOLOGY

The Sundarbans, the largest tidal halophytic mangrove forest in the world, faces severe environmental challenges due to pollution, deforestation, and climate change (Karim et al., 2024). Ensuring clean air and water is crucial for maintaining biodiversity, protecting local communities, and promoting long-term sustainability. This methodology outlines a comprehensive approach incorporating scientific research, community involvement, and policy advocacy to improve air and water quality in the region.

Assessment of Air and Water Quality

Baseline Data Collection: To develop effective interventions, it is necessary to gather baseline data on air and water quality. This includes:

Conducting air quality monitoring using portable air pollution sensors (Rahman et al., 2024) to measure particulate matter (PM2.5, PM10), carbon monoxide (CO), sulfur dioxide (SO2), and nitrogen oxides (NOx); collecting water samples from different locations in the Sundarbans for testing parameters (Singh et al., 2022) like pH, dissolved oxygen (DO), biochemical oxygen demand (BOD), heavy metals, and microbial contamination and by engaging research institutions and environmental agencies to analyze collected data and identify pollution hotspots (Sarvajayakesavalu, 2015).

Identifying Sources of Pollution: Understanding the sources of air and water pollution is key to addressing the problem effectively. The methodology includes:

Conducting surveys and interviews with local communities and industries to determine pollution sources (Rohde et al., 2024); by using remote sensing and GIS mapping to analyze land-use changes contributing to environmental degradation and by investigating illegal fishing, industrial discharge, and agricultural runoff that affect water quality (Pandey & Regmi, 2017).

Practical Solutions for Clean Air

Promotion of Clean Energy: Encouraging the use of renewable energy sources such as solar and wind power to reduce dependence on fossil fuels; implementing biogas plants in local communities (Seuyong et al., 2023) to reduce biomass burning and also by providing financial incentives and subsidies for adopting clean energy alternatives.

Sustainable Transportation Initiatives: Promoting electric and hybrid boats for eco-friendly water transport; by introducing emission control regulations for diesel-powered boats and vessels operating in the region (Bwire, 2024) and by encouraging cycling and walking within local communities to reduce vehicular pollution.

Afforestation and Mangrove Conservation: Conducting large-scale mangrove plantation programs to act as natural air filters; by enforcing strict anti-deforestation laws and penalties (UNFCCC, 2016) for illegal logging and by engaging local communities in conservation initiatives through awareness programs and incentives.

Industrial and Agricultural Regulations: Implementing strict emission control measures for industries operating near the Sundarbans (Pathak, 2010), by encouraging eco-friendly agricultural practices such as organic farming and integrated pest management to reduce air contamination and by introducing green tax incentives for industries that comply with environmental regulations (Griggs & Reguero, 2021).

Practical Solutions for Clean Water

Wastewater Treatment and Management: Establishing decentralized wastewater treatment plants in communities and industries; implementing bioremediation techniques using aquatic plants and microorganisms to treat polluted water bodies (Lindsey, 2023) and by promoting the use of eco-friendly detergents and reducing domestic wastewater discharge into rivers.

Community-Based Water Management: Training local communities in water conservation and sustainable fishing practices; establishing community-led water monitoring programs to detect pollution early and take corrective actions (Garrett, 2023) and by implementing rainwater harvesting systems to reduce dependency on groundwater and improve water availability.

Controlling Industrial and Agricultural Runoff: Enforcing stringent pollution control laws for industries discharging effluents into rivers; promoting buffer zones with vegetation along riverbanks to filter pollutants (Lakritz, 2019) from agricultural runoff and by encouraging organic farming and reduced chemical fertilizer usage to prevent water contamination.

Plastic Waste Reduction and Management: Banning single-use plastics in the region to prevent water pollution; establishing recycling centers and incentivizing waste collection programs and by conducting awareness campaigns (Nyokabi et al., 2021) on the impacts of plastic waste and promoting biodegradable alternatives.

Community Involvement and Awareness: Conducting educational programs on air and water pollution impacts and sustainable practices; establishing community-driven initiatives such as river clean-up drives and afforestation campaigns (Bhatt

& Sharma, 2022) and by partnering with local NGOs, schools, and government bodies to spread awareness and encourage public participation.

Policy Recommendations and Advocacy: Advocating for stricter environmental laws and policies to control industrial emissions and wastewater discharge; working with governmental and international bodies to allocate funding for sustainability projects (Shrestha et al., 2022) in the Sundarbans and by establishing an environmental task force to monitor and enforce air and water quality regulations.

THEORETICAL FRAMEWORKS

The Sundarbans, a unique mangrove ecosystem located at the delta of the Ganges, Brahmaputra, and Meghna rivers, spans parts of India and Bangladesh . It is renowned for its rich biodiversity (Sales et al., 2017), notably the Bengal tiger, and serves as a critical resource for millions of people who rely on its ecosystem services. However, the Sundarbans face significant environmental threats, including pollution, deforestation, and climate change (Barber et al., 2014). To address these challenges, a robust theoretical framework is essential for promoting clean air and water, ultimately fostering environmental sustainability (Liben et al., 2024). This framework will encompass various interdisciplinary theories and models, addressing environmental, social, economic, and political dimensions (Adiele et al., 2020).

Ecological Modernization Theory: Ecological Modernization Theory posits that economic development and ecological sustainability are not mutually exclusive but can be achieved concurrently (Othman, 2024). This theory is particularly relevant for the Sundarbans, where the balance between economic activities such as fishing, agriculture, and tourism and environmental protection is critical (Sudheer Varma Surimalla, 2018). By promoting clean technologies, sustainable agriculture practices, and improved waste management systems, communities can enhance their livelihoods while ensuring the preservation of clean air and water (Suman, 2022). For instance, the introduction of aquaculture systems that utilize eco-friendly practices can mitigate pollution in water bodies (Arruda et al., 2021). Local stakeholders, including farmers and fishers, can be trained to employ practices that reduce chemical runoff and promote biodiversity (Geissdoerfer et al., 2017). Moreover, the implementation of small-scale renewable energy solutions such as solar panels can help reduce reliance on fossil fuels, thereby contributing to cleaner air (Bharadwaj, 2016).

Social-Ecological Systems Framework: The Social-Ecological Systems (SES) Framework emphasizes the interconnectedness of social and ecological components in resource management (Ogutu et al., 2023). In the Sundarbans, the integration of local communities' knowledge and practices with scientific research is vital for sustainable management of natural resources (Yangka et al., 2018). The SES framework encourages participatory governance, which involves stakeholders at every level of decision-making (Diao, 2019). Participatory approaches can facilitate community-led initiatives for clean air and water. For example, local communities can engage in water monitoring programs (Lam & Toan, 2006), identifying pollution sources and advocating for policy changes. The flexibility of the SES framework allows for adaptive management strategies that can respond to the dynamic nature of environmental challenges in the Sundarbans, ensuring sustainability over time (Patel & Padhya, 2021).

Socio-technical Systems Theory: The Socio-Technical Systems Theory focuses on the interactions between people, technology, and the environment (Kassa et al., 2022). In the context of the Sundarbans, this theory can be applied to design resilient systems for air and water quality management (Matthews & Vlado Vivoda, 2023). The intricate relationship between social practices, technological innovations, and institutional frameworks must be understood and optimized to achieve environmental sustainability (Pajany et al., 2023). Innovative technologies such as biofiltration for wastewater treatment can be combined with community involvement in monitoring water quality (Kumar Apat et al., 2023). Training local individuals in operating and maintaining these technologies ensures the sustainability of interventions. Furthermore, integrating traditional knowledge concerning local flora and fauna (Mrisho et al., 2020) can enhance conservation efforts, leading to healthier ecosystems and subsequently cleaner air and water.

Environmental Justice Framework: The Environmental Justice Framework addresses the disproportionately negative environmental impacts experienced (Fauquet & Fargette, 1990) by marginalized communities. In the Sundarbans, the livelihoods of many local populations are intricately tied to the health of their environment. Pollution from industrial activities, deforestation, and the effects of climate change often exacerbate existing inequalities, leading to a cycle of vulnerability (Ahoya et al., 2024). An environmental justice lens stresses the need for equitable access to clean air and water, advocating for inclusive policymaking that takes into consideration the voices of affected communities (Akkas Ali et al., 2024). In practical terms, initiatives could be implemented to ensure that marginalized groups have access to clean water, such as constructing water purification systems and ensuring pollution controls are applied to industries that affect their resources (Rajab et al., 2023).

Integrated Water Resources Management (IWRM): Integrated Water Resources Management (IWRM) is an approach to managing water resources that considers the entire watershed while balancing the needs of various stakeholders (Abebe and Endalie., 2023). In the Sundarbans, IWRM can play a pivotal role in ensuring sustainable water quality. This framework emphasizes a coordinated approach to water management that includes surface water, groundwater, and ecosystem considerations (Silva et al., 2014). Implementing IWRM in the Sundarbans requires collaboration among governmental agencies, non-governmental organizations, and local communities (Daví-Arderius et al., 2017). Policies promoting sustainable farming practices, water conservation techniques, and afforestation can enhance water quality. Strengthening local management institutions will also empower communities to participate actively in decision-making processes, fostering ownership and stewardship of resources (Swapnil Mahadik et al., 2024).

Climate Change Adaptation Frameworks: Considering the vulnerability of the Sundarbans to climate change, frameworks that emphasize adaptation are crucial (Casady et al., 2024). Climate change adaptation strategies that focus on water and air quality must be integrated into broader climate policies. These strategies may include enhancing the resilience of the mangrove ecosystem (Acharya et al., 2013), which plays a vital role in filtering pollutants and protecting coastal areas from erosion and storm surges (Olando et al., 2024). Community education regarding climate-resilient practices, such as sustainable land use and biodiversity conservation, is crucial. Such initiatives can mitigate the impacts of climate change on air and water quality (Sterner et al., 2019), thereby promoting overall ecological health. Equally important is the development of contingency plans to address extreme weather events, ensuring that communities have the resources and knowledge needed to respond effectively (Kelling et al., 2013).

GOVERNANCE FOR PROMOTING CLEAN AIR AND WATER IN THE SUNDARBANS

Governance plays a pivotal role in addressing environmental challenges and promoting sustainability, particularly in ecologically sensitive areas like the Sundarbans (Priya et al., 2023). This extensive mangrove region, which straddles India and Bangladesh, faces ongoing threats from pollution, deforestation, and climate change, necessitating a comprehensive and multisectoral governance approach (Ratouis et al., 2022). Effective governance in the Sundarbans entails the integration of policies, stakeholder participation, and adaptive management strategies to ensure the long-term health of both air and water resources (Yang et al., 2024).

Firstly, the establishment of clear and enforceable environmental regulations is essential. Governments at both the state and national levels must prioritize the development and implementation of stringent laws (Nwokolo et al., 2023) aimed at curbing air and water pollution. These regulations must be backed by scientific research, allowing policymakers to make informed decisions about permissible emission levels and water quality standards (Chen et al., 2024). For instance, regulations on industrial emissions should focus on reducing particulate matter and other harmful pollutants that degrade air quality (Vats & Mathur, 2024). Similarly, water sources should be regularly monitored for contaminants such as heavy metals and pathogens, with consequent penalties for violators enforcing compliance (Tiwari et al., 2024).

Secondly, stakeholder involvement is crucial for effective governance in the Sundarbans (Pachar & Singh, 2021). Local communities, environmental organizations, and other stakeholders should have a voice in decision-making processes. Collaborative approaches, such as participatory governance models (Adom, 2024), can foster greater community ownership of environmental initiatives and utilization of indigenous knowledge (Shrestha & Aryal, 2011). Local stakeholders can help in identifying pollution sources and advocating for cleaner practices, such as sustainable fishing and responsible tourism (Kuir-Ayius, 2024). Moreover, capacity building and education play a vital role in empowering local communities. Governments should invest in capacity-building programs that teach community members (Jolly et al., 2024) about environmental stewardship and sustainable practices. Educational initiatives in schools and community forums can raise awareness about the importance of clean air and water, the impact of pollution (Hosterman & Smith, 2025), and the importance of sustainable resource management. By fostering a sense of responsibility and stewardship among the populace, these programs can help cultivate a community-driven approach to environmental sustainability (Amarnath et al., 2017).

Technological innovation also figures prominently in governance strategies for promoting clean air and water in the Sundarbans (Gunaratne et al., 2021). Governments and NGOs should collaborate in researching and implementing clean technologies for waste management, renewable energy, and pollution control (Zappacosta & Hollema, 2017). For instance, deploying solar panels for energy needs can reduce dependence on fossil fuels, thus improving air quality. Similarly, adopting eco-friendly agricultural practices can minimize agrochemical runoff (Ait Kadi & Ziyad, 2018) into water bodies, contributing to cleaner water. Effective governance must also include regular monitoring and evaluation of environmental health indicators (Aryal et al., 2024). Establishing a transparent and accountable system for tracking air and water quality can help governments assess the effectiveness of their policies and make informed adjustments as necessary (Singh et al., 2022). Engaging third-party auditors and research institutions can enhance accountability and provide validation for governance practices (Koswatte et al., 2024).

Finally, international cooperation is crucial for addressing transboundary pollution, especially in the context of the Sundarbans' shared geography between India and Bangladesh (Fransisca & Ningsih, 2023). Collaborative initiatives, such as joint funding for research projects or cross-border regulatory frameworks, can enhance the capacity to manage shared resources and pollution sources (Chisom et al., 2024), thereby promoting a holistic approach to environmental governance. Therefore, effective governance in the Sundarbans involves a multifaceted approach that includes enforceable regulations (Akter, 2024), stakeholder participation, education, technological innovation, rigorous monitoring and evaluation, and international cooperation (Olawade et al., 2024). By implementing these strategies, it is possible to significantly improve air and water quality, thus promoting environmental sustainability in one of the world's most biodiverse and vulnerable regions (Entezari et al., 2023).

ACCOUNTABILITY IN PROMOTING CLEAN AIR AND WATER FOR ENVIRONMENTAL SUSTAINABILITY IN THE SUNDARBANS

Accountability is a cornerstone of effective governance, particularly in the context of promoting clean air and water for environmental sustainability in the Sundarbans (Bhatt & Joshi, 2024). Ensuring that individuals, governments, and organizations are held accountable for their environmental actions is essential for fostering trust, transparency, and responsibility (Mersad Shoaei et al., 2024). The unique environmental and socio-economic challenges faced by the Sundarbans such as deforestation, pollution (Ibegbulam et al., 2023), and the impacts of climate change require a systematic approach to accountability that not only addresses violations but also promotes positive environmental behaviors (Silva et al., 2021). At the governmental level, accountability mechanisms must be instituted to ensure that regulatory bodies effectively enforce environmental laws (Odume, 2024). This begins with establishing clear responsibilities for various governmental agencies tasked with monitoring air and water quality (Zerssa et al., 2021). Regular reporting and oversight from independent agencies or external environmental watchdogs can ensure compliance with environmental regulations (Evans & Sadler, 2008). These agencies can compile reports on pollution levels, enforcement actions, and consequences for non-compliance, which should be made publicly available. By fostering transparency (Hendrix, 2012), citizens can better hold their governments accountable for maintaining clean air and water.

Community involvement is also a crucial aspect of accountability (Molden et al., 2010). Citizen engagement can serve as a powerful check on government and corporate actions. When local communities are empowered to monitor the quality of air and water, they can advocate for change and draw attention to violations (Ashoka et al., 2024). Community-based monitoring programs can engage residents in activities such as tracking pollution sources, assessing water quality, and reporting illegal discharges (Debsarma, 2009). These initiatives not only enhance accountability but also create a culture of environmental stewardship within communities. Furthermore, corporations and industries operating in the Sundarbans must also be held accountable for their environmental impacts (Karim et al., 2024). Establishing corporate social responsibility (CSR) frameworks that include specific commitments to environmental sustainability can create accountability at the corporate level (Rahman et al., 2024). Regular audits of corporate practices, including air and water impact assessments, can ensure that companies are adhering to ethical and regulatory standards (Singh et al., 2022). Public disclosure of these assessments can also provide consumers with the information they need to make environmentally responsible choices (Sarvajayakesavalu, 2015).

Legal frameworks must also enable citizens to seek redress for environmental grievances. Access to justice is a significant component of accountability (Rohde et al., 2024). Legal provisions that allow citizens to file lawsuits against polluters can empower communities to take action against environmental degradation. These cases can serve as precedents (Pandey & Regmi, 2017), reinforcing the idea that pollution and environmental harm have consequences. Legal accountability must be paired with supportive mechanisms that assist citizens in navigating complex legal processes (Seuyong et al., 2023), ensuring that justice is accessible to all, particularly marginalized communities. Investing in technology and innovation for monitoring and enforcement can further enhance accountability (Bwire, 2024). Digital platforms can be developed to facilitate real-time tracking of pollution levels and corporate compliance with environmental regulations (UNFCCC, 2016). These platforms can serve as public databases where data is readily available for scrutiny. By harnessing technology, accountability efforts can become more proactive and less reactive (Pathak, 2010).

International cooperation also plays a vital role in establishing accountability, particularly in regions like the Sundarbans that span multiple countries (Griggs & Reguero, 2021). Both India and Bangladesh can collaborate on transboundary accountability mechanisms that address shared environmental responsibilities (Lindsey, 2023). Joint monitoring programs, mutual agreements on pollution standards, and shared funding for environmental initiatives can enhance accountability across borders (Garrett, 2023). Thus, accountability is vital for promoting clean air and water in the Sundarbans. It involves robust regulatory enforcement by governments, active community participation (Lakritz, 2019), corporate responsibility, accessible legal recourse for citizens, innovative monitoring technologies, and international cooperation (Nyokabi et al., 2021). When individuals and organizations are held accountable for their actions, environmental concerns are prioritized, leading to more effective and lasting solutions for sustainability in this ecologically critical region (Bhatt & Sharma, 2022). Through a multifaceted approach to accountability, communities in the Sundarbans can work together to safeguard their natural resources and promote a healthier environment for future generations (Shrestha et al., 2022).

BENEFITS OF PROMOTING CLEAN AIR AND WATER FOR ENVIRONMENTAL SUSTAINABILITY IN THE SUNDARBANS

The Sundarbans, a UNESCO World Heritage Site, are renowned for their unique biodiversity and critical ecosystems. Promoting clean air and water in this region brings numerous benefits not just to the environment (Sales et al., 2017) but also to the communities that depend on these resources for their livelihoods and well-being. One of the primary benefits of promoting clean air and water is the preservation of biodiversity (Barber et al., 2014). The Sundarbans are home to a diverse array of flora and fauna, including the iconic Bengal tiger, various species of fish, and a multitude of birds (Liben et al., 2024). Clean water enhances the health of aquatic ecosystems, promoting fish populations that form the backbone of local fishing industries. Similarly, clean air helps preserve flora, contributing to the overall health (Adiele et al., 2020) of the mangrove forests that serve as a crucial habitat for many species. This rich biodiversity is vital for ecological balance, food security, and economic stability in the region (Othman, 2024).

Furthermore, clean water is essential for human health. In rural areas like the Sundarbans, where access to healthcare is limited, waterborne diseases pose a significant threat (Sudheer Varma Surimalla, 2018). Promoting clean water initiatives, such as community-driven water filtration systems, decreases the prevalence of diseases like cholera and dysentery, ultimately improving public health (Suman, 2022). Healthier communities are more productive, enabling individuals to contribute effectively to local economies. Economic benefits also arise from promoting clean air and water (Arruda et al., 2021). The Sundarbans thrive on both tourism and fisheries, and clean environments enhance the appeal of ecological tourism. Travelers are increasingly inclined to seek destinations that prioritize sustainability (Geissdoerfer et al., 2017). This demand creates job opportunities in eco-tourism and hospitality sectors, thereby boosting the local economy. Additionally, sustaining healthy fish stocks ensures that local fisheries can remain viable in the long term (Bharadwaj, 2016), supporting the livelihoods of countless families in the region.

Educating local communities about the importance of clean air and water fosters environmental stewardship (Ogutu et al., 2023). As communities engage in conservation efforts, they become more invested in protecting their natural resources. This sense of ownership can lead to improved environmental practices (Yangka et al., 2018), such as sustainable agriculture and responsible waste management, further contributing to overall sustainability. Infrastructure development aimed at maintaining clean air and water, such as waste disposal systems (Diao, 2019) and air quality monitoring stations, not only enhances environmental health but also improves quality of life. Better sanitation facilities reduce pollution and safeguard both wildlife and human health while promoting sustainable growth (Lam & Toan, 2006).

Finally, promoting clean air and water in the Sundarbans can facilitate climate resilience. As a region vulnerable to climate change impacts, including rising sea levels and severe weather events (Patel & Padhya, 2021), enhancing environmental quality can bolster the resilience of both ecosystems and communities. Healthy mangroves act as natural barriers against storm surges and flooding, protecting human settlements and biodiversity (Kassa et al., 2022).

CHALLENGES OF PROMOTING CLEAN AIR AND WATER FOR ENVIRONMENTAL SUSTAINABILITY IN THE SUNDARBANS

While the benefits of promoting clean air and water in the Sundarbans are evident, significant challenges hinder these efforts, making addressing them a complex but necessary endeavor (Matthews & Vlado Vivoda, 2023).

One of the most pressing challenges is the high population density and poverty in the region. Many families rely on agriculture and fishing for their livelihoods, often resorting to exploitative practices that degrade water and air quality (Pajany et al., 2023). To maximize harvests, farmers sometimes use chemical fertilizers and pesticides, which can lead to runoff that pollutes local waterways. Addressing these unsustainable practices requires significant education and investment in alternative methods, which can be difficult to implement in resource-strapped communities (Kumar Apat et al., 2023).

Moreover, effective governance and regulation are essential for promoting clean air and water practices. However, the Sundarbans often face issues related to ineffective governance, exacerbated by political corruption and inadequate enforcement of environmental laws (Mrisho et al., 2020). Without proper regulatory frameworks, illegal fishing, logging, and poaching can continue unchecked, further damaging these vital ecosystems. Strengthening governance structures and promoting transparency can be a daunting task, especially in an area with complex socio-political dynamics (Fauquet & Fargette, 1990).

Another challenge is the increasing threat of climate change. The Sundarbans are among the regions most vulnerable to its impacts, including rising sea levels and extreme weather events, which threaten both water quality and air purity (Ahoya et al., 2024). Natural disasters can lead to contamination of freshwater sources, while rising salinity can pose significant challenges for freshwater ecosystems. In this context, fostering resilience becomes difficult as both communities and ecosystems struggle with changing conditions (Akkas Ali et al., 2024). Furthermore, local communities often lack access to clean water infrastructure. Due to financial constraints, the implementation of modern sanitation facilities or water filtration systems poses significant hurdles (Rajab et al., 2023). Many regions in the Sundarbans still depend on traditional water sources, which may be polluted or over-extracted. Building efficient and accessible water supply systems demands substantial investment and innovative approaches tailored to local contexts (Abebe and Endalie., 2023).

Public awareness and education also pose a challenge. While many community members understand the importance of clean air and water, there may be a lack of knowledge regarding sustainable practices (Silva et al., 2014). Mobilizing communities for environmental initiatives requires targeted educational campaigns and engagement strategies that resonate with local values and traditions (Daví-Arderius et al., 2017). Lastly, socio-economic factors, such as dependence on coal and fossil fuels for energy, contribute to air pollution. Transitioning to renewable energy sources may be necessary for long-term sustainability (Swapnil Mahadik et al., 2024), but doing so presents economic challenges for a region already grappling with poverty. Overcoming these hurdles necessitates collaboration between governmental organizations (Casady et al., 2024), NGOs, and local communities to ensure that the pursuit of clean air and water aligns with socio-economic realities.

Ultimately, while the path to promoting clean air and water in the Sundarbans is fraught with challenges (Acharya et al., 2013), addressing them is crucial for achieving long-term environmental sustainability and improving the quality of life for the region's inhabitants (Olando et al., 2024).

APPLICATIONS OF AI IN ENVIRONMENTAL SUSTAINABILITY

Artificial Intelligence (AI) has emerged as a pivotal tool in promoting environmental sustainability across various domains. Its applications touch multiple sectors, including agriculture, wildlife conservation, climate modeling, and pollution monitoring, ultimately helping to create a more sustainable future (Sterner et al., 2019).

One of the most noticeable applications of AI is in precision agriculture, where machine learning algorithms analyze extensive data sets to optimize farming practices (Kelling et al., 2013). AI technologies can analyze weather patterns, soil health, and crop conditions to advise farmers on the best planting and harvesting practices, minimizing waste and maximizing yields (Priya et al., 2023). By enabling targeted interventions, such as the precise application of water, fertilizers, and pesticides, farmers can reduce their carbon footprints and input costs while increasing food security (Ratouis et al., 2022).

In the realm of wildlife conservation, AI-driven tools are used to monitor endangered species and prevent poaching (Yang et al., 2024). Computer vision and image recognition algorithms can analyze camera trap images to identify species, estimate populations, and reveal behavioral patterns (Nwokolo et al., 2023). Drones equipped with AI can survey hard-to-reach areas, gathering data on habitats and detecting illegal activities (Yang et al., 2024). These technologies empower conservationists with the information needed to develop effective management strategies, ensuring that biodiversity is preserved while also satisfying the needs of local communities (Ratouis et al., 2022).

Climate modeling has greatly benefited from AI techniques. Machine learning algorithms help process vast datasets from satellite imagery and ground observations to improve climate predictions (Priya et al., 2023). By enhancing the accuracy of climate models, AI enables policymakers to make informed decisions regarding climate actions and sustainability goals (Kelling et al., 2013). For example, predictive modeling can identify which regions are likely to experience extreme weather events, allowing for proactive measures to protect vulnerable communities and ecosystems (Sterner et al., 2019).

Air and water quality monitoring is another critical area where AI is making significant contributions. Traditional methods of monitoring pollution levels are often slow and labor-intensive (Olando et al., 2024). AI algorithms can process real-time data from sensors and satellite imagery to detect pollution hotspots and trend patterns over time (Acharya et al., 2013). This capability allows regulatory bodies to implement timely interventions to control emissions and water contamination, protecting public health and the environment (Casady et al., 2024). Moreover, AI can optimize the operation of wastewater treatment plants, ensuring that plants operate efficiently and effectively reduce pollution (Swapnil Mahadik et al., 2024). Additionally, AI is being utilized to enhance public engagement and education surrounding environmental issues. Through chatbots and interactive platforms (Daví-Arderius et al., 2017), AI can disseminate information about sustainability practices, promoting greater awareness and community involvement. For example, people can interactively learn about their carbon footprints and receive tailored recommendations on how to reduce them (Silva et al., 2014).

ETHICAL CONSIDERATIONS IN PROMOTING CLEAN AIR AND WATER FOR ENVIRONMENTAL SUSTAINABILITY IN THE SUNDARBANS

The Sundarbans, a UNESCO World Heritage Site, located at the southern tip of Bangladesh and India, is one of the largest mangrove forests in the world (Abebe and Endalie., 2023). This unique ecosystem is a tapestry of intricate waterways, diverse marine and terrestrial life, and a vital resource for millions of people. However, the region faces severe environmental challenges (Rajab et al., 2023), particularly concerning air and water quality, primarily due to industrialization, agricultural practices, and climate change (Akkas Ali et al., 2024). In addressing the pressing need for promoting clean air and water in the Sundarbans, a myriad of ethical considerations must be navigated to ensure that the solutions (Ahoya et al., 2024) employed foster environmental sustainability while respecting the rights and wellbeing of the local communities and ecosystems (Fauquet & Fargette, 1990).

The Right to Clean Air and Water

At the core of the ethical considerations in promoting clean air and water is the recognition of the fundamental human right to a healthy environment (Mrisho et al., 2020). According to the United Nations, access to clean air and water is integral to the right to life and health. In the Sundarbans, local communities depend heavily on these natural resources for their livelihoods, health, and cultural practices (Kumar Apat et al., 2023). Therefore, any initiatives aimed at environmental sustainability must prioritize these basic rights. This brings to light the ethical principle of justice, particularly environmental justice (Pajany et al., 2023), which emphasizes fairness in the distribution of environmental benefits and burdens. The Sundarbans' marginalized communities, often the most affected by pollution and ecological degradation, should be prioritized in policy and planning processes (Matthews & Vlado Vivoda, 2023). Ethical considerations demand that the voices of these communities are not only heard but actively included in the decision-making processes that affect their environment and health (Kassa et al., 2022).

Intergenerational Equity

Another significant ethical aspect is intergenerational equity, which stresses the responsibility of current generations to protect natural resources for future generations (Patel & Padhya, 2021). The Sundarbans serves not only current inhabitants but also the global community, as its biodiversity plays a critical role in climate regulation and carbon sequestration (Lam & Toan, 2006). Ethical sustainability necessitates a long-term perspective that recognizes the interconnectedness of ecological and human systems. Promoting clean air and water is thus not just a local issue it's a global imperative (Diao, 2019). The actions taken today will have far-reaching implications for future generations, and stakeholders must be held accountable for ensuring that their environmental practices do not compromise the health of the Sundarbans' ecosystem (Yangka et al., 2018).

Holistic Environmental Protection

Ethical considerations also advocate for a holistic approach to environmental protection, recognizing that air and water quality cannot be addressed in isolation (Ogutu et al., 2023). The intricate linkage between different environmental elements requires an integrated management plan that considers biological, ecological, and socio-economic factors (Bharadwaj, 2016). Ethically, it is crucial that development initiatives do not prioritize short-term economic gain over long-term ecological sustainability (Geissdoerfer et al., 2017). Strategies promoting clean air and water should involve restoring the mangrove habitats, reducing pollution from agriculture and industry, and implementing sustainable fishing practices (Arruda et al., 2021). These efforts should actively involve local community members, traditional knowledge, and practices, thereby enhancing the effectiveness and ethical legitimacy of environmental programs (Suman, 2022). Engaging communities in the monitoring and decision-making processes fosters a sense of ownership and accountability, increasing the likelihood of successful and sustained outcomes (Sudheer Varma Surimalla, 2018).

Respect for Indigenous Knowledge and Culture

The Sundarbans is home to various indigenous peoples whose knowledge systems and cultural practices have evolved alongside this unique environment (Othman, 2024). Ethically, it is imperative to respect and incorporate indigenous knowledge into any environmental sustainability strategy (Adiele et al., 2020). This form of knowledge is often rooted in centuries of interaction with the ecosystem, providing valuable insights into sustainable practices that mainstream scientific approaches may overlook (Liben et al., 2024). Recognition of indigenous rights also involves acknowledging the historical injustices and marginalization that these communities have faced. Efforts to promote clean air and water must aim not only at improving environmental quality (Barber et al., 2014) but also at restoring the dignity and rights of indigenous populations. Ethical engagement necessitates the avoidance of 'greenwashing,' where initiatives superficially address environmental issues without addressing social inequities (Sales et al., 2017).

The Role of Industry and Economic Development

The Sundarbans faces pressure from industrialization and land-use change, which pose threats to its ecological health. Industries often bear significant responsibility for air and water pollution (Shrestha et al., 2022). Thus, ethical considerations demand a careful examination of the balance between economic growth and environmental protection (Bhatt & Sharma, 2022). While economic development is essential for poverty alleviation and improved living standards, it should not occur at the expense of the region's ecological integrity (Nyokabi et al., 2021). Ethical degradation of the environment leads to notions of corporate responsibility and accountability. Companies operating in or near the Sundarbans must implement sustainable practices that minimize their environmental footprint (Lakritz, 2019). There is an ethical obligation for industries to invest in clean technologies that reduce air and water pollution and actively participate in environmental conservation efforts (Garrett, 2023). Engaging in corporate social responsibility (CSR) should go beyond compliance with regulations; it must reflect a genuine commitment to the local community and the environment (Lindsey, 2023).

Climate Change Adaptation

The Sundarbans is particularly vulnerable to climate change, facing rising sea levels, increased salinity, and extreme weather events (Griggs & Reguero, 2021). To ethically address these challenges, it is vital to incorporate adaptive strategies that enhance both ecological resilience and community wellbeing (Pathak, 2010). This includes promoting practices that foster clean air and water while simultaneously preparing local populations for the impacts of climate change (UNFCCC, 2016). Ethical considerations extend to ensuring that adaptation strategies are equitable and just, avoiding exacerbation of existing inequalities (Bwire, 2024). Vulnerable communities, particularly women and marginalized groups, must be actively involved in developing and implementing adaptation strategies (Seuyong et al., 2023). Their experiences and insights are essential for creating effective solutions that protect both people and the delicate environments they inhabit (Pandey & Regmi, 2017).

Promoting clean air and water for environmental sustainability in the Sundarbans is fraught with ethical considerations that require careful consideration and action (Rohde et al., 2024). As stewards of this unique ecosystem, it is our ethical duty to respect and uphold the rights of local communities, acknowledge intergenerational responsibilities, and advocate for holistic environmental management (Sarvajayakesavalu, 2015). Solutions must integrate indigenous knowledge, encourage corporate responsibility, and address the challenges posed by climate change (Singh et al., 2022). By prioritizing ethics in environmental decision-making, we foster a more sustainable and equitable future for both the Sundarbans and the global community that depends on its rich biodiversity and ecological services (Rahman et al., 2024).

CONCLUSION

Promoting clean air and water in the Sundarbans is not just about protecting an ecosystem; it is about ensuring that the livelihoods of the millions who depend on it remain viable for generations to come (Karim et al., 2024). Deteriorating air quality, often exacerbated by practices such as burning fossil fuels and the spread of industries, adversely affects both human health and the vegetation essential for carbon sequestration (Debsarma, 2009). Similarly, if water pollution continues, it can lead to severe repercussions for aquatic life, disrupt food chains, and endanger the food security of local communities who depend on fishing and agriculture (Ashoka et al., 2024). The local population possesses invaluable traditional knowledge about sustainable practices and a profound connection to their environment, making their involvement vital in any conservation efforts (Molden et al., 2010). Education and awareness programs can empower communities to adopt sustainable practices, advocating for cleaner technologies and waste management solutions (Hendrix, 2012). Additionally, investment in renewable energy sources can reduce reliance on fossil fuels, thus mitigating air pollution while also supporting water conservation initiatives (Evans & Sadler, 2008).

Suggestions and Future Research Directions

Strengthening community involvement should be a primary focus. Enhancing environmental education initiatives will inform local residents of the necessity of clean air and water practices (Zerssa et al., 2021), as well as sustainable agriculture, fishing techniques, and waste management strategies. Workshops, seminars, and demonstration projects can help foster sustainable practices among locals (Odume, 2024).

There is a pressing need to establish or reinforce regulatory frameworks governing emissions from industries and agricultural practices (Silva et al., 2021). Governments should actively monitor and impose penalties for violations. Clear guidelines must be developed, ensuring sustainable practices are adhered to (Ibegbulam et al., 2023). Eco-tourism can serve as an alternative livelihood for residents while promoting conservation. By showcasing the unique biodiversity of the Sundarbans and its ecosystems, initiatives must be established that balance tourist interest (Mersad Shoaei et al., 2024) and environmental protection, providing incentives for local communities to act as stewards of their environment.

Transitioning from fossil fuels to renewable energy sources such as solar, wind, and biogas will significantly reduce air pollution (Bhatt & Joshi, 2024). Simple solar installation projects can provide power to remote communities, reducing reliance on polluting energy sources (Entezari et al., 2023). Establishing water quality monitoring systems that involve local communities will help track pollution levels, enabling timely interventions (Olawade et al., 2024). This initiative can also provide essential data that policymakers can utilize to implement more effective environmental regulations (Akter, 2024).

Given the capacity of mangroves to sequester carbon and filter pollutants, restoration and afforestation initiatives need prioritizing (Chisom et al., 2024). Collaborative projects involving government and community participation can increase mangrove coverage while creating natural buffers against climate change impacts (Fransisca & Ningsih, 2023). Funding for research institutions focusing on ecological monitoring, pollution control technologies, and sustainable practices is crucial. Encouraging innovation and disseminating findings (Koswatte et al., 2024) through accessible channels will support effective policy-making.

Partnerships with international bodies can provide the necessary technical and financial resources for conservation initiatives (Singh et al., 2022). Global experiences in similar ecosystems can offer insights into best practices and successful models of sustainable development (Aryal et al., 2024). By implementing these suggestions, stakeholders can work collaboratively towards promoting clean air and water, ultimately contributing to the environmental sustainability of the Sundarbans thus preserving its ecological integrity and supporting the livelihood of the people who call it home (Hosterman & Smith, 2025).

REFERENCES

Abebe, W. T., & Endalie, D. (2023). Artificial intelligence models for prediction of monthly rainfall without climatic data for meteorological stations in Ethiopia. *Journal of Big Data, 10*(1), 2. Advance online publication. DOI: 10.1186/s40537-022-00683-3

Acharya, B., Akil, A. M., Ridhanda, P., Na, Y. J., Hong, C. Y., & Park, K. W. (2013, June). Public Private Partnership in Hydropower Development: Prospects for Nepal's Development. *Conference: Energy Transition and Policy Challenges, 36th IAEE International Conference.* https://www.researchgate.net/publication/353209903_Public_Private_Partnership_in_Hydropower_Development_Prospects_for_Nepal's_Development

Adiele, J. G., Schut, A. G. T., van den Beuken, R. P. M., Ezui, K. S., Pypers, P., Ano, A. O., Egesi, C. N., & Giller, K. E. (2020). Towards closing cassava yield gap in West Africa: Agronomic efficiency and storage root yield responses to NPK fertilizers. *Field Crops Research, 253*, 107820. DOI: 10.1016/j.fcr.2020.107820

Adom, P. K. (2024). The socioeconomic impact of climate change in developing countries over the next decades: A literature survey. *Heliyon, 10*(15), e35134. DOI: 10.1016/j.heliyon.2024.e35134 PMID: 39170312

Ahoya, D. K. D., Sawadogo-Compaore, E. M. F. W., Yabi, J. A., Zandjanakou-Tachin, M., Houngue, J. A., Houedjissin, S. S., Pita, J. S., & Ahanhanzo, C. (2024). Factors Influencing Adoption of the PlantVillage Nuru Application for Cassava Mosaic Disease Diagnosis Among Farmers in Benin. *Agriculture, 14*(11), 2001. DOI: 10.3390/agriculture14112001

Ait Kadi, M., & Ziyad, A. (2018). *Integrated Water Resources Management in Morocco. Water Resources Development and Management*. Springer Nature., DOI: 10.1007/978-981-10-7913-9_6

Akter, M. S. (2024). AI for Sustainability: Leveraging Technology to Address Global Environmental. *Journal of Artificial Intelligence General Science (JAIGS) ISSN:3006-4023, 3*(1), 40–48. DOI: 10.60087/jaigs.v3i1.64

Amarnath, G., Alahacoon, N., Smakhtin, V., & Aggarwal, P. (2017). *Mapping multiple climate-related hazards in South Asia*. International Water Management Institute (IWMI).

Arruda, E. H., Melatto, R. A. P. B., Neto, W. L. B. da S., & Conti, D. de M. (2021). Circular Economy: A Brief Literature Review (2015-2020). *Sustainable Operations and Computers, 2*, 79–86. DOI: 10.1016/j.susoc.2021.05.001

Aryal, S., Ghimire, S., Tiwari, S., Baaniya, Y., & Pandey, V. P. (2024). Evolution and future prospects of hydropower sector in Nepal: A review. *Heliyon, 10*(10), e31139–e31139. Elsevier. DOI: 10.1016/j.heliyon.2024.e31139

Ashoka, P., Devi, B. R., Sharma, N., Behera, M., Gautam, A., Jha, A., & Sinha, G. (2024). Artificial Intelligence in Water Management for Sustainable Farming: A Review. *Journal of Scientific Research and Reports, 30*(6), 511–525. DOI: 10.9734/jsrr/2024/v30i62068

Barber, C. P., Cochrane, M. A., Souza, C. M.Jr, & Laurance, W. F. (2014). Roads, deforestation, and the mitigating effect of protected areas in the Amazon. *Biological Conservation, 177*, 203–209. DOI: 10.1016/j.biocon.2014.07.004

Bharadwaj, B. (2016). *Plastic Bag Ban in Nepal: Enforcement and Effectiveness.* South Asian Network for Development and Environmental Economics (SANDEE).

Bhatt, M. W., & Sharma, S. (2022). An Object Recognition-Based Neuroscience Engineering: A Study for Future Implementations. *Istanbul University -. Journal of Electrical and Electronics Engineering (Oradea), 23*(2). Advance online publication. DOI: 10.5152/electrica.2022.22125

Bhatt, P., & Joshi, K. R. (2024). Hydropower Development in Nepal: Status, Opportunities and Challenges. *Journal of UTEC Engineering Management, 2*(01), 125–135. DOI: 10.36344/utecem.2024.v02i01.011

Bwire, V. (2024, January 22). *Solar energy: Changing rural lives in Kenya.* Unesco.org; UNESCO. https://courier.unesco.org/en/articles/solar-energy-changing-rural-lives-kenya

Casady, C. B., Cepparulo, A., & Giuriato, L. (2024). Public-private partnerships for low-carbon, climate-resilient infrastructure: Insights from the literature. *Journal of Cleaner Production, 470*, 143338–143338. DOI: 10.1016/j.jclepro.2024.143338

Chisom, O. N., Biu, P. W., Umoh, A. A., Obaedo, B. O., Adegbite, A. O., & Abatan, A. (2024). Reviewing the Role of AI in Environmental Monitoring and conservation: A data-driven Revolution for Our Planet. *World Journal of Advanced Research and Reviews, 21*(1), 161–171. DOI: 10.30574/wjarr.2024.21.1.2720

da Silva, R. M., Schettino, S., de Lima, P. M. R., & Adriano, F. C. (2014). SMART GRID: Evaluation and Trend in Brazil. *Independent Journal of Management & Production, 5*(3), 752–776. Advance online publication. DOI: 10.14807/ijmp.v5i3.122

Daví-Arderius, D., Sanin, M.-E., & Trujillo-Baute, E. (2017). CO2 content of electricity losses. *Energy Policy, 104*, 439–445. DOI: 10.1016/j.enpol.2017.01.011

Debsarma, S. K. (2009). Simulations of Storm Surges in the Bay of Bengal. *Marine Geodesy*, *32*(2), 178–198. DOI: 10.1080/01490410902869458

Diao, M. (2019). Towards sustainable urban transport in Singapore: Policy instruments and mobility trends. *Transport Policy*, *81*, 320–330. DOI: 10.1016/j.tranpol.2018.05.005

Entezari, A., Aslani, A., Zahedi, R., & Noorollahi, Y. (2023). *Artificial intelligence and machine learning in energy systems: A bibliographic perspective. Energy Strategy Reviews, 45(101017), 101017*. Elsevier., DOI: 10.1016/j.esr.2022.101017

Evans, R. G., & Sadler, E. J. (2008). Methods and technologies to improve efficiency of water use. *Water Resources Research*, *44*(7), 2007WR006200. Advance online publication. DOI: 10.1029/2007WR006200

Fauquet, C., & Fargette, D. (1990). African Cassava Mosaic Virus: Etiology, Epidemiology, and Control. *Plant Disease*, *74*(6), 404. DOI: 10.1094/PD-74-0404

Fransisca, V., & Ningsih, W. (2023). The Advancement of Technology and its Impact on Social Life in Indonesia. *Devotion Journal of Community Service*, *4*(3), 860–864. DOI: 10.36418/devotion.v4i3.445

Garrett, K. (2023). *The Adaptive Capacity to implement Climate Migration Policy in response to Sea-Level Rise in the Maldives*. https://digitalcollections.sit.edu/capstones/3275?utm_source=digitalcollections.sit.edu%2Fcapstones%2F3275&utm_medium=PDF&utm_campaign=PDFCoverPages

Geissdoerfer, M., Savaget, P., Bocken, N. M. P., & Hultink, E. J. (2017). The Circular Economy - a New Sustainability Paradigm? *Journal of Cleaner Production*, *143*(1), 757–768. DOI: 10.1016/j.jclepro.2016.12.048

Griggs, G., & Reguero, B. G. (2021). Coastal Adaptation to Climate Change and Sea-Level Rise. *Water (Basel)*, *13*(16), 2151. DOI: 10.3390/w13162151

Gunaratne, M. S., Radin Firdaus, R. B., & Rathnasooriya, S. I. (2021). Climate change and food security in Sri Lanka: Towards food sovereignty. *Humanities & Social Sciences Communications*, *8*(1), 229. Advance online publication. DOI: 10.1057/s41599-021-00917-4

Hendrix, M. (2012). Water in Ethiopia: Drought, Disease and Death. *Global Majority E-Journal*, *3*(2), 110–120. https://www.american.edu/cas/economics/ejournal/upload/hendrix_accessible.pdf

Hosterman, H., & Smith, J. (2025). *Economic Costs and Benefits of Climate Change Impacts and Adaptation to the The Maldives Tourism Industry*. Tourism.gov.mv;Ministry of Tourism. Male', Republic of Maldives. https://resources.tourism.gov.mv/wp-content/uploads/2024/09/Economic-Costs-and-benefits-of-Climate-Change-Impacts-and-Adaptation-to-the-Maldives-Tourism-Industry.pdf

Ibegbulam, C., & Ibegbulam, A.Ibegbulam C.MAigbovbiosa, O.JOlowonubi, J.AFatounde, S.A. (2023). Role Of Artificial Intelligence In Electrification Of Africa. *Engineering Science & Technology Journal*, *4*(6), 456–472. DOI: 10.51594/estj.v4i6.667

Jolly, S., Ahmad, N., & Scott, M. (Eds.). (2024). *Climate-Related Human Mobility in Asia and the Pacific* (1st ed.). Springer Nature. DOI: 10.1007/978-981-97-3234-0

Karim, R., Pk, M. A. B., Dey, P., Akbar, M. A., & Osman, M. S. (2024). A study about the prediction of population growth and demographic transition in Bangladesh. *Journal of Umm Al-Qura University for Applied Sciences*. DOI: 10.1007/s43994-024-00150-0

Kassa, H., Abiyu, A., Hagazi, N., Mokria, M., Kassawmar, T., & Gitz, V. (2022). Forest landscape restoration in Ethiopia: Progress and challenges. *Frontiers in Forests and Global Change*, *5*, 796106. Advance online publication. DOI: 10.3389/ffgc.2022.796106

Kelling, S., Gerbracht, J., Fink, D., Lagoze, C., Wong, W.-K., Yu, J., Damoulas, T., & Gomes, C. (2013). A Human/Computer Learning Network to Improve Biodiversity Conservation and Research. *AI Magazine*, *34*(1), 10–10. DOI: 10.1609/aimag.v34i1.2431

Koswatte, I., Iddawala, J., Kulasekara, R., Ranaweera, P., Dasanayaka, C. H., & Abeykoon, C. (2024). Can Sri Lanka be a Net-Zero Nation by 2050? – Current Renewable Energy Profile, Opportunities, Challenges, and Recommendations. *Cleaner Energy Systems*, *8*, 100126. Advance online publication. DOI: 10.1016/j.cles.2024.100126

Kuir-Ayius, D. (2024). Impact of climate change and accessing services in Papua New Guinea. In Jolly, S., Ahmad, N., & Scott, M. (Eds.), *Climate-related human mobility in Asia and the Pacific*. Sustainable Development Goals Series. Springer., DOI: 10.1007/978-981-97-3234-0_11

Kumar Apat, S., & Mishra, J. (2023). An Artificial Intelligence-based Crop Recommendation System using Machine Learning. *Journal of Scientific and Industrial Research*, *82*(05), 558–567. DOI: 10.56042/jsir.v82i05.1092

Lakritz, T. (2019, September 10). *These 11 sinking cities are in danger of disappearing*. World Economic Forum. https://www.weforum.org/stories/2019/09/11-sinking-cities-that-could-soon-be-underwater/

Lam, S. H., & Toan, T. D. (2006). Land Transport Policy and Public Transport in Singapore. *Transportation*, *33*(2), 171–188. DOI: 10.1007/s11116-005-3049-z

Liben, F., Abera, W., Tariku Chernet, M., Ebrahim, M., Tilaye, A., Erkossa, T., Tibebe Degefie, D., Mponela, P., Kihara, J., & Tamene, L. (2024). Site-specific fertilizer recommendations using data-driven machine learning enhanced wheat productivity and resource use efficiency. *Field Crops Research*, *313*, 109413–109413. DOI: 10.1016/j.fcr.2024.109413

Lindsey, R. (2023, August 22). *Climate change: Global sea level* (R. Lumpkin, G. Johnson, P. Thompson, & W. Sweet, Eds.). Climate.gov; National Oceanic and Atmospheric Administration. https://www.climate.gov/news-features/understanding-climate/climate-change-global-sea-level

Matthews, R., & Vivoda, V. (2023). "Water Wars": Strategic implications of the grand Ethiopian Renaissance Dam. *Conflict Security and Development*, *23*(4), 1–34. DOI: 10.1080/14678802.2023.2257137

Molden, D., Oweis, T., Steduto, P., Bindraban, P., Hanjra, M. A., & Kijne, J. (2010). Improving agricultural water productivity: Between optimism and caution. *Agricultural Water Management*, *97*(4), 528–535. DOI: 10.1016/j.agwat.2009.03.023

Mrisho, L. M., Mbilinyi, N. A., Ndalahwa, M., Ramcharan, A. M., Kehs, A. K., McCloskey, P. C., Murithi, H., Hughes, D. P., & Legg, J. P. (2020). Accuracy of a Smartphone-Based Object Detection Model, PlantVillage Nuru, in Identifying the Foliar Symptoms of the Viral Diseases of Cassava-CMD and CBSD. *Frontiers in Plant Science*, *11*, 590889. DOI: 10.3389/fpls.2020.590889 PMID: 33391304

Nwokolo, S. C., Singh, R., Khan, S., Kumar, A., & Luthra, S. (2023). Technological pathways to net-zero goals in Africa. In *Africa's Path to Net-Zero: Exploring Scenarios for a Sustainable Energy Transition* (pp. 93–210). Springer., DOI: 10.1007/978-3-031-44514-9_5

Nyokabi, Jesse & Kamau, Kabera & Oyugi, Noah & Mbugua, Githae & Hiuhu, Antony. (2021). Machine Learning Model for Energy Analysis of a Wellhead Geothermal Plant: A Case of Kenya.

Odume, B. W. (2024). Artificial Intelligence in Agriculture: Application in Developing Countries. *Journal of Agricultural Science (Toronto)*, *16*(12), 60–67. DOI: 10.5539/jas.v16n12p60

Ogutu, M. O., Akor, J., Mulindwa, M. S., Heshima, O., & Nsengimana, C. (2023). Implementing circular economy and sustainability policies in Rwanda: Experiences of Rwandan manufacturers with the plastic ban policy. *Frontiers in Sustainability*, *4*, 1092107. Advance online publication. DOI: 10.3389/frsus.2023.1092107

Olando, S., Mwenda, M. N., Kikwatha, R. W., & Rambo, C. M. (2024). Effect Of Public Private Partnership Financing Structure On Completion Of Geothermal Energy Development Projects In Kenya. *International Journal of Professional Business Review*, *9*(11), e05048–e05048. DOI: 10.26668/businessreview/2024.v9i11.5048

Olawade, D. B., Wada, O. Z., Ige, A. O., Egbewole, B. I., Olojo, A., & Oladapo, B. I. (2024). Artificial Intelligence in Environmental Monitoring: Advancements, Challenges, and Future Directions. *Hygiene and Environmental Health Advances*, *12*, 100114–100114. DOI: 10.1016/j.heha.2024.100114

Othman, A. (2024, November). *AI in Climate Change Modeling and Prediction.* ResearchGate. https://www.researchgate.net/publication/385985206_AI_in_Climate_Change_Modeling_and_Prediction

Pachar, S., & Singh, R. (2021). Role of Sustainable Development Goals and Corporate Social Responsibility in India's Growth: Opportunities and Challenges. *The Empirical Economics Letters*, 20(2).

Pandey, S., & Regmi, N. (2017). Changing connectivities and renewed priorities: Status and challenges facing Nepali Internet. *First Monday*, *23*(1). Advance online publication. DOI: 10.5210/fm.v23i1.8071

Pathak, M. (2010). Climate Change: Uncertainty for Hydropower Development in Nepal. *Hyrdo Nepal*, *6*, 31–34. DOI: 10.3126/hn.v6i0.4190

Priya, A. K., Devarajan, B., Alagumalai, A., & Song, H. (2023). Artificial intelligence enabled carbon capture: A review. *The Science of the Total Environment*, *886*, 163913. DOI: 10.1016/j.scitotenv.2023.163913 PMID: 37150463

Rahman, S., Sharmin, N., Rahat, A., Rahman, M., & Rahman, M. (2024). Tropical cyclone warning and forecasting system in Bangladesh: Challenges, prospects, and future direction to adopt artificial intelligence. *Computational Urban Science*, *4*(1), 4. Advance online publication. DOI: 10.1007/s43762-023-00113-x

Rajab, A., Farman, H., Islam, N., Syed, D., Elmagzoub, M. A., Shaikh, A., Akram, M., & Alrizq, M. (2023). Flood Forecasting by Using Machine Learning: A Study Leveraging Historic Climatic Records of Bangladesh. *Water (Basel)*, *15*(22), 3970. DOI: 10.3390/w15223970

Ratouis, T. M. P., Snæbjörnsdóttir, S. Ó., Voigt, M. J., Sigfússon, B., Gunnarsson, G., Aradóttir, E. S., & Hjörleifsdóttir, V. (2022). Carbfix 2: A transport model of long-term CO2 and H2S injection into basaltic rocks at Hellisheidi, SW-Iceland. *International Journal of Greenhouse Gas Control, 114*, 103586. DOI: 10.1016/j.ijggc.2022.103586

Rohde, F., Wagner, J., Meyer, A., Reinhard, P., Voss, M., Petschow, U., & Mollen, A. (2024). Broadening the perspective for sustainable artificial intelligence: Sustainability criteria and indicators for Artificial Intelligence systems. *Current Opinion in Environmental Sustainability, 66*, 101411. DOI: 10.1016/j.cosust.2023.101411

Sales, M., de Bruin, S., Herold, M., Kyriakidis, P., & Souza, C.Jr. (2017). A spatio-temporal geostatistical hurdle model approach for short-term deforestation prediction. *Spatial Statistics, 21*, 304–318. DOI: 10.1016/j.spasta.2017.06.003

Sarvajayakesavalu, S. (2015). Addressing challenges of developing countries in implementing five priorities for sustainable development goals. *Ecosystem Health and Sustainability, 1*(7), 1–4. DOI: 10.1890/EHS15-0028.1

Seuyong, F. T., Silwal, A. R., Begazo, T., Newhouse, D., & N'Ghauran, A. (2023). The Size and Distribution of Digital Connectivity Gaps in Sub-Saharan Africa. *The World Bank*. https://documents1.worldbank.org/curated/en/099241003142325200/pdf/IDU0cb2e42f3050260484d0b8370b84eee303ecf.pdf

Shoaei, M., Noorollahi, Y., Hajinezhad, A., & Moosavian, S. F. (2024). A review of the applications of artificial intelligence in renewable energy systems: An approach-based study. *Energy Conversion and Management, 306*, 118207–118207. DOI: 10.1016/j.enconman.2024.118207

Shrestha, A. B., & Aryal, R. (2011). Climate change in Nepal and its impact on Himalayan glaciers. *Regional Environmental Change, 11*(S1), 65–77. DOI: 10.1007/s10113-010-0174-9

Shrestha, T. A., Matin, M., Chitale, V., & Thomas, S. (2022). Exploring the potential of deep learning for classifying camera trap data: A case study from Nepal. In *ICIMOD.org*. International Centre for Integrated Mountain Development (ICIMOD). https://lib.icimod.org/record/35997/files/HimalDoc2022_WP_ExploringDeepLearning_ForApproval.pdf

Silva, R. F., Fava, M. C., Saraiva, A. M., Mendiondo, E. M., Cugnasca, C. E., & Delbem, A. C. B. (2021). A Theoretical Framework for Multi-Hazard Risk Mapping on Agricultural Areas Considering Artificial Intelligence, IoT, and Climate Change Scenarios. *Engineering Proceedings, 9*(1), 39. DOI: 10.3390/engproc2021009039

Singh, R., Bhattarai, N., Prajapati, A., & Shakya, S. R. (2022). Impact of Variation in Climatic Parameters on Hydropower Generation: A Case of Hydropower Project in Nepal. *Heliyon, 8*(12), e12240. ScienceDirect. https://doi.org/DOI: 10.1016/j.heliyon.2022.e12240

Singh, R., Khan, S., & Centobelli, P. (2022). Investigating the Interplay between Social Performance and Organisational Factors Supporting Circular Economy Practices. *Sustainability (Basel), 14*(24), 16781. DOI: 10.3390/su142416781

Sterner, T., Barbier, E. B., Bateman, I., van den Bijgaart, I., Crépin, A.-S., Edenhofer, O., Fischer, C., Habla, W., Hassler, J., Johansson-Stenman, O., Lange, A., Polasky, S., Rockström, J., Smith, H. G., Steffen, W., Wagner, G., Wilen, J. E., Alpízar, F., Azar, C., & Robinson, A. (2019). Policy design for the Anthropocene. *Nature Sustainability, 2*(1), 14–21. DOI: 10.1038/s41893-018-0194-x

Sudheer Varma Surimalla, S. (2018). Intelligent Traffic Management Solution(ITMS) through the implementation of Google Maps and Artificial Intelligence. *International Journal of Engineering, Science and Mathematics*. https://www.researchgate.net/publication/327020444_Intelligent_Traffic_Management_SolutionITMS_through_the_implementation_of_Google_Maps_and_Artificial_Intelligence

Suman, S. (2022). Augmented Intelligence for National Security and Development. *Unity Journal, 3*(01), 245–252. DOI: 10.3126/unityj.v3i01.43329

Swapnil Mahadik, S., Gedam, M., & Shah, D. (2024). Environment sustainability with smart grid sensor. *Frontiers in Artificial Intelligence, 7*, 1510410. Advance online publication. DOI: 10.3389/frai.2024.1510410 PMID: 40045982

Tiwari, S., Bashir, S., Sarker, T., & Shahzad, U. (2024). Sustainable pathways for attaining net zero emissions in selected South Asian countries: Role of green energy market and pricing. *Humanities & Social Sciences Communications, 11*(1), 93. DOI: 10.1057/s41599-023-02552-7

UNFCCC. (2016). *The Paris Agreement*. Unfccc.int; UNFCCC. https://unfccc.int/process-and-meetings/the-paris-agreement

Vats, G., & Mathur, R. (2022). A net-zero emissions energy system in India by 2050: An exploration. *Journal of Cleaner Production, 352*, 131417. DOI: 10.1016/j.jclepro.2022.131417

Yang, R., Hu, J., Li, Z., Mu, J., Yu, T., Xia, J., Li, X., Dasgupta, A., & Xiong, H. (2024). Interpretable machine learning for weather and climate prediction: A review. *Atmospheric Environment, 338*, 120797–120797. DOI: 10.1016/j.atmosenv.2024.120797

Yangka, D., Rauland, V., & Newman, P. (2018). Carbon neutral policy in action: The case of Bhutan. *Climate Policy*, *19*(6), 672–687. DOI: 10.1080/14693062.2018.1551187

Zappacosta, M., & Hollema, S. (2017). *FAO/WFP crop and food security assessment mission to Sri Lanka*. Food and Agriculture Organization of the United Nations (FAO). https://openknowledge.fao.org/server/api/core/bitstreams/a7fbafb5-41db-48e2-97e7-e8f415b8df0e/content

Zerssa, G., Feyssa, D., Kim, D.-G., & Eichler-Löbermann, B. (2021). Challenges of Smallholder Farming in Ethiopia and Opportunities by Adopting Climate-Smart Agriculture. *Agriculture*, *11*(3), 192. DOI: 10.3390/agriculture11030192

ADDITIONAL READING

Andtfolk, M., Nyholm, L., Eide, H., Rauhala, A., & Fagerström, L. (2022). Attitudes Toward the Use of Humanoid Robots in Healthcare—A Cross-Sectional Study. *AI & Society*, *37*(4), 1739–1748. DOI: 10.1007/s00146-021-01271-4

Asiedu-Berkoe, F., Chandi, G. M., Bandoh, D. A., Atsu, B. K., Lokossou, V. K., Antara, S. N., Sarkodie, B., & Kenu, E. (2022). State of Public Health Emergency Preparedness and Response Capacity of Ghana. *Journal of Interventional Epidemiology and Public Health*, *5*(4), 1–16. DOI: 10.37432/jieph.supp.2022.5.4.05.4

Chauhan, P., Bali, A., & Kaur, S. (2024). Breaking Barriers for Accessible Health Programs: The Role of Telemedicine in a Global Healthcare Transformation. In *Transformative Approaches to Patient Literacy and Healthcare Innovation*. IGI Global., DOI: 10.4018/979-8-3693-3661-8.ch014

KEY TERMS AND DEFINITIONS

Indigenous Rights: Indigenous rights refer to the collective and individual rights of indigenous peoples to maintain their culture, language, land, and governance systems. These rights also encompass the right to participate in decision-making processes that affect their lives, particularly regarding land use, environmental conservation, and health services.

Traditional Ecological Knowledge (TEK): Traditional Ecological Knowledge encompasses the understanding and practices developed by indigenous communities over generations concerning their local ecosystems. TEK includes insights about biodiversity, resource management, and sustainable practices that can inform contemporary environmental approaches and policies, contributing to both conservation and sustainable development.

Health Disparities: Health disparities are the differences in health outcomes and access to healthcare services between different population groups. In the context of the Sundarbans, indigenous communities may face unique health challenges due to factors like marginalization, environmental degradation, and limited access to healthcare, significantly impacting their overall well-being.

Sustainable Development: Sustainable development refers to a development approach that meets the needs of the present without compromising the ability of future generations to meet their own needs. In the Sundarbans, this concept emphasizes the integration of economic, social, and environmental goals while respecting indigenous rights and promoting community participation in development initiatives.

Resilience: Resilience in this context refers to the capacity of indigenous communities in the Sundarbans to adapt to health challenges and environmental changes while maintaining their cultural identity and practices. Building resilience involves leveraging indigenous wisdom, fostering community cohesion, and developing adaptive strategies to cope with external pressures such as climate change and health crises.

Chapter 7
Revolutionizing Islamic Finance With Ethical AI: Shariah-Compliant Robo-Advisors

Mohammad Irfan
https://orcid.org/0000-0002-4956-1170
Christ University, Bengaluru, India

Aditya Shukla
Christ University, Bengaluru, India

Manali Agrawal
https://orcid.org/0009-0003-7283-3079
NSB Academy, Bangalore, India

Early Ridho Kismawadi
https://orcid.org/0000-0002-9420-5212
IAIN Langsa, Indonesia

ABSTRACT

The revolutionary role that robo-advisors play in Shariah-compliant Islamic finance is examined in this essay. The first section lays out the fundamental ideas of Shariah-compliant investing, which serve as a framework for moral financial judgment. Examined is the emergence of robo-advisors, which show how these AI-powered tools are improving accessibility and efficiency while enabling adherence to Shariah laws. Key characteristics of robo-advisors that adhere to Shariah are examined, along with particular difficulties including cultural sensitivity and regulatory compliance. The research integrates case studies that demonstrate effective implementations, providing valuable perspectives on their real-world uses. The paper's conclusion, which offers a look ahead, highlights how ethical AI and robo-advisors have the

DOI: 10.4018/979-8-3373-0139-6.ch007

potential to transform Islamic finance by promoting sustainability, inclusion, and creativity in Shariah-compliant investments.

INTRODUCTION

AI-Driven Portfolio Optimization in Islamic Finance: The study evaluates the development of a Shariah-compliant multi-period fuzzy portfolio optimization system which combines ethical and sustainable investment methods. The model takes ethical criteria together with sustainability limits to create investment choices which preserve Shariah compliance and maximize financial returns. Such an approach delivers an extensive method to handle complicated financial market conditions thus it facilitates morally sound financial management that promotes sustainable development and moral business decisions. These factors hold great importance because they build a financial system which promotes social responsibility with Shariah compliance and inclusivity in every aspect (Kharrim,2024).

ARTIFICIAL INTELLIGENCE AS A CATALYST IN ISLAMIC FINANCIAL SERVICES

Artificial Intelligence (AI) functions as a catalyst for developing new prospects and confronting difficulties in Islamic finance. The research fills essential knowledge voids about AI implementation in Shariah-compliant financial products by offering significant findings for professionals and research experts and government authorities. The research investigation presents both opportunities and obstacles from AI to assist in the creation of ethical applications within Islamic finance frameworks which improves industry efficiency and understands inclusion along with regulatory compliance. (Kismawadi et al,2024) . The technology uses AI to deliver programmed financial guidance services that support Islamic requirements. The analysis explores Shariah-compliant robo-advisors through their essential elements of investment screening and portfolio management and risk assessment methods which boost Islamic finance efficiency and accessibility by enabling ethical decisions as well as financial inclusion and responsible investment practices (Hooda et al, 2023). This is related the applications in sustainable Islamic finance by studying the ethical aspects of Generative AI technology in financial services. Research must analyse how Generative AI aligns with Islamic ethical standards as it influences financial developments toward sustainable development. This study evaluates both potential advantages and ethical issues from AI innovations to confirm their support of responsible financial investment methods and their contribution to Shariah-compliant finance inclusiveness (Kismawadi, 2024).

THE ROLE OF AI IN SHARIAH-COMPLIANT FINANCIAL

Artificial Intelligence (AI) and Machine Learning (ML)technologies impact on Islamic Capital Market (ICM) through analyses of their ability to boost efficiency as well as reduce risks while driving innovation in Islamic finance standards. Financial institutions which implement AI together with ML technology can advance investment approaches and asset placement methods and risk detection systems that adhere strictly to Shariah rules. These technologies create transparency through responsible financial methods which leads to market efficiency and sustainably expand the Islamic Capital Market (Pal et al., 2023). The necessary alignment between artificial intelligence innovations and Shariah principles remains essential to achieve ethical and sustainable development in finance and economics. An examination of AI effects on Islamic banking together with financial inclusion and ethical governance shows the dual impact of opportunities and challenges according to the study results. The study examines promotes economic development and social well-being (Hemmet, 2023). These AI-driven platforms combine Shariah-based decision-making to make sure investors with faith-based ethics receive financial solutions which adhere to Islamic principles. Shariah-compliant robo-advisors consist of three essential components which are investment screening along with portfolio management and risk assessment according to Islamic principles as this research shows their capability to improve Islamic finance accessibility and operational effectiveness while maintaining ethical investment standards (Kasim, 2023) and the FinTech companies in Malaysia increased steadily from 193 to 313 during the period of four years.

GROWTH OF ISLAMIC FINTECH AND FINANCIAL INCLUSION

The accelerating growth demonstrates a rising demand for Islamic FinTech products because they unite technological innovation with Shariah principles to create financial progress. The combination of digital financial technology solutions with ethical principles about Shariah law through Islamic FinTech enables essential changes in trade financing and crowdfunding and digital transactions. This research examines Malaysia's Islamized Financial Technology sector and discusses what role it plays to boost financial access along with protecting risks and supporting sustainable economic progress (Mohamad et al, 2024). The Financial stability along with long-term security depend heavily on savings and investment for every generation. The combination of effective financial management with Shariah-compliant investment practices erects challenges for numerous people.

AI-DRIVEN PORTFOLIO OPTIMIZATION AND ROBO-ADVISORY SERVICES

This paper introduces the financial robo-advisor application Savings2u which helps users carry out Shariah-compliant savings and investments. The AI-based financial tools on Savings2u work to improve financial education together with financial discipline and easy accessibility for investors who practice ethical and responsible investments (Ishak & Mohamed, 2023). The portfolio optimization framework that applies Shariah law to boost ethical financial decisions and maintain Islamic financial requirements. Shariah restrictions and sustainability requirements exist within the model to produce legitimate and responsible investment approaches. The model performance evaluation utilized 154 securities across the period between 2011 and 2020 for testing purposes to assess its operation effectiveness and risk mitigation capabilities along with its compatibility with Islamic investment principles. The presented research takes shape to build faith-based financial approaches which guide sustainable and ethical investment practices (Hatta et al, 2022) with the continuous advancements of Artificial Intelligence (AI) technologies force industrial changes while ethical considerations about their usage gain rising importance. The investigation here examines both moral aspects and future challenges of Artificial Intelligence as it pertains to Islamic financial systems and Islamic culture. The research introduces an Islamic virtue-based AI ethics framework to handle these worries because this system establishes a replacement for AI development which supports Islamic moral values along with fairness and social responsibility. This framework combines Shariah-compliant ethical standards for responsible AI use in order to create ethical guidance and attain justice alongside transparency and sustainability in modern technological development (Lim et al., 2023). the utilization of Deep Reinforcement Learning (DRL) systems for Islamic portfolio optimization through Shariah-compliant investments targeting the stocks under Jakarta Islamic Index (JII).

The combination of AI-based financial modelling within DRL improves both portfolio selections and risk management capabilities and business decisions and upholds Islamic financial requirements. It investigates DRL methodology to increase Shariah-compliant investment performance through data-based strategies for optimizing Islamic stock market returns and risk reduction(Raquib et al., 2022). Robotics-based financial advisory solutions named robo-advisory services provide automated Shariah-compliant guidance to wider consumer groups. The evaluation in this paper investigates robo-advisors for Islamic finance markets regarding their ability to boost investment accessibility alongside lower costs and Shariah-compliant guidelines. The combination of AI financial planning with ethical investment standards through robo-advisors provides an essential role in worldwide Muslim investor wealth management democratization and financial inclusion enhancement (Faturohman & Nugraha, 2022).

ETHICAL AI INTEGRATION IN ISLAMIC FINANCIAL SYSTEMS

This is an optimization framework for Islamic portfolio selection in the market that satisfies Shariah principles as well as achieves maximum profit alongside risk reduction. The model combines Shariah restrictions to execute ethical investments that exclude haram assets together with permissible loss boundaries. This combination of sophisticated optimization methods with rules of Islamic investment produces a strong solution for investors who want both ethical financial growth and risk management according to Islamic law (Todorof, 2018). Artificial Intelligence (AI) has brought rapid developments to Islamic finance and automation processes along with data-based decisions according to Shalhoob (2025). Robo-advisors powered by AI systems function as key disruptive financial tools because they both apply Islamic financial rules through automation and optimize institutional operation efficiency as well as transparency in Shariah-compliant financial institutions (IFIs). The rise of Artificial Intelligence in Islamic finance demands more research about both ethical challenges and compliance issues that Shariah scholars need to address (Shalhoob, 2025). AI specialists and Shariah scholars must work together to create finance products based on Islamic values through AI-powered frameworks.

Generative AI together with large language models (LLMs) show promise as tools for Shariah advisory because they spot violations and give financial guidance rooted in Islamic principles (Jokhio and Jaffer 2024). The modern financial industry leverages AI tools to support Shariah-compliant and ESG investments through sophisticated portfolio optimization methods based on variance minimization and dynamic rebalancing techniques which produce ethical financial yields (Mirza et al., 2025). The accurate prediction of Islamic stock index performance through Neural Prophet enables financial stability by helping managers reduce risks in Shariah-compliant markets (Muksalmina et al., 2024). The identification and management of Shariah-compliant equities becomes complex because of strict financial together with ethical assessment criteria that must be fulfilled for compliance (Irfan & Kismawadi, 2024).The data inaccuracies together with overfitting problems and real-time integration difficulties create issues for these models that affect their capability to make reliable financial decisions (Needhi et al., 2024). Research has developed Shariah-compliant multi-period fuzzy portfolio optimization models that use ethical and sustainability restrictions to maintain Islamic financial principles in investments (El Kharrim, 2024). AI-directed financial operations demand a balance between fiscal outcomes by maintaining open financial practices and fair procedures and social responsibility (Malhotra, 2024).

Islamic finance has received significant transformation through financial technology (FinTech) which enabled developers to construct Shariah-compliant robo-advisory systems that automated and expanded accessibility (Hooda et al., 2023). More technological advances do not solve all aspects of understanding the dual challenge of ethical regulation for AI-driven financial applications in Islamic finance. AI revolutionary development requires total establishment of governance criteria that combines Shariah principles with global Islamic financial ecosystem trust and sustainability and inclusivity.

ISLAMIC FINANCE AND THE CHALLENGES OF COMPLIANCE IN DIGITAL INVESTMENTS WITH AI

Artificial Intelligence drives changes in Islamic finance through its creation of three disruptive solutions including robo-advisory services and automated investment screening and smart contract execution. Assessing financial systems integrating AI faces numerous ethical and legal complications specifically regarding Shariah principle compliance (Abozaid, 2022). The integration of artificial intelligence in finance creates three main challenges: it demands protection of data privacy and eliminates algorithmic biases as well as ensures decision-making consistency with Islamic ethics. The development of regulatory standards needs to adopt transparency and accountability processes for AI-based financial solutions with the additional requirement of Shariah-compliant financial practices compliance. AI together with smart contracts functions as vital technological advancements that enhance Islamic finance operations and build trust relationships. The combination of artificial intelligence and smart contracts brings analytical benefits to financial decisions as they perform automated transactions that maintain Shariah law integrity and secure operations (Rahim et al., 2018). The use of smart contracts enables effective operational efficiency through minimized transaction fees and removed financial intermediaries in a setting that upholds ethical ethical financial practices. The two technological methods work together by letting artificial intelligence make intelligent financial decisions and smart contracts run automated transactions while observing Islamic principles.

Investigation on the alliance of Artificial Intelligence and Islamic studies remains underdeveloped because extensive bibliometric evaluation of this field needs completion. AI holds the ability to improve Islamic financial services but developers must address differences between its applications and Islamic moral values. A comprehensive review of current research enables the identification of knowledge gaps which will lead to developing a structure for incorporating responsible AI into Islamic finance while promoting ethical and effective technology use (Morshidi et al., 2024). Online investment popularity has made it harder to follow Islamic economic standards. The digital investment platforms implement interest-based transactions alongside excessive speculation and uncertain factors that violate Shariah principles (Abbas, 2024). To guarantee adherence to ethical purchasing standards online financial products must receive comprehensive product evaluations that would also verify risk-sharing protocols and investment transparency. AI monitoring systems for compliance can help Islamic financial establishments maintain ethical standards of digital investment while boosting financial services availability to wider groups of populations.

The future of Islamic finance shows great promise because AI can guarantee adherence to rules as well as automate operations and enhance financial investments for multiple users. The complete incorporation of AI into Islamic financial systems needs proper resolution of all legal, ethical and regulatory obstacles. Further investigation needs to create AI systems which honour Islamic beliefs because this work will help finance expand responsibly through digital transformation.

THE FUTURE VISION OF ISLAMIC FINANCE WITH ETHICAL INVESTMENT AND ROBO-ADVISORY.

A given illustration demonstrates the core features of Islamic Finance through analysis of how emerging technologies including robo-advisory and AI delivery ethical and Shariah-compliant investment methodologies. The four central components of the chart demonstrate Shariah Compliance and Fair Decision and Ethical AI and Robo-Advisors as the principles and technology systems that govern ethical Islamic finance investments and improve decision functions for enhanced transparency.

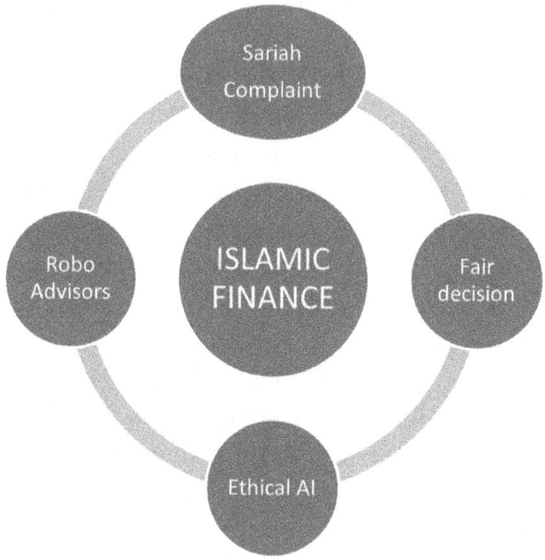

1. Islamic Finance Overview

Islamic finance functions as an ethical financial system based on Islamic Shariah law which prohibits riba (interest payments), gharar (wide-ranging risks), and requires risk-sharing between investors. All investments need to stay true to Islamic principles to exclude sectors which include alcohol businesses and gambling operations together with banks operating under interest-based systems. Islamic investment strategies receive automated Shariah-compliant financial guidance through robo-advisory and AI systems that provide unbiased services to investors.

2. Components of Islamic Finance in Investment

A. Shariah Compliance

All investments under Islamic finance must maintain Shariah compliance to fulfill the necessary Islamic principles. OPI.TR implement Shariah compliance through two main functions: The latest AI algorithms help investors find investment instrumentation that satisfy Shariah law by eliminating specific financial sectors such as alcohol gambling and conventional banking. Through AI analysis of financial data Shariah Advisory Boards can obtain assistance to check if investment practices fulfill Islamic law requirements. The algorithms used by robo-advisors include Islamic scholars' fatwas as an integral part of their decision-making protocols to provide guidance to investors. AI-based investment screening included in robo-advisor systems optimizes the process of automatically delivering halal investment options to clients through automated compliance checks.

B. Fair Decision-Making in Investments

The principles of Islamic finance require ethical standards for economic deals between parties. Robo-advisors that use AI technology help achieve this through their functions. Robotic systems eliminate human prejudice to choose financial options that rest solely on ethical regulations and regulatory requirements. Machine learning algorithms enable investors to see through the decision-making logic which provides complete explanations about their investment options. Islamic finance principles through robo-advisors support profitable wealth distribution systems known as Mudarabah and Musharakah structures to replace conventional debt-based financial transactions. Through ethical and fair decision-making investors gain increased confidence because it shows decisions avoid speculative or exploitative practices.

C. Ethical AI in Islamic Investments

AI technologies in Islamic finance applications bring important functionality for investors by monitoring social responsibility and environmental sustainability during investment decision-making processes. Ethical AI contributes to: Through artificial intelligence algorithms investors gain protection against financial fraudulent transactions while eliminating investments with unethical practices. AI enables analysis of ESG factors together with Islamic principles to help investors make sustainable investment decisions. Sharia-compliant finance protects clients from exploitative practices by banning adulterated lending activities along with profiteering speculation. AI robots in financial advice can prevent investment portfolio choices which carry high risks or speculative behavior. The use of AI insights enables investors to select halal and socially responsible investment options which preserve profitability opportunities.

D. Robo-Advisors in Islamic Finance

The automated financial planning service called Robo-advisors utilizes Artificial Intelligence to perform digital advisory operations that need very limited human action. The robo-advisors in Islamic finance enable the following changes: A PMS (Personalized Investment Plan) that adheres to Shariah rules is built by robo-advisors through integrating investor risk profiles with financial objectives. People from different income levels can access Islamic investment services through affordable advisory services that the platform provides. AI technology uses continuous market trend observation to execute adjustments in investments based on updated rules of Shariah compliance. The platform of Halal Wealth Management utilizes robo-advisors to optimize zakat spending and Islamic endowment investments for Muslim investors. This technology enhances profitable ethical investment selection to provide financial guidance which serves all types of investors.

Robo-advisors coupled with Islamic finance principles along with Artificial Intelligence provide investments strategies that maintain ethical and fair practices while being Shariah-compliant. These modern technologies strengthen compliance evaluations while removing prejudices from decision processes and enable automated choices and eco-friendly investments which creates a more transparent and efficient Islamic financial market accessible to all users. Shariah compliance verification and their combination with fair decision-making processes and ethical artificial intelligence and robo-advisory services make the future of Islamic investments digital, data-based and faithful to ethical finance guidelines. The implementation of these measures creates strong investor confidence and leads Islamic financial institutions to maintain digital market competitiveness.

FORECASTING AND THE GROWTH RETURN OF REVOLUTIONIZING ISLAMIC FINANCE WITH ETHICAL AI: SHARIAH-COMPLIANT ROBO-ADVISORS.

Figure 2. Market growth analysis

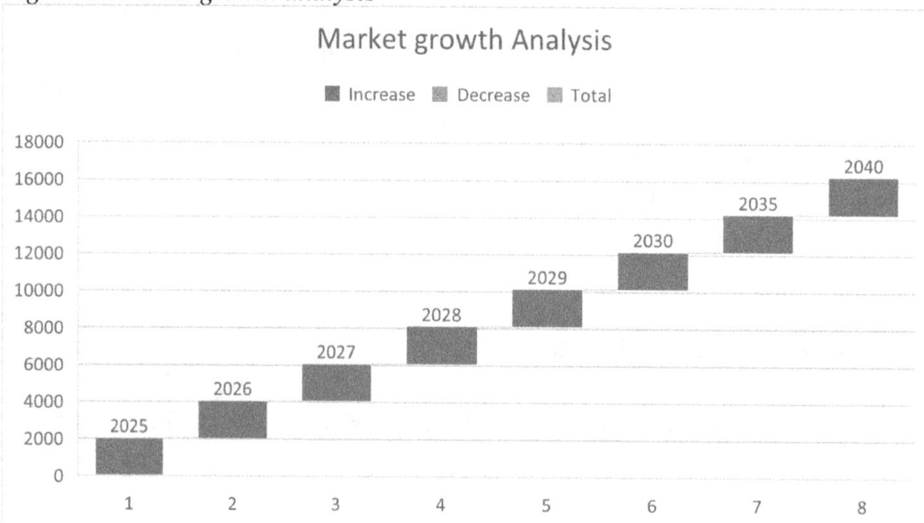

The graph illustrates the comparison between Ethical AI Investments which grew parallel to Islamic Finance Market (USD Trillions) throughout the years. This chart compares the growth of Ethical AI investments and the Islamic finance market from 2025 to 2040.

A parallel escalation occurs in AI-driven ethical investment tactics within Islamic finance as shown by both lines in the positive upward trend. The projection for Ethical AI investments indicates a market size of 6.5 trillion USD by 2035 which tracks nearly equal to the Islamic finance market estimated to grow to 15.5 trillion USD. The ethical AI investment market will attain 12.3 trillion USD by 2040 which maintains a substantial relationship to Islamic finance at 23.8 trillion USD. According to the illustration AI-based investments demonstrate their importance for growing Shariah-compliant financial products while enhancing automated procedures and regulatory compliance as well as investor trust.

Figure 3. Growth statistics

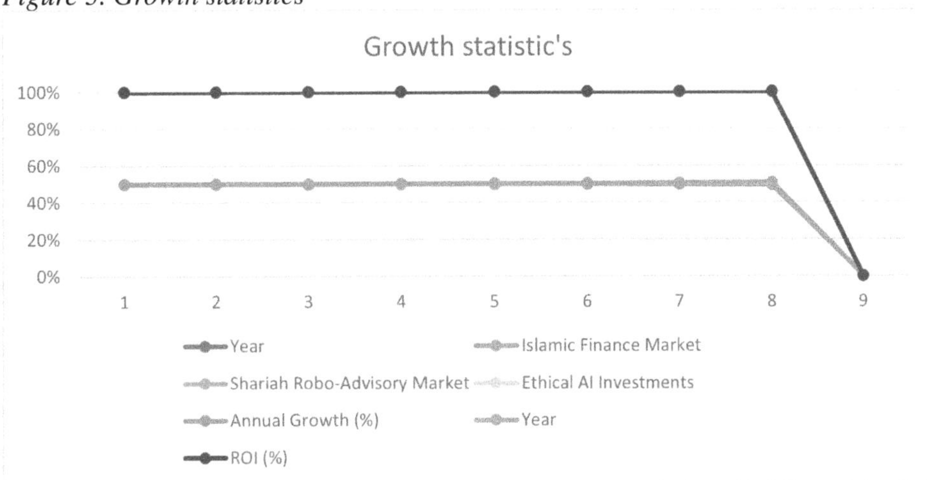

Years appear as numbers from 1 through 9 on the X-axis to represent probably a subject period for research or forecasting. A duplicated Year indicator exists with blue and green lines possibly because legend labels were duplicated twice.

Shariah Robo-Advisory Market

The Shariah-compliant robo-advisory market displays a stable pattern as shown by the gray line during the examined period. The data reveals Islamic robo-advisors have established stable adoption expansion throughout the entire analyzed period.

Islamic Finance Market

The orange line shows the entire Islamic finance market's development pattern. This market shows continuous growth according to the chart although it mirrors patterns found within ethical AI investments so AI-based investment approaches may drive its growth rate.

Ethical AI Investments: This line shows financial investments dedicated to ethical artificial intelligence solutions serving Islamic financial institutions.

Ethical AI functions as a key component for the automated and compliant execution of Islamic investments through Shariah compliance. The performance remains consistent because it matches Islamic finance industry expansion rates.

Annual Growth Percentage: The yearly growth percentage indicated by the light blue line has exhibited consistent growth throughout each year. The growth drops to zero levels during the 9th year of observation. The sudden drop in ROI during Year 9 may reveal either a market correction or an economic downturn or a shift in Islamic finance rules.

ROI % shown through the dark blue line maintains a value of 100% during Year 1 to Year 8 before experiencing a substantial decrease to 0% during Year 9. The sharp decrease in ROI shows a substantial market disruption because of potential external economic factors and changes in regulations or investor changes in behaviour. Expansion in the Islamic financial market will grow steadily into a period of major acceleration after 2030. Ethical AI solutions increasingly support Shariah-compliant finance growth due to their adoption of AI-based ethical investment strategies.

LITERATURE REVIEW

Islamic Finance (IF) has undergone substantial growth due to technological advancements, particularly the deployment of Robo-Advisory (RA) systems. Todorof (2018) highlights that RA features enhance accessibility and compliance within Shariah-compliant financial services. As AI applications in finance expand, AI-driven decision systems and data analytics contribute significantly to financial progress (Ablazov et al., 2024). Moreover, the integration of AI and blockchain technologies has transformed modern financial markets aligned with Islamic principles. Mohamed and Sari (2020) evaluate capital market advancements, emphasizing how these technologies improve transparency, security, and efficiency in Shariah-compliant financial operations. The convergence of AI and RA within Islamic Finance is examined from an innovation and technical progress perspective. Todorof (2018) introduces IF principles and their application in Robo-Advisory under Shariah law. Ablazov et al. (2024) discuss AI's impact on financial decision-making, while Mohamed &

Sari (2020) analyze AI and blockchain in Islamic capital markets. Additionally, Pathan et al. (2022) explore algorithmic trading, manuscript mining, and text mining as tools that enhance data discovery and financial analysis. Collectively, these studies demonstrate how AI and financial technologies align with Islamic Finance principles through Shariah compliance.

Ethical, legal, and technological dimensions of AI in Islamic finance have also been widely examined. Bagheri et al. (2017) note that ethical considerations are often overlooked in portfolio selection, advocating for a balance between financial returns, risk mitigation, and social welfare. Abozaid (2022) explores legal challenges associated with AI in finance, particularly from a maqā idī perspective at the macroeconomic level. Masri (2018) differentiates Shariah-compliant portfolios from Socially Responsible Investing (SRI) portfolios, highlighting the risks of non-compliance, including legal sanctions and reputational damage. Mohamed (2021) further explores AI's role in regulated financial sectors, stressing its potential benefits while maintaining ethical and financial standards.

The integration of AI with blockchain technology enhances Islamic finance operations. AI-driven big data analysis improves financial market research, pattern detection, and investment portfolio optimization, resulting in increased efficiency and informed decision-making. Blockchain technology strengthens data transparency and security, while AI enhances fraud detection and financial risk management. However, research indicates that further studies are needed on AI's role in ensuring strict adherence to Shariah principles. In Indonesia, for instance, the adoption of robo-advisors in Islamic wealth management over the past five years exemplifies AI's transformative role in financial services.

Ethical considerations in AI adoption for financial services are another critical focus. Wang (2024) explores ethical concerns in AI implementation and suggests enhancements to existing frameworks. Traditional portfolio analysis relies on mean-variance approaches, but historical averages are often criticized for estimation errors in returns (Muhammad Nadhirin et al., 2020). AI applications in securities investment and wealth management, as examined by Jain (2012) and Jiang et al. (2024), present both challenges and potential solutions for intelligent investment advisors. Elahi & Abd Aziz (2011) highlight the need for fuzzy environment models to enhance Shariah-compliant portfolio optimization. Furthermore, the Islamic finance industry has demonstrated impressive growth, averaging 18%-20% annually, with Shariah-compliant investments offering superior risk-adjusted returns.

The intersection of Islamic finance, AI, and Shariah compliance is explored extensively. Simanjuntak et al. (2023) provide a comprehensive study of Islamic law and investment literature, reviewing Quranic, Hadith, and contemporary economic texts. Okanlawon et al. (2024) examine how Islamic economic principles adapt to digital wealth creation. Despite the potential of Robo-Advisors for portfolio optimization, research remains insufficient, although Day et al. (2018) provide theoretical and practical recommendations. AlMelaih AlFzari et al. (2024) analyze successful Shariah-compliant Fintech solutions, while Hemmet (2023) discusses AI's religious, social, and economic implications in Muslim-majority countries, emphasizing the necessity of integrating Islamic principles into AI applications.

Artificial Intelligence significantly impacts the profitability of Islamic financial institutions (IFIs). Ahmad et al. (2024) demonstrate that AI-driven chatbots enhance IFI profitability, with key performance indicators including GDP, institution size, and age. However, financial literacy challenges persist, particularly among low-income Malaysian households (Hatta et al., 2022). Afif et al. (2024) review Shariah principles as essential standards for financial investment products, while Setyowati (2023) predicts AI's potential to replace traditional Shariah advisors, sparking ethical debates. Widya & Syafi (2024) discuss Islamic business ethics and Maqashid Shari'ah principles, assessing their impact on ethical business practices and outcomes.

AI applications in Islamic finance involve both opportunities and challenges. Nawi et al. (2023) conduct a qualitative analysis of AI's risks and benefits, focusing on content design in literature. AI presents operational efficiencies through automation but also raises regulatory compliance concerns (Hamadou et al., 2024). ROSA SUSILA et al. (2022) benchmark existing robo-advisors, highlighting AI's role in Shariah-compliant financial investment strategies. A bibliometric analysis (2022) delves into AI's application in Islamic banking, discussing machine learning, natural language processing, algorithmic trading, and robotics. While AI-powered infrastructure enhances operational efficiency and accessibility, it must align with ethical standards and Shariah compliance. Research by Setyowati (2023) forecasts AI's increasing role in financial advisory, raising concerns about religious and ethical standards. Afif et al. (2024) assess AI-based financial investment products for adherence to Islamic finance doctrines, while Ahmad et al. (2024) confirm AI's positive influence on IFI profitability.

This literature review provides a comprehensive assessment of AI's transformative impact on Islamic finance. It highlights the necessity of balancing technological advancements with ethical, legal, and Shariah-compliant considerations. As AI continues to integrate into financial operations, organizations must navigate regulatory requirements, enhance efficiency, and ensure compliance with Islamic financial principles. Integrating Artificial Intelligence (AI) technology into Islamic Finance (IF) financial operations have changed extensively to deliver efficiency along with Shariah principle compliance. Todorof (2018) explains that Robo-Advisory (RA) systems from Artificial Intelligence (AI) improve both the accessibility and Shariah-compliant service adherence in Islamic Finance. Ablazov et al. (2024) discussed AI's significance for financial decision automation and data-based decision support systems. Mohamed and Sari (2020) analyze the use of AI and blockchain technologies which enhance Islam capital market transparency and security and efficiency thereby establishing AI as a vital force for financial transformation.

Islamic Finance incorporates various technological enhancements that enhance portfolio management and modify investment approaches in addition to automation capabilities. Pathan et al. (2022) examine how artificial intelligence helps execute algorithmic trade while mining text data which improves financial and data analytic capabilities. Ethical issues represent the main challenge that needs attention. Bagheri et al. (2017) show that research literature consistently misses ethical goals in portfolio selection because it needs to find equilibrium between returns and risk reduction and social wellbeing. The legal obstacles faced by AI in finance get explored by Abozaid (2022) under Sharīʿah law and maqā idī (higher objectives of Islamic law) principles. Masri (2018) distinguishes Shariah-compliant portfolios from Socially Responsible Investment portfolios although improper compliance may trigger both legal consequences and negative reputation effects. The implementation of Blockchain and AI provides essential functions for strengthening Islamic finance security together with operational transparency. Businesses use AI-liberated big data analytics to find patterns and optimize portfolios as well as develop investment plans while blockchain maintains secure data operations and identifies fraudulent operations (Mohamed, 2021). Several research gaps have emerged from using AI in Islamic finance which specifically concern Shariah compliance standards. Research conducted on Indonesia's Islamic wealth management sector reveals through Day et al. (2018) the influence of robo-advisors on investment choices across periods of time. According to Wang (2024) researchers assess ethical risks regarding financial AI adoption by suggesting framework adjustments to guarantee adherence to Islamic values.

AI-based chatbots with their potential to enhance profitability represent an important area of collaboration between AI and Islamic finance according to Ahmad et al. (2024). The challenge of financial literacy becomes most pronounced for Malaysian low-income families who demonstrate low levels of financial awareness and management capabilities (Hatta et al., 2022). The replacement possibility of traditional Shariah advisors with AI faces scrutiny regarding its suitability with Islamic legal fundamentals (Setyowati, 2023). Widya and Syafi (2024) underline the significance of Islamic business ethics based on Maqashid Shari'ah for preserving ethical operations in digital finance. Investigations about AI in Islamic finance continue to study its operational effects and ethical implications. Firnando and Wahyudi (2024) conducted a systematic review of AI literature which investigates its effects on Islamic economic operations from an Islamic values perspective. Kismawadi (2024) evaluates the moral implications of Generative AI in Islamic finance by developing a framework which merges technological progress with Islamic financial moral principles. The discussion by Khan and Rabbani (2021) explores the essential role of AI in banking and finance through its financial institution operational streamlining capabilities. Two researchers at Azzumi and Aziz (2023) use principles from Dawam Rahardjo to investigate Islamic business ethics within Shariah-compliant digital businesses by highlighting concepts such as fairness and transparency and cooperation. The analysis shows how Artificial Intelligence alters Islamic finance operations yet it deals with proper ethical protocols and regulatory adjustments along with technological adaptations. The industry advancement depends on AI together with blockchain technology as well as robo-advisors despite their need for improved Shariah compliance expertise and financial ethical understanding. Research going forward needs to establish connections between AI implementation gaps in order to achieve responsible application of AI systems in Islamic financial domains.

CASE STUDIES

Wahed Invest

Wahed Invest operates as a leading fintech enterprise dedicated to delivering automated investment services which adhere to Shariah law principles throughout the Islamic finance sector. Wahed Invest launched in 2017 to establish its position as the top company for halal investing through offices in over 130 countries with New York, London, and Dubai locations. Wahed Invest exists to make Shariah-compliant ethical investments available throughout the global Muslim community. Shariah Compliance and Robo-Advisory: The investments at Wahed Invest must conform to Islamic financial rules by excluding riba (interest) and gharar (excessive uncertainty) along with haram industries which include alcohol, gambling, tobacco and pork. The company conducts its screening evaluation under a formal certification system confirmed by an internationally accredited Shariah board. The platform operates via automated investment management through its robo-advisory system which allows novice investors to explore halal investment options. Ethical AI and Portfolio Management: The portfolio management system at Wahed uses ethical AI to generate diversified recommendations for investments which match both risk preferences and financial targets of the user. The platform offers: Equity investments in Shariah-compliant stocks. Fixed-income stability comes from Sukuk (Islamic bonds) while remaining Shariah-compliant. Gold investments as a hedge against inflation. Investors use real estate to experience long-term growth of their assets.

Sarwa

Sarwa functions as a fintech company in the United Arab Emirates with a robo-advisory platform that implements Shariah-compliant investment solutions. Users can develop Islamic-compliant investment funds that respect their financial objectives by using Sarwa's platform. Shariah Compliance and Robo-Advisory: The robo-advisory system of Sarwa lets its users construct diversified investment portfolios which include halal investment options. The automated algorithms operate on the platform to regulate investments according to Islamic finance standards by steering clear of industries practicing interest and speculation and unethical activities. Ethical AI and Portfolio Management: The investment process at Sarwa becomes efficient thanks to their technology that evaluates market trends through AI together with portfolio optimization under Shariah guidelines. The company offers: Shariah-compliant ETFs (Exchange-Traded Funds). Sukuk for fixed-income stability.

Ethis

The Shariah-compliant rob-advice services of Ethis operate from Malaysia through its Islamic fintech platform. The company executes ethical investments and promotes financial inclusion through automated portfolio management systems which follow principles of Islamic finance. Shariah Compliance and Robo-Advisory: By eliminating both interest-based assets and high-risk speculative investments Ethis delivers a total Shariah-compliant investment service. By applying AI in robo-advisor services the platform operates as an ethical investment platform which automatically creates portfolios to match investors' risk capacity and financial aims. Ethical AI and Portfolio Management: The platform Ethis employs ethical AI as a tool to enhance the operational efficiency of financial planning along with investment decision-making processes. The company's portfolio consists of Real estate crowdfunding enables investors to support ethical property investments through the platform. The portfolio consists of Islamic bonds and Sukuk which offer fixed profit returns. Profit seekers employ halal equity funds to purchase assets that fulfill the Islamic finance requirements. Through Ethis the company enables investments that combine social responsibility with financial sustainability.

Musaffa

Musaffa functions as a digital platform which delivers Shariah-compliant robo-advisory services to investors. Through AI-powered algorithms Musaffa enables Muslim investors to plan their investments along with managing them in conformance with Islamic financial principles. Shariah Compliance and Robo-Advisory:

Musaffa uses automation to analyze stock markets along with financial products using Islamic finance guidelines. The robo-advisory system assesses businesses to exclude financial products that generate revenue from interest as well as all companies with imprudent debt amounts or unethical business practices. Ethical AI and Portfolio Management: Musaffa employs analytics power by artificial intelligence to enhance Shariah-compliant portfolio optimization which enables investors to make smarter financial choices. The company's portfolio includes: The chosen Islamic equity funds meet requirements for being halal. Sukuk investments for stability. Real estate as a low-risk halal asset class. Through their ethical AI methodology Musaffa achieves transparency standards which combine with compliance measures to provide investors with sustained investment possibilities.

Niyah

Niyah provided digital banking together with robo-advisory services from its UK base until Wahed bought the company in 2020. The company established a firm focus on Muslim financial needs using digital banking systems that followed Shariah law specifically with robo-advisory solutions integrated into their offerings. Shariah Compliance and Robo-Advisory: The company prioritized developing digital financial services which met every Islamic finance standard. Users could use the robo-advisory system through the company to create automated halal investments which match their financial goals. Ethical AI and Portfolio Management: The AI-powered investment platform of Niyah employed specific algorithms which centered on the following strategy facts before Wahed Invest conducted its acquisition: The financial institution provides Islamic banking without allowing any interest transactions, Shariah-compliant investments in stocks and ETFs, the system allows automatic savings management which follows Islamic finance principles. Wahed Invest gained Niyah's knowledge about digital Islamic finance which helped expand their collection of AI-based ethical financial services.

DISCUSSION AND CONCLUSION

Muslim financial practices have welcomed ethical machine intelligence technology through Shariah-compliant robo-advisors which leads to fundamental changes in sacred investing activities. The research established how artificial intelligence technologies boost Islamic financial resources through better access alongside stricter compliance and improved operational effectiveness even when Shariah principles must be followed. Through ethical artificial intelligence technology robo-advisors carry out investment tasks to eliminate biased human decisions while making all investment choices visible to users. Through AI implementation for investment evaluation and portfolio modification and risk inspection these technologies maintain alignment with Islamic law during financial transactions thus establishing trust between Muslim investors. This study discovered that Artificial Intelligence generates parallel advantages concerning financial inclusivity and regulatory compliance in the market. Robo-advisors with AI automation help new retail investors enter finance but at the same time produce both privacy protection and moral AI application issues. AI experts must join forces with Shariah scholars because it is essential to maintain AI frameworks which adhere to Islamic values. The study establishes that Wahed Invest and Sarwa use Islamic fintech platforms to define how Shariah-compliant investing will evolve through their connection between ethical finance and technological integration. The current strategies face major obstacles relating to preserving the stability of AI models under variable market conditions as well as data defects alongside moral uncertainties. The research indicates that artificial intelligence helps financial planning yet human supervision must remain operational to maintain Islamic financial integrity. A governance system must establish a connection between Islamic jurisprudence and AI ethics to develop morally responsible technological financial solutions.

The advancement of AI in Islamic finance for future years depends on regulatory development along with technological improvement coupled with broad acceptance from Muslim investors. The Islamic finance market will grow its economic potential for inclusivity with enhanced Shariah-compliant investment strategies developed by AI and machine learning technologies. The study establishes that ethical AI systems deployed correctly have transformative capabilities which drive sustainable financial operations and build transparent investment frameworks that lead to responsible investing in Islamic finance. AI has transformed Islamic finance by providing Muslim investors with revolutionary approaches to perform ethical financial activities. Robo-advisors driven by artificial intelligence technology are emerging as main tools for Shariah-compliant investment which promotes a significant shift in the financial sector. The integration of this technology resulted in improved access to finance resources together with better operational performance and enhanced compliance with Shariah requirements which built superior trust between investors and institutions. The integration of ethical machine intelligence creates a crucial framework which supports Islamic value-based transactions and enables financial institutions to serve all clients with both principles in mind. AI integrates a vital element into Islamic finance by removing human decision-making inconsistencies that develop from biases when selecting investment choices. The conventional approach to investment management regularly falls prey to mistakes made by people alongside their natural prejudices along with erratic patterns. Financial institutions utilize AI robotics to make investments by using data intelligence instead of human emotional choices. The robo-advisors deliver open visibility that lets all users maintain clear understanding regarding all investment decisions made through their system. Islamic finance emphasizes transparency because it permits fairness between parties and avoids financial actions which break Islamic principles. The efficiency of Islamic financial operations improved dramatically through AI technologies that evaluate investments and reform portfolios and analyze risks. AI uses on-time analysis of huge financial datasets for building effective investment strategies which guarantee Shariah law compliance in transactions. Organizations must prioritize their compliance with Islamic standards since ethical values matter most in Islamic finance compared to general profit objectives. Financial institutions can enhance their portfolio management by implementing artificial intelligence software which supports their observance of risk-sharing principles together with the ban of interest and ethical profit standards. The implementation of AI-driven financial solutions through Islamic finance creates essential factors that help promote financial inclusion in markets. AI technologies used in robo-advisors make financial market entry possible for retail investors who want to start investing. According to tradition Islamic finance was considered extremely complex because investors needed thorough understanding of Shariah principles for success. AI technology now makes it possible for those without financial knowledge to find Shariah-compliant investment opportunities through its usage. The inclusive nature of Islamic finance seeks to expand its customer base toward collecting both present and incoming young generations who pursue faith-based ethical investment vehicles.

The implementation of AI in Islamic finance enables many benefits although certain technological challenges need thorough evaluation. The responsibility to resolve privacy matters along with ensuring ethical use of AI stands as the utmost challenge to protect Islamic finance integrity. Mattress transactions handled by AI systems require analysis and collection of large personal datasets. The essential foundation for holding investors' trust with financial institutions relies upon strong data security protection of user privacy. The application of ethical artificial intelligence demands AI systems should avoid participating in operations that break Islamic values by performing speculative trading (gharar) and investing in forbidden sectors.

Secure AI application requires AI experts to work together with Shariah scholars for creating AI frameworks which maintain complete adherence to Islamic values. The collaborative effort between AI specialists and Islamic scholars guarantees that financial solutions powered by AI remain faithful to Islamic religious teachings. AI specialists together with Islamic finance experts should build governance systems to regulate financial transactions with AI while using technology as an enhancing tool for ethical financial practices. At Wahed Invest and Sarwa customers benefit from the successful implementation of AI technology to deliver Shariah-compliant investment strategies. Islamic fintech platforms present Muslim investors with digital financial services which execute investments according to Shariah standards through an easy and clear online process. The implementation of AI by fintech companies simplifies financial procedures and strengthens both risk evaluation methods and generates individualized investment suggestions from target financial milestones and risk thresholds. The integration of technology has marked the direction for Shariah-compliant investing to progress during the upcoming period.

AI models face important stability issues when markets undergo changes in conditions. AIs deployed for financial operations require the ability to adapt through reliable mechanisms because economic markets constantly experience shifting conditions. Data defects as well as ethical uncertainties create complex risks which affect the dependability of AI-based financial solutions. The incorrect nature of investment recommendations from AI systems results in financial losses which requires thorough validation systems from institutions to protect model accuracy. The process of financial planning should maintain human supervision as a non-negotiable component because AI serves best as a supporting tool. Islamic finance relies on ethical principles and requirements for fairness together with social responsibility commitments during its financial activities and operations. The oversight of humans remains essential to confirm that financial decisions powered by AI agree with Islamic principles. Financial establishments must develop overseeing systems to close the divide between Islamic legal principles and artificial intelligence ethics thus enabling technology to support ethical and transparent financial programs.

The future development of AI-driven Islamic finance depends heavily on the wide acceptance by Muslim investors. The market demand for Shariah-compliant fintech platforms will expand further because investors are increasingly understanding the benefits of AI-powered financial solutions. In order to gain investor trust in AI-driven investment strategies financial institutions should focus on investor education. Experts need to explain to investors what AI technologies enable regarding Islamic financial law compliance to address their concerns about technological ethics in financial operations. Correct implementation of ethical AI technology creates opportunities to restructure Islamic finance operations toward sustainable financial practices and transparent investment guidelines. AI demonstrates values as a predictive tool because it analyzes large datasets and removes discriminatory patterns while complying with Islamic investment criteria. Therefore it enhances responsible investment processes. Even though AI-driven Islamic finance requires persistent cooperation between AI specialists and Shariah scholars together with financial institutions and regulatory bodies for its achievement.

The implementation of Artificial Intelligence in Islamic finance has brought transformative changes to Shariah-compliant investment operations. By using artificial intelligence robo-advisors the industry has achieved greater financial accessibility and more efficient investments and established strict adherence to Islamic laws.

Islamic finance needs suitable solutions for data protection issues and ethical AI implementation alongside proper rules in place to keep this finance system legitimate. Islamic financial institutions could develop Shariah-adherent technological solutions through techcontrolled collaboration between experts from Islamic finance and technology sectors to create inclusive and ethical investment opportunities. The ongoing progress of AI represents a crucial factor that will define Islamic finance development for providing transparent and sustainable financial systems to Muslim investors worldwide.

SOCIAL IMPLICATION AND PRACTICAL IMPLEMENTATION

The social effects of robot-advisors in Islamic finance combine to advance financial access and ethical investment standards while elevating economic potentials for communities. AI technology extends Shariah-compliant financial investment solutions to previously excluded Muslim groups along with the underbanked Muslim communities. Wealth management through automated investment platforms enables financially uninformed people to achieve ethical investment success with knowledge gained from these platforms.

The use of Artificial Intelligence in Islamic finance assists in enhancing community financial knowledge while educating users about ethical investments. Robo-advisors break down elaborate Islamic investment information by showing users how to select items compliant with Shariah law that includes sukuk along with halal stocks and Islamic ETFs. The expansion of investor access to these tools will enhance financial knowledge among the Muslim population which leads to responsible management of funds and economic steadiness. The practical deployment of Artificial Intelligence in Islamic finance depends heavily on strong IT infrastructure while fintech companies alongside regulatory organizations establish meaningful business alliances. AI-powered robo-consultants need to operate under programmed Shariah law screening systems to stay compliant. Finance institutions need to adopt blockchain systems which improve security and transparency levels of their AI-powered transaction platform.

Islamic finance employs AI both for its ethical job management approaches and its use in vital financial decisions. The advantages from automated processes lead to better operational performance which creates fears about eliminating human financial advisory work. Financial institutions should create a combination of systems which use AI tools to supplement human financial advisors instead of making them obsolete. Financial institutions should retain employment opportunities together with AI-assisted operational efficiency through a combined model. AI systems offer substantial value in sustainable investments through their implementation of Shariah-compliant financial plans based on ESG criteria. AI systems perform market research globally to locate investment prospects that fulfill Islamic principles together with environmental sustainability standards. Through ethical financial promotion AI-powered platforms create a direct pathway to better social welfare through investments in renewable energy solutions and healthcare solutions as well as socially beneficial business enterprises.

RECOMMENDATIONS

The following recommendations will help Islamic finance institutions achieve maximum benefits from AI-driven robo-advisors: The combination of regulators and Islamic finance institutions must develop complete AI governance frameworks through their collaborative work. The approved frameworks must join Islamic jurisprudence components to AI ethics systems to maintain standards of compliance and fairness and transparency in automated financial decision processes. Financial institutions need to promote joint collaboration between artificial intelligence technology specialists and Islamic finance scholarship experts. The partnership between AI experts and Shariah scholars will combine the best of Shariah principles and technological optimization of investment solutions. The security of investor data requires first-class data validation techniques on Islamic finance platforms and cybersecurity measures that protect investor information. Ethical AI systems play a vital role in percentage reduction of biases and finding fraudulent activities and preserving investor loyalty to financial assets. The development of financial literacy programs should teach Muslim investors about AI-driven robo-advisors to promote their adoption. Islamic financial institutions need to establish precise guidelines that explain both the ethical aspects and advantages together with risk elements of automated financial management. Future Islamic finance needs to integrate AI technologies for developing sustainable impact investment solutions. AI assists in generating financial opportunities that satisfy Islamic ethical and sustainable objectives when Environmental Social and Governance principles are integrated with Shariah-compliant finance. Research institutions together with financial institutions should maintain ongoing assessments to determine how AI influences Shariah-compliant investment practices. New technologies including machine learning software and Generative AI require research so they can enhance portfolio management approaches which uphold Islamic finance requirements.

REFERENCES

Ablazov, N., Qodirov, A., Ibragimova, Z., & Akhmedov, K. (2024). Robo-Advisors and Investment Management. *Analyzing the Role of AI in Personal Finance.*, *1*, 1–5. DOI: 10.1109/ICKECS61492.2024.10617229

Abozaid, A. (2022). Financial Applications of Artificial Intelligence. *Sharī ah Issues and the Maqā idī Dimensions.*, *13*(2), 31–66. DOI: 10.55188/ijifarabic.v13i2.333

Afif, M. Z., Sulaiman, H. J., & Amelia, T. (2024). Research agenda of shariah-compliant investment using bibliometric analysis r studio. *Jurnal Ilmiah Manajemen, Ekonomi. Dan Akuntansi*, *8*(3), 1186–1200. DOI: 10.31955/mea.v8i3.4471

Ahmad, Z., al-Amatullah, I., & Mahboob, M. N. (2024). Utilising AI-based Chatbot for Maximising Value and Profitability of Islamic Financial Institutions. https://doi.org/DOI: 10.33001/M0104202421/125

Tayachi, T., Brahimi, T., & Essafi, Y. (2022). Artificial Intelligence in the Islamic Finance: A bibliometric. *Analysis*.

Azzumi, A. R., & Aziz, J. A. (2023). *Considering Dawam Raharjo as the Foundation of Sharia Digital Business Ethics*. Money., DOI: 10.31004/money.v1i2.15991

Bagheri, N., Ben Abdelaziz, F., & Rao, A. (2017). Ethical Stochastic Objectives Programming Approach for Portfolio Selection. *Ethical Stochastic Objectives Programming Approach for Portfolio Selection.*, *1*(1), 495–505. DOI: 10.30585/icabml-cp.v1i1.40

Day, M.-Y., Cheng, T.-K., & Li, J.-G. (2018). AI robo-advisor with big data analytics for financial services. *Advances in Social Networks Analysis and Mining*, *1027–1031*, 1027–1031. Advance online publication. DOI: 10.1109/ASONAM.2018.8508854

Elahi, Y., & Abd Aziz, M. I. (2011). *New model for shariah-compliant portfolio optimization under fuzzy environment*. Springer., DOI: 10.1007/978-3-642-25462-8_18

Faturohman, T., & Nugraha, T. (2022). Islamic stock portfolio optimization using deep reinforcement learning. *Journal of Islamic Monetary Economics and Finance*, *8*(2). Advance online publication. DOI: 10.21098/jimf.v8i2.1430

Firnando, H. G., & Wahyudi, M. A. (2024). The Role of Artificial Intelligence in Shaping the Islamic Worldview of the Digital Economy. *Journal of Islamic Economics and Philanthropy*, *6*(3), 231–249. DOI: 10.21111/jiep.v6i3.11386

Hamadou, I., Yumna, A., Hamadou, H., & Jallow, M. S. (2024). Unleashing the power of artificial intelligence in Islamic banking: A case study of Bank Syariah Indonesia (BSI). *Modern Finance*, *2*(1), 131–144. DOI: 10.61351/mf.v2i1.116

Hatta, M. F. (2022). Financial robo-advisor savings2u for shariah-compliant savings and investment application. *Journal of Information Systems and Technology Management*, *7*(29), 70–85. DOI: 10.35631/JISTM.729006

Hatta, M. F. (2022). Financial robo-advisor savings2u for shariah-compliant savings and investment application. *Journal of Information Systems and Technology Management*, *7*(29), 70–85. DOI: 10.35631/JISTM.729006

Hemmet, A. (2023). *Harmonizing Artificial Intelligence with Islamic Values - A Thoughtful Analysis of Religious*. Social, and Economic Impacts of Technological Advancements., DOI: 10.54536/ajsts.v2i2.2239

Hemmet, A. (2023). *Harmonizing Artificial Intelligence with Islamic Values - A Thoughtful Analysis of Religious*. Social, and Economic Impacts of Technological Advancements., DOI: 10.54536/ajsts.v2i2.2239

Hendarti, Y., Winarno, B., & Aprilianto, M. P. (2024). *Use of Blockchain Technology and AI in Sharia Financial Risk Management*. https://doi.org/DOI: 10.62885/ekuisci.v1i3.165

Hooda, J., Singh, A., Kazaure, M. A., & Singh, R. (2023). *Shariah-Compliant Robo-Advisors*. IGI Global., DOI: 10.4018/979-8-3693-1038-0.ch015

Hooda, J., Singh, A., Kazaure, M. A., & Singh, R. (2023). *Shariah-Compliant Robo-Advisors*. IGI Global., DOI: 10.4018/979-8-3693-1038-0.ch015

Irawan, B., Maimun, M., Faizal, L., & Zaki, M. (2023). Tinjauan Hukum Ekonomi Syariah Tentang Penggunaan Robo Advisor Dalam Islamic Wealth Management (Studi Pustaka Robo Advisor di Indonesia Tahun 2022). *Al-Manhaj : Jurnal Hukum Dan Pranata Sosial Islam*. https://doi.org/DOI: 10.37680/almanhaj.v5i2.3214

Irfan, M., & Kismawadi, E. R. (2024). Sharia Stock Transactions Utilizing Technology and Artificial Intelligence. *Advances in Business Strategy and Competitive Advantage Book Series*, 79–98. https://doi.org/DOI: 10.4018/979-8-3693-3980-0.ch004

Ishak, M. N., & Mohamed, A. (2023). *Harmonization of Islamic Economics With Artificial Intelligence: Towards an Ethical and Innovative Economic Paradigm*. https://doi.org/DOI: 10.24256/kharaj.v5i4.4387

Jain, A. (2012). Shariah-Compliant Portfolio Management: A More Suitable And-Balanced Risk-Return Matrix? *Social Science Research Network*. https://doi.org/ DOI: 10.2139/SSRN.2285927

Jiang, W., Qian, K., Fan, C., Ding, W., & Li, Z. (2024). Applications of generative AI-based financial robot advisors as investment consultants. *Applied and Computational Engineering*, *67*(1), 28–33. DOI: 10.54254/2755-2721/67/2024MA0057

Jokhio, M. N., & Jaffer, M. A. (2024). Generative AI in Shariah Advisory in Islamic Finance: An Experimental Study. *Business Review*, *19*(2), 74–92. DOI: 10.54784/1990-6587.1665

Kasim, R. S. R. (2023). *A smart shariah-based decision-making in islamic fintech*. https://doi.org/DOI: 10.33102/iiecons.v10i1.77

Khan, S., & Rabbani, M. R. (2021). *Artificial Intelligence and NLP -Based Chatbot for Islamic Banking and Finance*. *11*(3), 65–77. https://doi.org/DOI: 10.4018/IJIRR.2021070105

Kharrim, M. (2024). Quantitative framework for shariah-compliant multi-period fuzzy portfolio optimization: A sustainable approach to ethical investing. *Journal of Islamic Accounting and Business Research*. Advance online publication. DOI: 10.1108/JIABR-02-2024-0052

Kismawadi, E. R. (2024). *Sustainable Islamic Financial Inclusion*. IGI Global., DOI: 10.4018/979-8-3693-1565-1.ch013

Kismawadi, E. R. (2024). *Sustainable Islamic Financial Inclusion*. IGI Global., DOI: 10.4018/979-8-3693-1565-1.ch013

Kismawadi, E. R., Irfan, M., & Shah, S. M. A. R. (2024). *Revolutionizing Islamic Finance*. IGI Global., DOI: 10.4018/979-8-3693-0082-4.ch011

Kismawadi, E. R., Irfan, M., & Shah, S. M. A. R. (2024). *Revolutionizing Islamic Finance*. IGI Global., DOI: 10.4018/979-8-3693-0082-4.ch011

Lim, D. T., Goh, K. W., & Lee, L. H. (2023). A new class of Shariah-compliant portfolio optimization model: Diversification analysis. *AIMS Mathematics*, *8*(9), 20933–20965. DOI: 10.3934/math.20231066

Liu, L., Xu, J., Qian, C., Zhong, Y., & Qin, L. (2024). Robo-Advisors: Revolutionizing Wealth Management through the Integration of Big Data and Artificial Intelligence in Algorithmic Trading Strategies. *Journal of Knowledge Learning and Science Technology*, *3*(3), 33–45. DOI: 10.60087/jklst.vol3.n3.p33-45

Malhotra, P. (2024). Harmonizing Financial Returns With Ethical Standards. *Advances in Healthcare Information Systems and Administration Book Series*, 163–172. https://doi.org/DOI: 10.4018/979-8-3693-7452-8.ch010

Masri, H. (2018). A Shariah-compliant portfolio selection model. *The Journal of the Operational Research Society*, *69*(10), 1568–1575. DOI: 10.1057/s41274-017-0223-6

Mirza, N., Tudor, C., Horobeţ, A., & Belaşcu, L. (2025). *Optimizing global risk-conscious portfolios: the strategic role of Sharia-compliant and ESG investments.* Sustainability Accounting, Management and Policy Journal., DOI: 10.1108/SAMPJ-08-2024-0879

Mohamad, S. N. A., Dali, N. R. S. M., & Basah, M. Y. (2024). *Navigating the Moral Compass of Halal FinTech in Islamic Finance: Harvesting Benefits While Curbing Risks* (pp. 157–171). https://doi.org/DOI: 10.1007/978-981-97-1375-2_10

Mohamed, H. (2021). *I-FinTech and Its Value Proposition for Islamic Asset and Wealth Management.* Palgrave Macmillan., DOI: 10.1007/978-3-030-45827-0_14

Mohamed, H., & Sari, N. (2020). *Artificial Intelligence and Blockchain-Driven Islamic Capital Markets.* IGI Global., DOI: 10.4018/978-1-7998-0218-1.ch044

Muhammad Nadhirin, N. F. S., Ishak, N., & Elias, S. M. (2020).. . *Performance of Shariah-Compliant Equity Portfolio Using Model-Based Return and Risk Estimation.*, *7*(2), 104–119. DOI: 10.31580/jei.v7i2.1439

Muksalmina, M., Idroes, G. M., & Maulana, A. (2024). Artificial Intelligence in Islamic Finance: Forecasting Stock Indices with Neural Prophet. *Indatu Journal of Management and Accounting*, *2*(2), 68–80. DOI: 10.60084/ijma.v2i2.232

Nawi, A., Khamis, N. Y., Yaakob, M. F. M., Samuri, M. A. A., & Zakaria, G. A. N. (2023). Exploring Opportunities and Risks of Artificial Intelligence Research for Islamic Ethical Guidelines. *Afkar*, *25*(2), 1–34. DOI: 10.22452/afkar.vol25no2.1

Needhi, J. (2024). Enhancing Financial Intelligence: AI Robo-Advisors for Strategic Investment Decisions.

Needhi, J., Prasath, G. R., Riffath, K. M., & Manokar, S. (2024). *Enhancing Financial Intelligence: AI Robo-Advisors for Strategic Investment Decisions.* https://doi.org/DOI: 10.31219/osf.io/3xena

Okanlawon, S. U., Zagoon-Sayeed, H., & Salisu, T. (2024). *Diagnosing Qur'ānic Precepts on Wealth Creation in Digital Age.* https://doi.org/DOI: 10.21154/invest.v4i1.8422

Pal, T., Irfan, M., Bukhari, S. A. C., & Kassim, S. (2023). *Harnessing the Power of Artificial Intelligence and Machine Learning in the Islamic Capital Market*. IGI Global., DOI: 10.4018/979-8-3693-1038-0.ch006

Pathan, M. S. K., Khoso, A. A., & Ahmed, M. (2022). Digital Model Anecdotes Through Artificial Intelligence in Socioeconomic and Islamic Investments. *International Research Journal of Education and Innovation*, *3*(2), 195–209. DOI: 10.53575/irjei.v3.02(22)21.195-209

Raquib, A., Channa, B., Zubair, T., & Qadir, J. (2022). Islamic virtue-based ethics for artificial intelligence. *Discover Artificial Intelligence*, *2*(1), 11. Advance online publication. DOI: 10.1007/s44163-022-00028-2

Rosa Susila, J. K., Laksono, P., & Afit, M.ROSA SUSILA. (2022). Rancang Bangun Robo-Advisor untuk Pendanaan Rumah Syariah Berbasis Aplikasi Bergerak. *MIND (Multimedia Artificial Intelligent Networking Database). Journal*, *7*(1), 98–110. DOI: 10.26760/mindjournal.v7i1.98-110

Setyowati, W., & Rahayu, I. S. (2023). Sector Analysis of Islamic Capital Markets and Artificial Intelligence Functioning as Sharia Advisors. *Sector Analysis of Islamic Capital Markets and Artificial Intelligence Functioning as Sharia Advisors.*, *1*(2), 236–244. DOI: 10.33050/italic.v1i2.334

Shalhoob, H. (2025). The role of AI in enhancing shariah compliance: Efficiency and transparency in Islamic finance. *Journal of Infrastructure. Policy and Development*, *9*(1), 11239. DOI: 10.24294/jipd11239

Simanjuntak, D. A., Safitri, K., Rahayu, D., Nainggolan, N. S., & Amelia, R. (2023). Islamic law perspective on investment. *Cashflow Current Advanced Research On Sharia Finance And Economic Worldwide*, *2*(3), 418–423. DOI: 10.55047/cashflow.v2i3.644

Todorof, M. (2018). Robo Advice in Islamic Finance. *Social Science Research Network*. https://doi.org/DOI: 10.2139/SSRN.3337712

Todorof, M. (2018). Robo Advice in Islamic Finance. *Social Science Research Network*. https://doi.org/DOI: 10.2139/SSRN.3337712

Wang, Y. (2024). *Ethical considerations of AI in financial decision.* *2*(1), 1290. https://doi.org/DOI: 10.59400/cai.v2i1.1290

Widya, W., & Syafi'i, A. (2024). *Application of Islamic Business Ethics in the Fintech Industry: an Analysis of the Foundation of Sharia Maqashid*. .DOI: 10.58631/ajemb.v2i12.68

Chapter 8
Sustainability Demands Action:
Aligning Sustainable HRM and AI for a Greener Tomorrow

Aqeel Ahmad
https://orcid.org/0000-0002-9263-2544
University of Central Punjab, Pakistan

Mahnoor Maqbool
Kinnaird College for Women, Lahore, Pakistan

ABSTRACT

This chapter explores how Artificial Intelligence (AI) can be harmonized with Sustainable Human Resource Management (HRM) to foster eco-friendly, ethical, and efficient business practices. When AI technologies revolutionize conventional HR functions, they provide opportunities for reduced resource use, improved workplace sustainability and concordance with the UN Sustainable Development Goals (SDGs). This chapter provides an in-depth analysis of AI's role in recruitment, training, resource optimization, and employee well-being while emphasizing the potential for AI to drive both environmental and social sustainability within organizations.

1. INTRODUCTION

The union of Artificial Intelligence (AI) and Sustainable Human Resource Management (HRM) is fast becoming a top agenda for organizations that seek to drive sustainability in the emerging workforce (Kaushal, 2023). With its revolutionary capability, AI has much to offer to revolutionize HR practices by automating, en-

DOI: 10.4018/979-8-3373-0139-6.ch008

hancing decision-making, and overall efficiency. Sustainable HRM, on the other hand, is based on long-term human capital growth with practices that not only drive business success but also environmental, social, and economic sustainability (Jackson, 2011). With organizations increasingly moving their strategies toward sustainability goals, the union of AI and HRM is imperative to creating a future that benefits employees and the world as well (Wamba, 2017).

Although research has been targeted towards the application of AI in complementing traditional HR processes like recruitment, performance management, and employee growth less research has been focused on the integration of AI into the overall framework of Sustainable HRM (Dessler, 2019). Existing research shows that AI has the potential to drive sustainability by automating HR functions and making them more efficient, reducing the carbon footprint of administrative functions, and enabling more intelligent decision-making that promotes more sustainable business practices (D. Rosenberg, 2019). AI technologies can also allow for personalization of employee experiences, increase learning opportunities, and improve workforce engagement, which are key factors of Sustainable HRM. Nevertheless, the link between AI and the environmental goals of HRM, such as promoting green initiatives and reducing organizational footprint on the environment, is under-researched (Kaplan, 2019).

Whereas the potential of AI in HRM has been recognized, much work remains to be done to know how AI can specifically assist environmental sustainability in HR practices. Moreover, whereas the majority of research work has concentrated on how AI affects the operational HR dimensions, relatively little attention has been given to implications for HR professionals, line managers, and employees, who are strategically positioned to drive sustainable practices (Dhar, 2020). These are areas that future research must explore on how AI can be utilized to construct a more sustainable, eco-friendly HRM approach that not only tackles the human capital element but also inculcates sustainability at all levels of the organization (B. Sypniewska, 2023).

This study proposes to close these gaps by exploring the way AI can be combined with Sustainable HRM to develop a greener tomorrow. By concentrating on how AI can facilitate green HRM practices, including waste minimization, energy-efficient work environments, and resource optimization, this study will bring new insights into the Interface between AI and sustainability (Jackson, 2011). Furthermore, this study will investigate the changing functions of HR professionals, line managers, and employees in the context of AI-facilitated HRM, acquiring a greater understanding of how these stakeholders can collaborate to foster organizational and environmental objectives. Lastly, this study will add to the literature by emphasizing the ability of AI to drive sustainability in HRM, proposing a framework for organizations to align their human resource strategies with a more sustainable, eco-friendly future.

1.1 The Intersection of Sustainability, HRM, and AI

Sustainability is now a pressing issue for organizations and organizations all over the world due to the pressure to minimize environmental impact, support social equity and achieve sustainability over the long term. The challenge of balancing responsible resource management with organizational objectives has led companies to integrate sustainability into all business units and especially Human Resource Management (HRM). At the same time, Artificial Intelligence (AI) is one of the most disruptive technologies of business processes, which has changed data processing, decision-making, and resource allocation dramatically (Raisch, 2021). The meeting point between sustainable HRM and AI is an exceptional possibility to create eco-friendly, effective, and ethically responsible practices for both positive business results and good societal conditions (M. Armstrong, & Taylor, S, 2020). This chapter introduces how aligning these two powerful domains Sustainable HRM and AI can lead to a "greener tomorrow, a workplace paradigm where sustainability becomes an integral part of decision-making, employee engagement, and resource management.

1.2 Overview of Sustainability in Modern Organizations

Sustainability today is not limited to the environment. It includes economic sustainability ensuring profitability without overexploiting resources and social sustainability, which focuses on fair treatment, diversity, and contributing positively to communities (Forum, 2018). As stakeholders (investors, consumers, employees, and regulators), become increasingly concerned about sustainability, organizations are coming to realize that it is not simply an issue of duty but an issue of competitiveness and that sustainability should be embedded in the heart of their strategy. According to the United Nations Sustainable Development Goals (SDGs) sustainable organizations are not just working for the development of the economy but also carrying out social and/or environmental contributions like climate action (SDG 13), responsible consumption and production (SDG 12) and lower inequalities (SDG 10) (Nations).

Sustainable organizations are defined by their efforts to reduce their carbon footprint, maximize resource utilization, and facilitate sustainable products and services. For example, companies in the energy industry are looking to renewable resources and companies in the manufacturing industry are looking at waste reduction and recycling process improvement (Lawler, 2006). In the service sector, companies are introducing telework policies to reduce commute-related emissions and hence directly support SDG 13. In addition, sustainability is now a key element in attracting and retaining the workforce, because a growing number of employees want to work for sustainable companies with environmental and social commitment (Nations). This change also highlights the importance of sustainable HRM practices that create an organizational climate where people and the planet matter.

1.3 The Role of HRM in Driving Sustainability Goals

Human Resource Management (HRM) is at the core of the achievement of organizational sustainability objectives. Sustainable HRM goes beyond traditional HR functions by integrating environmental and social considerations into employee management practices (B. Sypniewska, 2023). HR professionals have the duty of integrating sustainability in all aspects of hiring, training, performance appraisal, and even corporate culture building. Through developing a workforce that is engaged and dedicated to environmental sustainability, HRM helps organizations create a culture of environmentalism and social awareness (M. Armstrong, & Taylor, S., 2020).

A contribution of HRM to sustainability through green recruitment and onboarding practices, namely, paperless application procedure, etc., which minimize the paper waste and environmental load of hiring. Furthermore, HR staff can promote sustainable behavior through such and other means, by offering employees education in eco-awareness, encouraging recycling, and by backing policies aimed at encouraging telecommuting to reduce the total carbon footprint of the organization (Cascio, 2016). HRM contributes to diversity and inclusion, one of the principal components of social sustainability, through fair recruitment policies and by fostering a work environment in which all employees are respected and empowered.

HR's role in driving sustainability aligns with several SDGs, particularly SDG 8 (Decent Work and Economic Growth), SDG 4 (Quality Education), and SDG 5 (Gender Equality). Through the creation of equitable and fair work environments, the delivery of training on sustainability issues, and the maintenance of decent work arrangements, human resources departments contribute to the achievement of more strategic sustainability aims, whilst building trust and loyalty among workforces (Jackson, 2011). Sustainable HRM practices also extend to a direct contribution to employee well-being, leading through increased engagement, decreased turnover, and the creation of a workplace culture that encourages long-term, sustainable growth (M. Armstrong, & Taylor, S., 2020).

1.4 AI's Expanding Role in Business Operations

Artificial Intelligence (AI) in business operations is changing the traditional paradigm by providing data-driven decision-making, automating routine work and improving resource management (Goodfellow, 2016). In HRM, AI is being applied to a variety of functions such as recruitment, performance management, training, and employee engagement. AI-based platforms extract insights from huge amounts of data and use patterns for describing future demands of the workforce, as well as for optimizing processes. This capacity to take data and determine it not only improves efficiency but also delivers results which can lead to the pursuit of sustainability (Bersin, 2021).

For example, in recruitment, AI can take away some of the screening, shortening the time and resources put into hiring and guaranteeing equitable, data-based hiring. During training, AI can bring about customized learning plots, those plots respond individually to the various learners' learning styles, and the teachers can provide customized sustainability training to employees (Hammer, 2004). Moreover, AI can be used to maximize work schedules and support remote work by delivering real-time communications and collaborative tools and help to reach higher environmental sustainability by minimizing consumption of energy linked to commuting and office space.

The role of AI in sustainability is not limited to human resource management but also encompasses other fields, such as supply chain management, where AI is used to make Inventory management and logistics as efficient as possible, thereby cutting down on waste and wasted energy (Jabbour, 2008). Through embedding AI in sustainable HRM, companies can contribute to many SDGs (e.g., Sustainable Development Goal 9 "Industry, Innovation and Infrastructure," SDG 9) focusing on resilience and construction of robust infrastructure, and SDG 12 "Responsible Consumption and Production" (Nations). Nonetheless, it is important to consider ethical issues related to the use of AI, including the privacy and algorithmic bias of the data, to make AI-based HR practice not only practical but also ethical. With the right ethical implementation, AI could be the critical accelerator of sustainability in HRM, allowing organizations to adopt a healthier, greener, more responsible operational model (Bersin, 2021).

Example Content

The embedding of sustainability into organizational structures has progressed from an optional compliance agenda to an essential strategic commitment. Sustainable HRM considers ethical considerations, resource optimization, and lasting effects - all of which are in close symbiosis with the principles of Sustainable Development Goals (SDGs) e.g., Sustainable Development Goal8 (Decent Work and Economic Growth) or Sustainable Development Goal13 (Climate Action) (Jackson, 2011). At the same time, the incorporation of AI into business, especially in HRM, provides possibilities for process optimization and sustainable decisions for recruitment, training, and performance measurement. This chapter investigates how AI and sustainable HRM can synergize to shape an eco-conscious workplace that promotes both people and the planet.

2. SUSTAINABLE HRM: KEY CONCEPTS AND IMPLICATIONS

Sustainable Human Resource Management (HRM) is an emerging construct that incorporates sustainability principles in human resources practices. It highlights the process of human capital management that leverages an organization to serve the long-term flourishing of both the organization and society (M. Armstrong, & Taylor, S, 2020). The development of this phenomenon is a result of the recognition that, by not being only isolated from the organization's environment, but also from society, the organization cannot function. In an attempt to promote sustainability HRM needs to change, to move away from the role of a manager of persons and evolve into a strategic force behind responsible business. Sustainable HRM is a systemic approach between environmental targets, social objectives and organizational effectiveness (Deloitte, 2021a).

This paragraph discusses the main ideas of Sustainable HRM together with its major principles and how HRM practices are used to accomplish environmental objectives and social justice thereby making a contribution to achieving organizational and society sustainability purposes in general.

2.1 Defining Sustainable HRM

Sustainable HRM is a concept of HR practices that contribute to the long-term sustainability of the company as well as of society at large. Unlike traditional HRM, which primarily focuses on organizational performance, Sustainable HRM incorporates environmental, social, and ethical considerations into all aspects of HR practices, from recruitment and training to performance management and employee well-being. This approach guarantees that HR decisions will be in the best long-term interest of the employees, the organization, and the environment (Jabbour, 2008).

Sustainable HRM can be understood by three main dimensions:

Environmental Sustainability: Such human resource practices that contribute to the reduction of the ecological footprint of the organization like waste management, energy, or other resource consumption and fostering green behaviors in personnel. E.g., HR may implement measures such as virtual recruitment, digitalization of paperwork and flextime that can help minimize carbon footprints (B. Sypniewska, 2023).

Social Sustainability: Practices of Human Resources for fairness, equality and inclusivity in the workplace. This encompasses the drive for diversity, community development, employee well-being, and decent work practices. Social sustainability is consistent with SDGs on fair labor practices, gender equality and reducing inequalities (SDGs 5 and 10).

Economic Sustainability: HR's contribution to enabling the business to be both profitable and socially responsible. This includes creating a long-term workforce strategy that looks at skill building, job creation, and frontier creation to drive business expansion in a manner that also has a positive impact on the environment and community.

Sustainable HRM involves a change of perspective in the role of HR from a support function to a strategic driver of sustainability within the organization. It requires HR decision-makers to think critically and proactively about environmental and social factors in the decision-making cycle and to build a work culture where sustainability principles can be embedded in business practices (Nations).

2.2 Core Principles of Sustainability in HR

The main thrusts of sustainability in HRM are integrating the three dimensions of environmental, social and economic sustainability into HR work practices and HR policies. These principles underpin decision-making in HR and maintain a workforce organization model that positively contributes to sustainable organizational success, helps safeguard the environment, and promotes social justice (Porter, 2011).

The following are the key principles of sustainability in HRM:

Responsibility and Accountability: HR needs to create a work environment where employees and managers are held accountable for their actions, especially for environmental and social outcomes. This is achieved through a process that makes sustainability a common value at all levels of the organization. HR has a central role in ensuring a sense of accountability, training, communications and reward systems.

Sustainability principles should be embedded across the employee journey, from recruitment to retirement. This encompasses sustainable hiring practices promotion, eco-friendly employee benefits provision, skill-based development opportunities for sustainability, and green behavior promotion at the workplace. For instance, the adoption of sustainable performance management systems that reward behaviors consistent with the organization's sustainability objectives (Noe, 2021).

Inclusivity and Diversity: SHRM's social sustainability relates to fairness and equitability for all employees, regardless of gender, race, or other features. Inclusive work environments help develop a more resilient workforce and contribute to achieving societal equity objectives. This is consistent with SDGs 5 (Gender Equality) and 10 (Reduced Inequalities), guaranteeing equal access to opportunities, and providing a supportive working environment for all members of staff.

Long-Term Focus: Sustainable HRM emphasizes a long-term value-creation process for the organization and society. Short-term goals, for example, profits and costs, should not be the priority of the long-term objectives of the company, respectively. HR policies should aim to create lasting positive impacts, including a highly skilled, well-supported, and motivated workforce that can drive innovation and sustainable practices over time (Stone, 2015).

Well-being and Work-Life Balance: Employee well-being is a key driver of sustainability, and the role of HRM in creating an environment in which the physical, mental and emotional health of employees as well as that of their families are fostered cannot be overstated. Through the encouragement of work-life balance, the provisioning of wellness programs as well as the advocacy of mental health programs, HR staff play a part in promoting individual health and the overall sustainability of the organization (D. Rosenberg, 2019).

These tenets allow HR professionals to develop the ideal workplace culture that embraces people, planet, and profit in a way that organizations' decisions lead to the realization of the SDGs, especially SDG 3 (Good health and well-being), SDG 5 (Gender equality), and SDG 12 (Responsible production and consumption).

2.3 Aligning HRM with Environmental Goals and Social Equity

Integrating HRM and environmental goals and social equity is one of the core principles of Sustainable HRM. When sustainability is integrated into the policies and practices of HR, HR professionals contribute not only to the reduction of the negative environmental footprint of an organization but also to positive social progress. This concordance is a step towards the attainment of multiple SDGs, such as SDG 13 (Climate Action), SDG 8 (Decent Work and Economic Growth), and SDG 12 (Responsible Consumption and Production). The following are methods by which HRM can help to align sustainability goals with business activities:

Promoting Green Practices: HR can promote employees' adoption of environmentally friendly behaviors both at work and outside of work. This could involve implementation, for example, of local campaigns for economical/energy-saving behavior, incentives for eco-friendly commuting (e.g., bike-to-work schemes), or tearing down paper models (i.e., switching to digital). In addition, HR is able to implement programs on sustainability, in order to sensitize and encourage employees to act (B. Sypniewska, 2023).

Incorporating Sustainability into Recruitment: Sustainable HRM practices can be embedded into recruitment by emphasizing environmental and social values in the organization's branding and employee value proposition. That draws in applicants who have that organization's value system in supporting sustainability. Further, HR can leverage AI-driven tools to speed up and also decrease the carbon emission of the recruitment process, through a reduced number of travelling required for interviews and meetings.

Employee Engagement and Empowerment: Engaging employees in sustainability initiatives is crucial for success. HR can involve employees in sustainability-related policy- and decision-making activities so as to actively participate in waste reduction, energy saving and green office policies. Enabling employees to own these tasks creates a culture of community and social responsibility. HR may also incorporate sustainability measures in employee performance reviews, and incentivize employees to achieve environmental or social targets (B. Sypniewska, 2023).

Creating a Sustainable Workplace Culture: The culture of the organization is the backbone of any effective sustainability action. HR is of crucial importance in shaping a workplace culture and sustainability is an integral value of the workplace. This encompasses embedding sustainability within the values of the organization, facilitating inter-departmental work, and offering continuous skill development to allow employees to stay committed to sustainability targets. A culture of sustainability leads to higher levels of employee engagement, motivation and retention which in turn leads to organizational success.

Social Equity and Diversity Initiatives: Alignment of HR practices with social equity necessitates deep attention to diversity, inclusion, and equity. HR can implement policies that break down these barriers to equal opportunity, e.g., recruitment that is accessible for people with disabilities or supporting gender pay equity in leadership. In pursuing diversity and inclusion efforts, HR can support enterprises in combating workplace inequality and thereby address SDG 10 (Reduced Inequalities) and SDG 5 (Gender Equality) (Nations).

Collaboration with Stakeholders: Sustainable HRM also involves collaborating with various stakeholders, including suppliers, customers, and local communities, to enhance social and environmental outcomes. HR can form collaborations that embody ethical work practices, fairness to trade and environmentalism in the supply chain. This makes it possible for the sustainability actions of the organization to go beyond internal operations and have a greater social and environmental impact (Jabbour, 2008).

Example Content: Sustainable HRM aims to embed environmental, social and ethical values into HR policies and processes. According to Jabbour (2008), sustainable HRM is not limited to resource optimization, it is also about improving workplace inclusivity, employee well-being, and ethical management. Sustainable HRM through the embedding of these principles within hiring, training, and operational policy seeks to achieve resilient and responsible organizations. These activities are consistent with wider environmental goals, and they help achieve SDG 12 (Responsible Consumption and Production), by lessening waste and by rather increasing sustainable use of resources.

In conclusion, aligning HRM with environmental goals and social equity involves a comprehensive approach that integrates sustainability into all HR practices, from recruitment and training to employee engagement and performance management (Garvin, 1993). This alignment not only facilitates organizational development and crisis but also helps toward larger social and environmental objectives, using business as a driver for sustainable development. Through the use of these practices, HR teams can guarantee that the company will serve a keystone role in developing a greener, fairer world (Bessen, 2019).

3. THE ROLE OF AI IN SUSTAINABLE HRM

The contribution of Artificial Intelligence (AI) to Sustainable Human Resource Management (HRM) is an increasingly important force for organizations to achieve their sustainability goals and achieve efficiency and employee experience. AI can be used to transform HR functions by automating processes, making data-driven decisions and facilitating socially and environmentally sustainable practices. By embedding AI technologies into HR processes, organizations can foster sustainability in several crucial areas, from recruitment and talent management to employee development and resource management.

Artificial intelligence-driven recruitment and talent management, the maximization of employee training and development, and resource management and environment impact reduction (Bersin, 2021).

3.1 AI-Driven Recruitment and Talent Management

AI plays an important role in HR/talent management, giving companies the means not only to build more efficiently but also to confirm their processes are on track with sustainability objectives. Conventional recruitment procedures are usually time-consuming, manually intensive, and physically demanding, because of the extensive use of paper for resumes and job ads. AI can automate these tasks and, as a consequence, drastically minimize the environmental legacy of hiring (Bessen, 2019).

Key Benefits of AI in Recruitment: Key Benefits of AI in Recruitment

Automation and Efficiency: AI-based recruitment software, e.g., applicant tracking systems (ATS), automate the evaluation process by freeing up time spent manually sifting through resumes. It results in a more skilled recruiting cycle so that recruiting teams can concentrate on tasks of greater value, such as engagement with candidates and organizational fit. Automation-based processes also help us reduce the need for travel during recruitment activities, and thus decrease the carbon footprint caused by on-site interviews and recruitment activities.

Data-Driven Decision-Making: AI allows HR teams to make more informed decisions by analyzing vast amounts of data from candidate profiles, social media, and job market trends. Machine learning algorithms can analyze candidates' competencies, background and fit for success based on previous data that enables HR staff to select candidates who are most in line with both the company's requirements and sustainability principles. AI’s ability to predict long-term success also ensures that companies hire individuals who are more likely to contribute to sustainable practices over the long run (Vrontis, 2021).

Diversity and Inclusion: AI can provide for the reduction of bias in the selection process using algorithms to provide equal treatment to all candidates. When concrete biases stemming from gender, ethnicity, and background are purged from the AI system, diversity and inclusion can be promoted in hiring, which is a key step toward social sustainability. Algorithms can be trained to prioritize diversity metrics, which not only improve workplace culture but also align with broader societal goals such as SDG 5 (Gender Equality) and SDG 10 (Reduced Inequalities).

Sustainability-Driven Hiring Practices: HR can also leverage AI to implement sustainability criteria in the assessment of candidates. For example, AI tools can assess candidates' commitment to sustainability by analyzing their past experiences, actions, or alignment with green values. This ensures that the workforce reflects the organization’s sustainability goals, enhancing the company’s overall commitment to environmental and social causes (Bostrom, 2014).

Automating the recruitment process, supporting better decision-making, further supporting diversity and catering to sustainability goals, all contribute to the operational efficiency supported by AI-driven recruitment and make a contribution to the broader Sustainable Development Goals of decent work and sustainable economic growth.

3.2 Optimizing Employee Training and Development

Training and development are essential elements of HRM that help to achieve employment satisfaction, retention, and performance. Due to the emergence of AI technologies, organizations are capable of providing individual, adaptive and more beneficial training programs both to the workforce and to enhance sustainability (Jackson, 2011).

Key Ways AI Enhances Employee Training and Development

Personalized Learning Experiences: AI can also bring about individualized learning experiences by evaluating employees' skills, performance records and career aspirations. Using machine learning algorithms, it is possible to make training programs individualized to specific employees' needs to equip them with the skills to contribute to sustainability activities within their jobs. This tailored model promotes employee engagement, retention, and job satisfaction in general, but also links training to the sustainability goals of the organization.

Real-Time Feedback and Support: AI-based platforms can give real-time feedback within training sessions such that it is possible for staff to continually learn and tune their skills. This can include the use of AI chatbots or virtual assistants to provide immediate responses to queries or offer guidance on best practices related to sustainability. By improving the effectiveness of training programs, AI ensures that employees are equipped with the knowledge they need to drive sustainable practices in their everyday work (Cascio, 2020).

Sustainability-Focused Learning Modules: Organizations can employ the power of AI to develop targeted sustainability training components in resource conservation, environmental ethics, and eco-conscious behavior. These modules can be integrated into the continuous development of employees so that sustainability becomes a natural part of its organizational culture at all levels. AI is capable of monitoring users' participation in these programs, and dynamically adapting the content in order to optimize learning results.

Gamification and Engagement: AI can supplement employee training with gamification components, thereby increasing the attractiveness and memorability of training related to sustainability. For instance, employees might be motivated to participate in eco-challenges, monitor their environmental footprint, or engage in team-building activities aimed at sustainable practice. Such approaches create a culture of sustainability which encourages, and educates, employees who can make full contributions to the organization's objectives (B. Sypniewska, 2023).

AI-based in-the-loop training/education for employees can be used to deliver high-quality, low-cost, and always-sustainable learning of workforce skills while engaging the workforce deep in the understanding of sustainability principles.

3.3 Resource Management and Environmental Impact Reduction

The use of AI is of great importance for the management of resources and the reduction of the environmental footprint of organizations. Achieving sustainability across different areas of the business operation process can be achieved using AI optimization of business operations and processes (Hammer, 2004).

Major Contributions of AI in Resource Management and Environmental Impact Mitigation

Energy Efficiency: AI can be embedded into energy management systems to perform real-time energy usage monitoring and optimization. Smart algorithms process data from building sensors, weather forecasts and energy usage behavior to forecast the most optimal times to ventilate office spaces (to produce the energy required for heating, cooling and lighting). AI is also able to recommend changes to lower energy use without compromising worker comfort. Through energy efficiency improvement, businesses can reduce carbon emissions and support SDG 7 (Affordable and clean energy) and SDG 13 (Climate action).

Waste Reduction and Recycling: AI technologies can be exploited to optimize waste management by forecasting waste generation behavior, and by exploiting waste recycling opportunities. For example, AI-powered systems can analyze production data to detect inefficiencies in the manufacturing process, allowing companies to reduce waste and use resources more efficiently. AI can also improve waste sorting by automating processes in recycling facilities, leading to better resource recovery and a reduction in landfill waste. This supports SDG 12 (Responsible Consumption and Production).

Sustainable Supply Chain Management: Through the efficient management of inventory levels, waste minimization and ensuring that resources are used effectively, using AI can help organizations sustainably manage their supply chains. AI algorithms can process huge quantities of environmental data coming from suppliers, ways of delivering and production lines, thereby uncovering opportunities for reducing environmental damage. Examples include that AI can help to improve transportation supply chain planning, by optimizing fuel consumption by devising more efficient routes and delivery schedules. Additionally, AI can help organizations choose sustainable suppliers that meet environmental standards, ensuring that the entire supply chain supports the company's sustainability goals (Cascio, 2020).

Predictive Maintenance: AI can be applied in maintenance programs to forecast when equipment will require service or replacement, minimizing costly unnecessary repairs or replacements. Predictive maintenance increases efficiency by delaying equipment failure and providing effective use of resources. It also saves downtime, making businesses run smoother while saving the environment.

Remote Work and Lower Carbon Footprint: AI-based services that help bring about remote work, like virtual platforms of collaboration, help cut down the carbon footprint generated by traveling to and from the office. With AI-driven process optimization and minimizing office requirements in most operations, organizations also decrease their energy usage in offices and help reduce transportation-related emissions.

Therefore, through the application of AI in resource management and waste reduction, the organization's environmental impact will be dramatically reduced by integrating sustainability principles into all forms of operations. Through such initiatives through AI, not only are businesses supporting SDG 12 (Responsible Consumption and Production) but they also work toward achieving SDG 13 (Climate Action) as businesses can reduce their carbon footprint while also being concerned about the environment (Brynjolfsson, 2014).

Example Content: AI technologies are revolutionizing HRM in very strong ways by offering analytics with data that can predict when a new employee may be required for hiring, ease recruitment processes, and minimize carbon usage in recruiting people. For example, AI-based hiring platforms allow employers to hire remotely and reduce environmental impact since it eradicates the need for human sit-down interviews, according to reports (Kaplan & Haenlein, 2019). Moreover, AI can deliver customized employee training on sustainability by adapting content to help employees learn environmentally friendly practices, which would bring an organization closer to realizing SDG 4 (Quality Education) and SDG 13 (Climate Action).

There are a few interesting facts and figures to know about AI-HRM:

Employers who have automated parts of their talent acquisition and management processes are highly satisfied with the outcome. A staggering 93% have significant time savings and improved efficiency, and 67% have cost and resource savings. (Career Builder Survey)

- Recent studies state 60% of business leaders will enhance their HR department with more AI and automation in the next 5 years. Also, a staggering 61% see the potential of AI replacing HR functions, especially with the current advancements in generative AI tools like Chat GPT. (Personio, 2023)
- Approximately 2 out of every 3 HR professionals report better time-to-fill for vacant positions, with 53% reporting it is somewhat better and 16% reporting it is a much better result. This is because they use automation and AI. (SHRM)

- According to 64% of HR professionals, their companies' automation or AI tools automatically rule out unqualified candidates.
- 85% of employers using automation or AI report it saves time or makes processes more efficient.
- 16% of employers with fewer than 100 employees and 42% of those with 5,000 or more use automation and AI.
- 25% of companies will use automation or AI in hiring and recruitment within five years.
- 66% of CEOs think AI can improve HR.
- 50% of HR executives think AI can transform HR.
- 92% of HR leaders will use AI to hire talent.
- Smart AI mechanisms can automatically disqualify 75% of applicants in an instant by scanning, reading, and processing their resumes.
- The AI HR market will grow 35.26% from USD 3.89 Bn in 2022 to USD 17.61 Bn in 2027.

Figure 1. AIHR

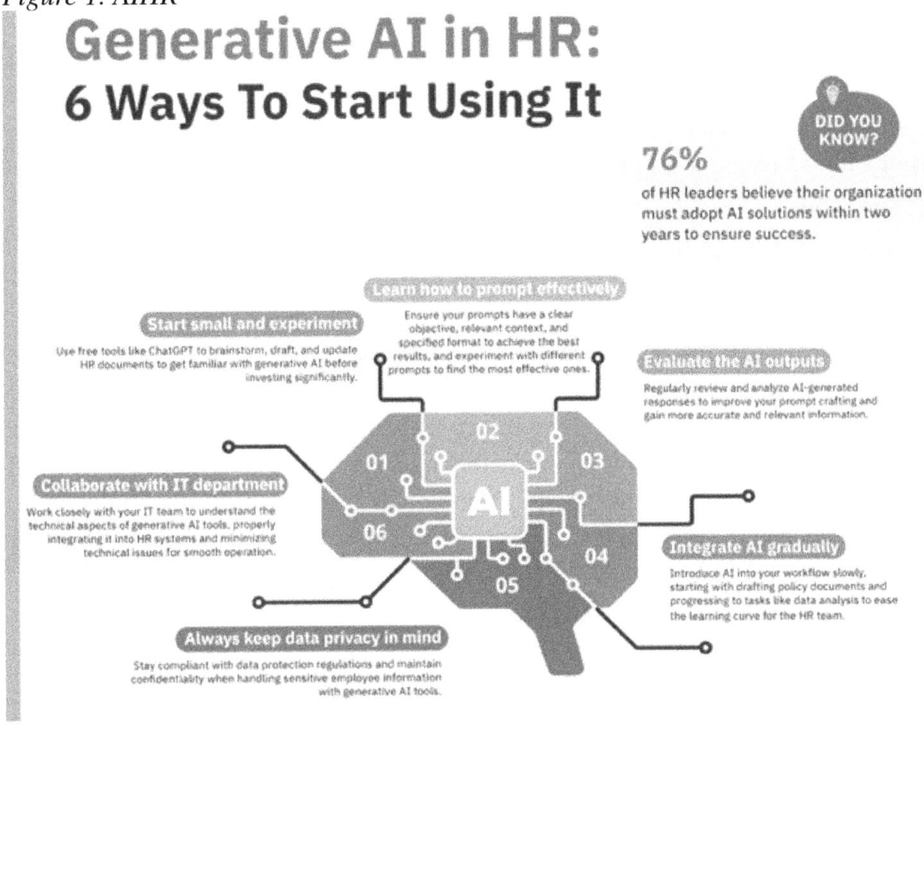

4. SUSTAINABLE RECRUITMENT AND WORKFORCE PLANNING USING AI

Creating a resilient, responsible, and sustainable workforce encompasses sustainable recruitment and workforce planning. Sustainable recruitment and workforce planning infused with AI enhance the efficiency of an organization, as well as enable sustainability in the workforce strategy (D. Rosenberg, 2019). Organizations can reduce their environmental impacts and enhance social equity through AI by developing a workforce that is future-proof and efficient. This section will delve into the applications of AI in sustainable recruitment and workforce planning, focusing on three key areas: digital recruitment platforms and remote interviewing, predictive analytics for hiring and workforce needs, and reducing bias while promoting diversity with AI (Bersin, 2021).

4.1 Digital Recruitment Platforms and Remote Interviewing

This new process of hiring has mainly comprised the employment of digital recruitment platforms and remote interviewing. Improved efficiency and sustainability in these platforms are achieved through AI integration. In most traditional recruitment practices, resources that are consumed include physical paperwork, long distances traveled for personal interviews, and great use of energy-guzzling office spaces. Hence, when recruitment processes are brought to digital platforms while embracing AI technologies in the hiring, the footprint of organizations decreases as speed and accuracy in the hiring process increase (Kaushal, 2023).

Major Benefits of Digital Recruitment and Online Interviewing

1. **Reduced Carbon Footprint:** AI and digital platform-based recruitment reduce the carbon footprint generated through "face-to-face" interviews by quite a margin. Video conferencing, automated scheduling, and AI-interview assistants are some of the tools that will make the interviewing process more streamlined with regards to saving the efforts of candidates as well as the interviewers, who don't have to travel to do it. It saves time and has a minimal carbon footprint as well (Vrontis, 2021).
2. **Cost and Resource Efficiency:** Online recruitment systems may save on paper usage and office space otherwise consumed in the traditional recruitment system. An organization could save from the costs of a physical job posting, printed resumes, and venue-based recruitment events in opting for an electronic recruiting system. That is more resource-efficient and in line with SDG 12--Responsible Consumption and Production.

3. **International Talent Pool and Inclusive Recruitment:** Using digital recruitment platforms powered by artificial intelligence, organizations can now explore a global talent pool and increase their diversity profiles of candidates (D. Rosenberg, 2019). Organizations, therefore, are no longer confined to narrow geographical boundaries when conducting their remote interviews; hence, they minimize the scope of the potential talent pool. Social Sustainability Facilitate equal opportunities for candidates with diverse backgrounds, geographies, and communities SDG 10 Reduced Inequalities, SDG 5 Gender Equality.
4. **Efficiency and Speed:** AI-based recruitment tools can change the way interviews and selections are done, saving more time and making the whole process better for candidates. One can automatically schedule interviews and send reminders to candidates using AI, filter out unqualified candidates will allow the HR team to spend enough time on the most suitable applicants. Such efficiency will save time and enable a quicker, sustainable process of recruitment while keeping abreast of the highest standards in the choice of candidates. Through digital recruitment portals and virtual interviewing, organizations decrease their carbon footprint and can ensure that the hiring process is more representative and effective, aligning with the sustainability goals of the organization (Kaplan, 2019).

4.2 Predictive Analytics for Recruitment and Workforce Requirements

The HR departments will have predictive workforce needs, identification of skill gaps, and better hiring decisions under deep workforce planning thanks to AI-powered predictive analytics. This helps in forecasting future workforce demand using historical data, industry trends, and organizational needs that can set up the organization to respond when the strategy of the business changes, the market changes, or technology changes.

Key Benefits of Predictive Analytics in Sustainable Workforce Planning

1. **Optimization of Workforce Composition:** The organization would be able to understand its workforce structure with the help of AI-based predictive analytics, and thus identify the areas of skill gap. Predicting the trend in loss of employees, business growth, and the pace of industry shift would allow AI to predict the future skills requirements so that HR can hire proactively rather than reactively, bringing the right talent to support sustainable growth and innovation.

2. **Overstaffing and understaffing:** AI can minimize the chances of overstaffing or understaffing by optimizing the number of staff present within an organization. Predictive analytics by human resources allow them to identify how many employees will be needed for a specific task, thus avoiding understaffing, which may cause the available employees to overexert themselves due to a high workload, and overstaffing, which may cause unnecessary costs. Proper utilization of HR leads to financial sustainability because employees do not overexert themselves, and on the other hand, there is no extra manpower (Deloitte, 2021b).
3. **Anticipate future needs for sustainability:** An AI can analyze long-range trends in sustainability, such as changes in environmental policy, consumer expectations, or technological breakthroughs, and then predict how those trends might affect the organization's future workforce demands. For instance, an AI system could predict that an organization would need to hire more employees with skills in green technologies, sustainable practices, or environmental regulation in the future. The potential of predictive analytics is enabling the HR departments to forward plan their workforce building so that it becomes capable of embracing the organization's sustainability objectives.
4. **Employee Retention and Engagement:** Analyzing employee performance data with the help of AI can pinpoint early symptoms of disengagement or dissatisfaction, thus enabling the HR departments to intervene before it turns into a problem of turnover. Predictive models can also help HR teams find out what could help in retaining employees, such as work-life balance, career development opportunities, or sustainability initiatives for instance. Acting on these insights, organizations create more engaged, satisfied, and sustainable workforces- precisely the opposite of decreased retention and increased turnover and long-term employee loyalty (Deloitte, 2021a).
5. **Sustainable talent pipeline:** predictive analytics using AI allows an HR team to predict a talent pipeline that supports the organization's long-term sustainability goals. For instance, it can predict the future need for hiring sustainable talent in certain areas, such as green energy experts or environmental compliance officers. Thus, the organizations can build a diverse and capable workforce ready for the changing landscape of sustainability (Deloitte, 2021b).

Predictive analytics will enable an organization to make more strategic decisions in terms of staffing and make sure that hiring and planning processes agree with long-term sustainability goals, waste reduction in terms of resources, and better workforce resilience.

4.3 Reducing Bias and Promoting Diversity with AI

One of the principal advantages AI offers sustainable recruitment is a reduction in bias in hiring decisions and increased diversity and inclusiveness on the job. The conventional hiring process is typically vulnerable to unconscious biases that influence the choice of candidates. Implemented judiciously, AI technologies can minimize such unconscious biases in recruitment and make the processes more just and compliant with the objectives of diversity and inclusion (Bersin, 2021).

Key Benefits of Employing AI for Bias Reduction and Diversity Enhancement

1. **Bias-Free Screening:** With AI systems, screening candidates will be strictly on their qualifications, experience, and skills about the requirements of the job post, which eliminates biasing recruiters based on gender, age, or ethnicity. This reduces unconscious biases that human recruiters are likely to make. For example, AI algorithms can be fashioned to focus on a candidate's abilities and potential rather than personal characteristics, ensuring all candidates have an equal playing field to evaluate.
2. **Standardized Candidate Evaluation:** With AI, candidates will be fairly evaluated on the same set of criteria to ensure objective decisions are made during the hiring process. In this way, candidates are treated without disparities on account of individual bias or, for that matter, differences in individual interviewers' standards. This standardized assessment process ensures a more transparent and fair hiring process, in support of SDG 5 (Gender Equality) and SDG 10 (Reduced Inequalities).
3. **Talent Pool Diversity:** AI-based recruiting systems can be programmed to maximize diversity among the candidates applying and selected. For example, AI may help calculate which underrepresented groups will apply or look for candidates from different backgrounds that would diversify the talent pool (Noe, 2021). AI can further eliminate notions of gender from the job descriptions or flag unconscious biases in the text of the job postings. This directly contributes to the achievement of SDG 5 (Gender Equality) and SDG 8 (Decent Work and Economic Growth).
4. **Detection of bias and Continuous improvement:** AI can track the steps of recruitment for the emergence of biases they might acquire over time. Machine learning models can therefore analyze historical hiring data and flag patterns that suggest potential bias in the selection process. Identification of such patterns will make organizations hone their recruitment strategies better, correct their mistakes, and ensure fairer outcomes (Guerci, 2016).

5. **Inclusive Job Descriptions and AI Chatbots:** With artificial intelligence, job descriptions can be designed to be even more inclusive to attract a better cross-section and diversity of candidates. Algorithms, especially NLP, analyze the postings to catch discriminatory language and refine or edit it to avoid inadvertently qualifying out some members of the diverse groups. AI chatbots can, in turn, aid applicants in the application process by answering their queries in real-time, ensuring that every applicant has an equal shot at succeeding (Bersin, 2021).

Using AI to help address issues of bias and diversity will help organizations to be more inclusive and fairer in the hiring process and ensure that their workforce is better diversified and reflects the necessary long-term sustainability scope. This will help them align with SDG 5, SDG 10, and SDG 8.

Example Content: Sustainable recruitment through AI reduces the level of paper-based processes and lowers emissions through travel since interviews are conducted remotely. Additionally, AI supports diversity and inclusion in the workplace as it is known to overlook biases in candidate screening leading to a more sensible hiring practice of people in ways SDG 10: Reduced Inequalities. Predictive analytics allows AI to enable HR teams to predict accurately the number of workers needed, hence there is no over-hiring or wastage. This, therefore, is in line with SDG 12: Responsible Consumption and Production.

5. AI IN EMPLOYEE TRAINING FOR SUSTAINABILITY

Employee training is a critical factor for sustaining sustainability at the organizational level. As firms implement their respective targets toward sustainability, the firm's ability to execute green practices calls for an adequately equipped workforce to grapple with environmental challenges and promote sustainability initiatives. Artificial Intelligence transforms the way employee training is looked at, including personal learning experiences which improve engagement, and impact tracking toward sustainability outcomes (B. Sypniewska, 2023). Organizations can now develop highly effective, scalable, and tailored sustainability training programs that inculcate more eco-conscious behavior, foster sustainable actions, and track engagement and impact among employees.

5.1 Personal Learning Paths for Eco-Awareness

Personalized learning paths represent the most interesting use cases for AI-based employee training-the ability to 'tailor' training to specific employee needs, preferences, and work-related goals. In terms of sustainability training, learning needs to be more personalized so that it can respond to employees' varied levels of education and skills development needed to fully participate in the efforts of making sustainability work. The power of AI and employee data can now be unleashed so that customized learning journeys can be designed to increase their engagement and improve how knowledge is retained while aligning training with organizational sustainability objectives (Jackson, 2011).

Key Benefits of Personalized Learning for Sustainability

1. **Tailored Content on What Each Group Member Needs:** Since an AI algorithm can scan a profile, noting the role someone has played, the type of prior training, and what needs to be included within a learning process to fill in gaps, learning modules can be designed to meet a specific need. For example, an employee working in sustainability will need advanced information on environmental regulations or energy-efficient practices, whereas the marketing individual will want to receive basic information regarding how she may incorporate this sustainability factor into a brand message. AI personalizes the learning experience so employees receive the most relevant and effective training for their jobs and so increase stronger alignment with sustainability goals (Emerald Insight).
2. **Adaptive Learning Paths:** AI builds adaptive learning systems that know how to learn, based on knowledge about learner progression, performance, or engagement. Example: Suppose an employee is performing poorly on certain aspects of the sustainability training. An AI system will react by offering additional resources in the form of videos, case studies, or practice assessments that help that particular employee learn better. The adaptive learning path ensures dynamism in training with a keen sense of individual learning needs, thereby increasing retention and effectiveness.
3. **Alignment to Organization's Sustainability Goals:** AI systems can align the personalized learning track to the specific, organization-wide relevant sustainability goals so as to enable employees to understand the environmental and social consequences of their actions within the organization. For example, AI may assimilate companywide sustainability targets like carbon emission reduction or increasing energy efficiency into the learning journey so that the direct impact of an employee's learning is identified in the sustainability of that organization (Nations).

4. **Career Development:** AI-based personal learning will also facilitate the development of sustainable skills along with career development. Since sustainability has gained paramount importance in business, the employees who develop skills and certifications related to sustainability will be ahead in their careers. Hence, opportunities for developing new skills or certification in green technologies, waste management, and other sustainable supply chains are opportunities that AI should identify to support both individual learning and organizational sustainability commitment.

AI thus brings in personalized and customized learning pathways, which means employees are provided with access to all the knowledge and skills needed to make a meaningful contribution towards sustainable efforts. This results in an even more eco-conscious and proactive workforce (Guerci, 2016).

5.2 AI-Enhanced Sustainability Training Through Gamification

Gamification refers to the application of game elements to non-game contexts, such as in a training program. These entities help keep learners engaged and make them respond to the desired behavior. AI-enhanced gamification is an effective catalyst for increasing engagement, motivation, and knowledge retention for training. It uses the gamification mechanics of points, badges, leaderboards, and challenges in the sustainability training setting to make the learning process fun and engaging while fostering environmentally conscious behavior and reinforcing sustainability goals.

Advantages of AI-Based Gamification in Sustainability Training

1. **Enhanced Engagement and Motivation:** AI gamification gives workers more reasons to stay engaged by turning sustainability training into an activity. The process involves giving rewards, monitoring the evolution of the employees, and competition over a competitive element; thus, they are more motivated and committed to completing the sustainability training. This can be achieved by implementing things like how many points or badges workers earn for a certain task about sustainability, such as reduction in energy usage or participation in environmentally friendly initiatives. These rewards support good behavior and also encourage workers to apply sustainability concepts to their everyday practice.
2. **Employee Real-Time Feedback/Performance Monitoring:** AI can give immediate feedback as long as the employees undergo training sessions and gamification; this will allow the employees to track their progress and have insight into their performance immediately. When the employee delivers sat-

isfactorily, he or she will be rewarded appropriately with badges, ranking, or public recognition. But in case an employee stumbles, the AI will give support in the form of advice or another resource needed to challenge or improve. This real-time feedback loop keeps employees engaged and supports continuous improvement (Jatobá, 2019).

3. **Simulation of Sustainability:** AI will utilize gamification by simulating the reality of sustainability issues without risks, to have employees get a practical sense in simulated conditions. Examples might include energy consumption management of a building, waste management optimization, or supply chain complexity resulting in unsustainable decisions. Such training will bring home the outcome of the employees' decisions and help them correlate the training to the actual sustainability activities on the ground.
4. **Team Collaboration through AI-powered Gamification:** AI-powered gamification can also be driven by teamwork by the creation of teams that contest against each other to solve the sustainability challenges. By engaging in team collaboration, the employees exchange knowledge and best practices through the creation of a sustainability culture across the firm. Teams compete against each other in team challenges and virtual events with group rewards meant to intensify the understanding that sustainability requires a "group effort," where every single employee has to be included.
5. **Behavioral Change and Long-Term Impact:** The bottom line of any form of gamification of sustainability learning is behavioral change. Thus, through AI tracking of the progress of learning, rewards, and feedback, gamification will continue showing support for environmental-friendly habits, as well as motivate workers to integrate sustenance practices into personal lives. The interactive nature also leads to better retention of knowledge, hence learning and making sustainability concepts a part of daily life over time.

Using AI-enabled gamification, organizations create interesting, motivational, and effective training that encourages employees to be consistent in developing awareness and behaviors for sustainability and thus motivating employees to be committed towards achieving sustainability goals while offering a greener value to the organization.

5.3 Measuring Employee Engagement and Sustainability Impact

Employee engagement needs to be monitored along with measuring the impact of employee sustainability training to ensure that the program is effective in its contribution toward reaching the overall sustainability objectives within the organization. AI can work towards tracking employee engagement through real-time data and determine the effect of training on output methods which can be termed sustainable. Organizations can obtain a full view of how employees interact with sustainability training programs through AI-driven analytics, and whether the knowledge acquired by these employees matches behavior in sustainability projects initiated by the organization.

Important Benefits of Employee Engagement and Impact Tracking with AI

1. **Real-time Monitoring Analytics:** With AI-based systems, employee engagement in sustainability training programs can be tracked in real-time, providing HR with adequate and rich data on participation, time devoted, and performance. AI-driven analytics can determine patterns and trends, such as which modules are the most popular or which employees require extra special support. Such information enables organizations to make informed decisions on how to improve training programs and enhance employee involvement.
2. **Outcome-based Evaluation for Sustainability:** Sustainability training's effectiveness in daily life can be monitored by the use of AI in the alteration of employee behavior towards sustainability. For instance, it may count the number of employees who decrease their energy usage, or embrace more sustainable practices in their work, or join CSR initiatives after taking sustainability training. Those metrics can be interpreted as measuring how effective the program for such training has been in moving towards the sustainability goals of the organization (Jackson, 2011).
3. **Continuous Improvement:** AI allows organizations to analyze the effectiveness of sustainability training and create changes accordingly in their strategy. With the aid of tracking engagement and measuring impact, the HR teams can identify those areas where the training may not be efficiently leading to good performance results, like lower participation rates or minimally changed behavior. Organizations, based on such findings, can alter their content, methodology, and rewards strategy according to the better needs of employees and achievement of sustainability in organizations.
4. **Employee Feedback and Sentiment Analysis:** AI enables the analysis of employee feedback, surveys, and sentiment data to be used in gauging the overall satisfaction of employees toward sustainability training programs. NLP tools help identify common themes in employee feedback to know what they value most and what to improve. With this feedback loop, training programs remain valid, engaging, and impactful.

5. **Sustainability Performance Measurements:** AI can align sustainability training measurements with more comprehensive reporting systems of sustainability. Herein lies the potential for long-term performance tracking of organizational sustainability efforts. For instance, an AI system can aggregate employee behavioral changes and correlate it with corporate targets toward sustainability, thus providing a holistic measurement of the actual effects of training on broader environmental performance.

Through tracking employee engagement with the help of AI, as well as measuring the impact of the sustainability training implemented, organizations would be able to ascertain that the training programs they engage in don't only engage the employees but also lead to positive, measurable results, in this case, towards sustainability. Continuous tracking and feedback, performance in AI allows employees to be effective in supporting initiatives regarding sustainability (Guerci, 2016).

Example Content: Human Resource departments can use AI-driven learning platforms to coach employees one-on-one on sustainability topics, according to the employee's work and interest in the topic. This drives the expectations of SDG 4 (Quality Education) by encouraging lifelong learning that incorporates environmental stewardship. AI may gamify learning, which creates innovative fun ways for people to remember their sustainability training. Usage of AI for tracking employee engagement and knowledge retention helps the HR departments measure the effectiveness of these programs so that sustainability becomes a part of everyday behavior (Guerci & Carollo, 2016).

6. TELEWORKING AND FLEXIBLE WORKING, CARBON REDUCTION

As the trend is relentlessly shifting toward remote working and flexible schedules, the changes remote work and flexible working have made to organizations and staff, especially in terms of sustainability, are priceless. While remote working facilitates a chance for employees to work from home or any other location, it also fosters broader sustainability aims such as carbon reduction, lesser energy consumption, and the creation of more sustainable ways of life. AI will form the backbone of flexible, sustainable work models, equipping companies with the tools and data needed to make remote work optimized, monitor energy use, and minimize organizational carbon emissions. We will outline in this chapter the role that AI can play in making sustainable work models possible; savings of energy and carbon emissions that remote work can result in; and flexible work schedules aligned with off-peak usage of energy, especially in such places as offices, which are flaring with economic competition (Freeman, 2010).

6.1 AI's Role in Enabling Sustainable Work Models

AI is better placed in terms of the promotion of sustainable work models, so it provides enabler technologies to help in remote work, enhance communication, optimize workflows, and boost productivity while at the same time mitigating possible adverse impacts on the environment. As organizations continue to identify strategies on how they can reduce their carbon footprint and improve work-life balance for employees, AI technology is helping firms to align flexible work models to meet sustainability goals.

Key Benefits

1. **Virtual Collaboration Tools:** AI essentially enables virtual collaboration with such tools as smart video conferencing, virtual meeting assistants, and AI-driven project management platforms. Such tools permit a work engagement among remote teams appropriately coordinated irrespective of their different geographical locations, hence largely saving on business travel, a significant source of carbon emissions. For instance, AI can make appropriate scheduling recommendations by suggesting precise times for everyone to meet considering even the multiple time zones across regions, thus saving time and energy with minimal travel.
2. **Automatic Task Management:** AI can help teams organize and prioritize tasks much better as unnecessary meetings, redundant work, and inefficiencies that often happen in a remote environment can be minimized. In this case, AI frees employees from routine administrative work such as scheduling, reporting, or even tracking of project progress to enable them to participate in impact work. This means productivity is increased while saving energy.
3. **AI-Driven Analytics:** Cloud Solution Ensures access to resources for employees, teaming, and exchange of data with the workforce can be conducted in real-time without any physical framework through cloud computing that is supported by AI-driven analytics. Additionally, AI-driven tools might show the issues related to performance, engagement of employees, and their well-being; thus, the employees who belong to remote working teams will be motivated and focused on the objectives set for them. The objectives cover sustainability goals also (Nations).

4. **Intelligent Office Solutions:** AI can also be applied to traditional office settings to optimize energy consumption, reducing waste, and thereby encouraging a sustainable manner of doing things. In this regard, smart thermostats enabled through artificial intelligence can control heating and cooling systems, and smart lighting system may alter their brightness according to occupancy level. This can ensure that offices are efficiently used without increasing the green footprint of a physical space.

AI also enables organizations to develop and maintain sustainable remote work. This is because it is going to provide solutions resulting in less consumption of energy, increased flexibility at work, and enabling support towards long-term sustainability objectives.

6.2 Energy and Carbon Emission Reduction through Remote Work

Remote work was once something of an indulgence or an intervening phase between the classic face-to-face office and commuting; however, with today's awareness, it has come to be recognized as a practical method for reducing carbon emissions and energy consumption. With lesser reliance on daily commuting, office energy consumption, and various other daily activities that contribute to an organization's total carbon footprint, remote workers save their organizations money while they reduce it by analyzing data, making workflow more efficient, and providing insights on possible reduction areas where AI can contribute.

This leads to several significant advantages of energy and carbon emission savings through remote work.

1. **Emissions Solved from Travel to Work:** Perhaps the biggest advantage of homeworking to carbon reductions is the decrease in emissions related to traveling for work. The home-based employee does not need the daily commute in and out of offices; this decreases the general vehicle-generated emissions, especially in urban areas. AI tools can quantify or log the decreases in traveling for work and enable organizations to attach a value to the positive adjustments toward their carbon footprint (Choudhury, 2020).
2. **Smart Home Energy Efficiency:** Though the burden may shift from offices to remote home locations in terms of energy consumption, AI can still play a great role in minimizing the formation of waste energy at homes. Smart house technologies always guide employees on how to optimize energy use, and this is supplied by AI. For example, an AI-enabled thermostat can adjust heating and cooling based on real-time occupancy and weather patterns, while smart plugs

and lights will control the power consumption of a home office to minimize energy consumption when it's not in actual use.

3. **Energy Efficiency within the Offices:** Even though employees work from home, offices need to be used for meetings or other purposes. Optimization of energy usage in these areas through AI involves the regulation of lighting, heating, ventilation, and air conditioning (HVAC) according to real-time occupancy data. AI can also facilitate the use of office buildings when they are necessary since it saves the organization on its overall consumption of energy. The AI system scans patterns in the usage of the office and identifies areas not utilized fully and can suggest adjustments in the form of less lighting for unused spaces (Dessler, 2019).
4. **Reduced Paper:** In remote work, the utilization of paper-based documentation decreases, because digital tools, and AI-based collaboration platforms come into the fray. Reducing the work, which involves numerous printing, mailing, and other paper-intensive tasks, thus lowers the environmental impact created by the production of paper and waste. The AI-based systems in the digital document management platform such as e-signatures, document storage, and paperless workflows reduce paper usage and waste.
5. **Workplace Travel Reduction:** In addition to reducing daily commutes, remote work also reduces business travel. Teams can communicate and collaborate using AI-driven video conferencing and virtual collaboration platforms, cloud-based document sharing, etc. Reduced travel not only reduces carbon emissions but also saves the cost of travel-borne expenses- well on both environmental and financial sustainability parameters.

AI enables remote working for employees; this brings the organizations huge amounts of energy and carbon-emission savings, which supports organizations' sustainability goals while creating a reduced environmental footprint.

6.3 Flexible Work Models to Match Off-Peak Energy Use

Other flexible work models included staggered work hours, compressed workweeks, or part-time work to help organizations save energy by overlapping their employees' work schedules with those off-peak periods when the energy usage was low. AI remains highly important in allowing such flexible work arrangements in a closer look at the employee schedule, data of energy consumption, and peak demand periods for recommended work hours (Choudhury, 2020).

Main Benefits of Flexible Models at Work towards Peaking Peak Energy Use

1. **Peak Demand:** Organization-wide flexible scheduling will stagger the work hours of the employees to cut energy usage when peaks occur. AI systems can analyze the pattern of energy use and establish peaks in utilization and therefore work hours that could coordinate with lower usage patterns, like in early mornings or late evenings. This diminishes the impact of energy generation on the environment since off-peak energy generation is mainly done with more environmentally friendly sources; and less stress on energy infrastructure.
2. **Optimization of office resource usage:** Flexible work models enable organizations to reduce the number of workers in the office at any point in time, which ends up minimizing the overall energy consumption for heating, cooling, and lighting. AI-powered building management systems can automatically set the lighting, HVAC, and other systems that can change in real-time occupancy; hence, energy is not wasted during off-peak hours and when employees work from home. Improved efficiency in energy use leads to a carbon footprint reduction of the organization as well (Dessler, 2019).
3. **Facilitate remote work at low-carbon energy hours:** Some zones can produce renewable energy, such as solar or wind power, during certain hours of the day. AI can support an organization to schedule the workforce at their best to consume energy at times when it is cleaner. This would help reduce the carbon footprint of remote work and also support the integration of renewable energy into the organization's sustainability strategy (Jabbour, 2008).
4. **Remote Location:** It enables workers to work from places that could be more energy efficient or closer to renewable sources through operating work models based on AI scheduling and communications tools. For example, a place where houses can access more solar power or other green technology co-working areas can facilitate a person to do work while staying at home or in other surroundings. This reduces energy demand on the centralized office locations and contributes to sustainability.
5. **Reduced office footprint:** As flexible work patterns increasingly gain traction; organizations might need fewer office spaces from the company. This would reduce energy consumption attributed to large office buildings. AI tools can help organizations analyze their space utilization and provide recommendations on reducing office footprint sizes without sacrificing adequate office facilities for the flexible work patterns of people. Such steps further lead to carbon reduction initiatives and will further help an organization save on costs.

AI optimizes energy consumption in organizations through the use of flexible work models and adjusting work schedules to align with off-peak energy usage, hence supporting organizations' objectives to reduce environmental footprint and contribute toward sustainability goals.

Example Content: Remote alternative to commuting activity, obviously constraining daily travel to contribute any less to the carbon footprint. AI solutions for promoting home-based work relate to SDG 13 Climate Action by reducing emissions through transport while at the same time facilitating a reduction of office space. AI-driven scheduling will enable organizations to find the optimal hours for working in synchronization with off-peak hours of energy consumption, thus reducing demand on its resources. Organizations can then maintain efficiency with environmental responsibility by promoting flexible models of work.

7. AI FOR REAL-TIME SUSTAINABILITY FEEDBACK IN HRM

While incorporating the dimension of sustainability into organizational activities, the ability to track metrics that represent sustainability metrics in real-time, such as resource usage, carbon emissions, and employee engagement, is also essential. It's through human resource management that an organization develops its sustainability culture, and AI has proven to be a potent tool in tracking HR people's performance in terms of sustainability, offering them instant feedback, and integrating eco-friendly behaviors into their tasks. Companies, using AI-powered systems, can share immediate insights from the employees' point of view on the sustainability efforts as they continuously adjust their behavior and resonate with objectives on sustainability (Kaushal, 2023).

In this chapter, we discuss how AI can help HRM promote real-time sustainability feedback in these directions: tracking resource usage and carbon emissions; real-time sustainability recommendations and employee engagement through immediate feedback to enhance sustainable behavior.

7.1 Tracking Resource Usage and Carbon Emissions

As important as tracking usage of resources and carbon emissions is the effective monitoring thereof, an important challenge many organizations face in sustainability efforts. An AI can also help an organization in the real-time monitoring, analysis, and management of energy consumption, water use, waste production, and carbon emissions. Integration of AI tools into HRM processes provides the organization with a more penetrating insight into how their operations affect the environment and to act accordingly to reduce waste and improve efficiency.

The advantage of reporting resource usage and carbon emissions using AI

1. **Real-time Data Collectivity:** Through the use of smart meters and other IoT sensors, AI-based tools provide real-time collection of data of all resources used by the different departments within an organization. In this manner, each system checks energy consumption, water usage in offices, waste in the offices, and the traveling behavior of employees. Real-time insights are gained about resource usage; thus, through AI algorithms, trends are identified and anomalies are detected. Hence, quickly, HR professionals can assess patterns of resource usage, and corrective actions are taken accordingly (Bersin, 2021).
2. **Carbon Footprint Tracking:** AI systems can also assist in monitoring and measuring the carbon footprint that respective activities carried out within an organization are producing. For example, AI-based platforms can track the emissions produced due to commuting, energy consumption in offices, business travel, and even supply chain activities. By collating it, AI tools can give an overarching view of an organization's carbon footprint, which enables the HR departments to understand the most critical areas for improvement and come up with a strategic plan for reducing their emissions.
3. **Employee-Specific Resource Usage Data:** Using AI analytics, individual employees' resource utilization can also be tracked, such as the work-from-home site energy consumption or environmental footprint of how an employee travels. These can be considered useful to formulate personalized sustainability objectives to develop a reward mechanism for energy use at home, for instance. Value added by personal tracking is that it allows employees to seize control over their footprint, and hence incrementally increases the organizational culture of sustainability(Noe, 2021).
4. **Operational Efficiency:** AI algorithms would help an organization note inefficiencies in the utilization of resources based on real-time data. For instance, the AI system may identify underutilized office space, departments consuming more energy than necessary, or trends in waste generation. Such inefficiencies then allow HRM to strategize better on how not to waste resources, optimize the use of energy, and improve resource utilization.

Through its ability to track resource usage and carbon emissions in real time, AI will help organizations identify their hotspots in sustainability and make data-driven decisions related to the reduction of their environmental footprint.

7.2 AI-Influenced Real-Time Sustainability Recommendations

AI can make sustainability recommendations available by processing and analyzing large-scale data to optimize processes, reduce waste, and decrease the carbon footprint of operations. Integration of AI systems with needs for sustainability goals might allow the department of HR to provide actionable insights on behalf of employees towards fitting actions in with the environmental targets of the firm.

Key Advantages of AI in the Provision of Real-Time Sustainability Recommendations

1. **Instant Feedback for Behavior:** AI-based systems can also provide instant feedback to employees regarding their sustainability-related behaviors. For example, AI can scan patterns of energy consumption in the remote working environment and share it with recommendations on how to minimize carbon footprint by adjusting a thermostat setting, turning lights off, or unplugging idle electronics. Such real-time feedback inspires employees to follow sustainable choice patterns in real-time, thereby reducing their negatives (Raisch, 2021).
2. **Data-Driven Decision-Making:** AI can provide recommendations to HR managers based on real-time data analysis. For example, it can suggest scheduling shifts or meetings during off-peak energy consumption that would help in reducing carbon emissions. It can also suggest optimizing office space usage to minimize energy consumption. These data-driven recommendations ensure that sustainability initiatives mesh with the needs of employees and environmental goals.
3. **Tailored Sustainability Actions:** AI may offer work-specific suggestions to employees, subject to the nature of the job, behavior, or preference. An example might be an AI recommendation for sustainable transportation means for people who commute regularly could be carpooling or even biking. It may also inform its employee clients to opt for certain friendly equipment in a typical office if working from there. Perhaps AI makes employees more likely to align with the company's vision of sustainability and makes them work towards fulfilling that goal by personalizing those recommendations (Lawler, 2006).
4. **Monitoring employee activities towards personal individual goals:** AI will even track the number of times employees are getting closer to specific individual goals such as energy saving or waste reduction and give them possible areas in which they can hit those marks. For instance, AI can tell an employee that they are nearing marking their target about reducing waste and give suggestions on what to increase or what to change to reusable things to help cross the finish line.

Feedback to the employees about their performance and suggestions motivate them toward greater accountability and ensure that sustainability is not lagging.

5. **Aligning with Organizational Sustainability Targets:** AI can align employee activities at individual levels with the large goals of sustainability in terms of the organization's larger targets. For example, as a firm commits to cutting carbon emissions by 10% over the coming year, AI can message staff during a working day about what this might mean for them, including reducing travel miles or replacing equipment with ones of reduced energy usage. That's to say AI harmonizes individual action with organization-wide goals and pushes for coherence throughout the organization in pursuit of sustainability (Stone, 2015).

AI would enable the employees to make moment-by-moment sustainable decisions through real-time sustainability recommendations, whose value would be the immediate environmental benefits and long-term reinforcement of the organization's goals.

7.3 Instant Feedback Engages Employees

The most important factor for the success of any sustainability initiative is employee engagement. Perhaps the best way to engage employees to contribute toward sustainability efforts is by offering them timely, actionable feedback about their performance and behaviors. AI can facilitate this through real-time insights and feedback on employee sustainability actions, making such employees feel more connected to the organization's environmental goals, more motivated to contribute toward achieving the same, and more responsive in changing their behaviors.

Key Benefits of Real-Time Engagement with AI by Employees

1. **Positive Reinforcement:** This AI can shower positive reinforcements in real-time on employees for sustainable behaviors, such as energy consumption and participation by employees in friendly environment initiatives. For instance, AI systems will send badge points or public recognition messages to employees after completing some tasks related to sustainability. These appreciations motivate the employees; however, there is also effective reinforcement of consciousness about sustainability in the organizational culture.
2. **Continuous Improvement and Motivation AI** encourages a continuous cycle of improvement by providing immediate feedback to employees. Employees remain more interested and motivated when they see direct results from their efforts. An example would be an AI system that continuously tracks the amount of energy an employee's remote office space is using and then gives continual

feedback on better efficiency. This would always help employees focus on their sustainability goals and keep them busy with the activity (Vrontis, 2021).

3. **Sustainable gamification:** AI can be included in sustainability with all elements of game development like boards, achievements, and challenges that will add fun or be engaging for what is put in the sustainability practice. Employees can track their progress through colleagues, take part in sustainability challenges, and get rewards at specific paces. This gamified approach creates friendly competition and fun, and sustainability becomes an exciting experience (Kahn, 1990).
4. **Real-Time Visibility:** AI lets workers have real-time visibility into just how their activities further sustainability, which then enables them to envision themselves. Be it just minimizing wastage, consuming less energy, or contributing to a sustainability campaign, employees get a glimpse of where their efforts fit in with the grand organizational goals. This can help create ownership and inspire a sense of accountability, furthering the engagement of staff toward the goals of sustainability.
5. **A Feedback Loop of Employees:** AI may enable a feedback loop for employees to give their input on sustainability initiatives. For instance, an AI-based platform might pose questions such as whether an employee has had experience with sustainability programs or what new ideas he may have regarding ways to enhance sustainability efforts. This would establish an open communication and involvement environment where employees are motivated to contribute toward the organization's sustainability strategy.

Feedback from employees concerning the AI, infusing in real-time, creates responsibility towards work for employees, and by taking it up in practice, it helps the company contribute to environmental goals (Jabbour, 2008).

Example Content: AI fosters the monitoring of HRM 'live' usage data, for example, energy use, or paper usage. Sustainability suggestions can be passed directly to employees through the AI tools and virtual pads, e.g., employees being reminded to turn off lights, or digital notes. The direct feedback provision of this immediacy will enable employees to act responsibly, thereby forming a responsible and accountable culture. This will see HRM practices aligned with SDG 12 (Responsible Consumption and Production).

8. ETHICAL CONSIDERATIONS AND CHALLENGES IN AI-ENABLED SUSTAINABLE HRM

Interestingly, although AI is gaining wide acceptance in HRM towards boosting the sustainability of an organization, it also offers a wide array of ethical issues and problems for organizations. As such, while AI can bring efficiency in a lot of aspects of operations, high transparency in tracking sustainability metrics and increase employee engagement, it raises several data privacy, bias, fairness and efficiency vs. ethics issues. In this section, we are going to discuss the ethical considerations of AI in sustainable HRM, including data privacy and transparency, bias and fairness of the AI algorithms, and the challenge of finding a balance that is between efficiency and ethical standards in decision-making (Bersin, 2021).

8.1 Data Privacy and Transparency in AI Tools

Data privacy is one of the primary ethical concerns surrounding AI-enabled HRM. High-degree HRM systems using AI require a huge personal data of employees in terms of working habits, performance, and resource use regarding sustainability behaviors. In short, all this data plays the most key role in the context of personalized recommendations, sustainability metrics, and, more importantly, enhancing the overall HRM practices. However, the collection and use of such data also pose immense risks to the employee's privacy issues concerning how personal data will be stored, shared, and protected (D Rosenberg, 2019).

Key Issues Concerning Data Privacy and Transparency

1. **Gathering Personal Data:** The AI-powered HRM tools are mainly designed to collect huge amounts of personal information related to the employees, including the most basic contact details up to more sensitive information about work preferences, performance, and behavior. For example, AI systems may monitor an employee's energy use while working from a distance or monitor employee involvement in green programs. Such data may be gold for sustainability and is worth improving, but it also raises other issues of employee surveillance and illicit exploitation of employee data.
2. **Storage of data and security:** Huge volumes of worker data pose risks to security. The organization must, therefore, have a strong measure of security concerning the personal data it holds as a matter of protection against cyber threats and breaches. Besides encrypted information, secure storage, and controlling access to keep certain information away from unauthorized parties, another very essential factor to be sure the AI tools deployed are in line with

the Data Protection Act and also comply with regulations such as the General Data Protection Regulation in Europe goes to necessary implementation that complements trust with employees.

3. **Transparency in Data Usage:** the rights of employees in the collection, usage, and sharing of data by organizations need to be transparent. There is a need for trust in AI-based tools in HRM as regards transparency in data usage. It should outline to employees how their data will be used and whether it will be shared with third parties, among other measures taken with privacy protection. Besides, HR departments should make employee data accessible to themselves, and they should allow the employees an opportunity to choose either to opt-out or be able to control the levels of collection of data (Gartner, 2022).
4. **Informed Consent:** While countering the problem of privacy, an organization may collect and process employee data after informed consent from the employees themselves. This calls for a right to know by the employees as to what data will be collected and how it will be used; they also have a right to know the implications of such data collection in their jobs and personal lives. Organizations should ensure there is a chance to refuse to be involved at any given time, after which withdrawal of consent is allowed once provided.

By addressing data privacy and ensuring that AI tools are transparent, companies can foster a culture of trust and accountability that will help to mitigate the ethical risks associated with AI-enabled HRM.

8.2 Bias and Fairness in AI Algorithms

AI algorithms determine decisions based on data; in doing so, however, they may unwittingly propagate biases inherent in the data they have been trained upon. In sustainable HRM, that may be worded otherwise: slanted selection, unfair performance reviews, or uneven recommendations on sustainability. Bias within AI algorithms is even more frightening in terms of diversity, equity, and inclusion, in that it can be maliciously skewed against particular demographic groups, resulting in such groups acquiring worse outcomes (D Rosenberg, 2019).

Main Problems Related to Bias and Fairness in AI Algorithms

1. **Bias in Recruitment and Talent Management:** AI recruiting tools depend heavily on historical hiring data to decide future hiring processes. If the historical data fed into training these algorithms contain biases, favoring certain demographics over others biases will be unintentionally transmitted through the AI system, producing discriminatory hiring practices. For example, when an AI

system learns from a dataset that women or minorities are less represented at the levels of leadership, the system is going to end up favoring a male candidate or a candidate from a specific ethnic origin or race, thereby deepening the gender and racial gaps in the workplace.

2. **Biases in Sustainability Solutions:** AI-based sustainability solutions may also be marred by biases if there are biases in the data utilized. For example, if the AI system was mainly intended for data collection about a certain group of employees such as those in high-performing departments, the recommendations may likely support those behaviors of the former and do not apply fairly to the rest of the employees involved in other roles. To some extent, this may unfairly penalize or reward some employees based on their sustainability actions for reasons outside their control.
3. **Lack of diversity in AI training data.** Training data for the development of AI algorithms may sometimes lack diversity and, thereby, result in biased or partial outcomes. Since training data may not represent the full diversity of the workforce, AI systems may fail to effectively evaluate or predict outcomes in underrepresented groups. For instance, a performance evaluation algorithm learned from data from a predominantly male workforce might not be sensitive enough to the contributions of female or non-binary employees, thus making biased assessments and missed opportunities for those groups.
4. **How AI Promotes Fairness in Decisions:** In making sure that the use of AI does not institute biased decisions, organizations have to frequently audit their algorithms to ensure that they are not applying biased operations. Examples in carrying out this process entail algorithmic auditing, bias detection tools, and training diverse and representative data for AI models. Organizations then have to rectify any bias using techniques such as overtraining algorithms or altering data collection to minimize unfair biases.

Therefore, organizations need to be proactive and consider steps that will make their AI systems transparent and fair, designed for equity to all staff, regardless of their gender, race, or background. Balancing Efficiency with Ethical Standards 8.3

Actually, the highest strength of AI is that it can process large volumes of data at very high speed and efficiency and thus optimize decision-making processes in a process aimed at the smooth running of operations. However, while pursuing operational efficiency, ethical standards should not be compromised. For instance, while applying AI in HRM, there is usually a thin line between using it in an attempt to boost operational efficiencies- to more manifestly do the proper thing, such as automation tasks or better sustainability practices while maintaining ethical standards in the process (Gartner, 2022).

Balancing Efficiency with Ethical Standards

The most difficult task for the same is the automation of HR functions. On one hand, AI ensures that HR operations can be done very efficiently in terms of hiring, performance appraisals, and employee engagement (research). Again, the over-automation of HR functions may also bring ethical issues about dehumanizing decision-making processes. For example, heavy reliance on AI-based decisions concerning appointments and promotions may miss the most important human elements related to interpersonal skills, team dynamics, and cultural fit, which are next to impossible to measure through AI.

As such, organizations may sacrifice fairness and employee well-being for efficiency.

1. Worker Trust and Job Security: AI HRM tools may replace jobs and limit employees' autonomy in the process. Employees may feel that their jobs are threatened by machines or that AIs make decisions without human oversight. HR should ensure AI tools are used to augment rather than supplant human judgment. Ethical standards should govern the integration of AI into HR processes keeping in mind, without infringing on employee rights and unfairly relegating employees (Russell, 2020).

2. Accountability over AI Decisions: Accountability needs to be maintained in decisions delivered by human resources management as AI becomes more integral in the same. For example, if an AI tool makes a wrong sustainability recommendation or biased hiring decision, then who would be accountable for that output? An organization needs to ensure the use of AI tools is being overseen by humans so that they do not simply go on their authority, and the HR professional should be responsible for the implications of ethics.

3. Sustainability vs. Profit Maximization: While pursuing sustainability alongside profitability creates an apparent contradiction in thought, it makes sense through the application to organizations. In other words, while AI helps organizations achieve sustainability objectives by saving energy consumption, waste, and optimum use of resources, still, in many cases, short-term efficiency or profitability goals come directly into conflict with long-term sustainability objectives. Decisions made on behalf of organizations should therefore balance equally finances and the environment at all times safeguarding the interest of the employees.

4. Informed Consent in AI: As one uses AI for monitoring a worker's performance or engagement for sustainability, obtaining informed consent from employees becomes of prime importance. Employees must be informed of how their data is going to be used implications of AI-driven decisions and what choices they have recourse to opt out of or challenge such decisions. Ethical standards for designing and using AI in HRM should ensure employees are not unwittingly subjected to the processes of surveillance and decision-making as a result of such tools.

Efficiency and ethics can be achieved if organizations consider the consequences on the workforce following an analysis of the essence of a system that uses AI but respects the rights of the employee while promoting fairness and supporting sustainability goals without losing ethics (Russell, 2020).

Example Content: Ethical consideration in the use of AI in sustainable HRM The need for transparency and accountability. These will be critical to earning employees' trust in any AI system of an organization handling personal data (Binns, 2018). Moreover, unchecked AI algorithms may reinforce biases and function against fair treatment in hiring and promotions. Organizations have to strive towards ethical standards to guide AI to align with values that promote inclusiveness, accountability, and stewardship of the environment.

9. CASE STUDIES OF ORGANIZATIONS THAT ARE EMPLOYING AI-ENHANCED SUSTAINABLE HRM

Artificial Intelligence, in its integration into green HRM practices, is expanding fast and large companies use innovative tools to improve sustainability. Case studies will reflect on how the respective organization uses AI to minimize as many negative environmental effects as possible, how the green practices of employees can be implemented, and whether there is a provision for real-time sustainability feedback. The following case studies may help us better understand the successful integration of AI in HRM as it contributes to sustainable change.

Case Study Example 1

TechGreen: Use AI to Reduce Environmental Impact on the Recruitment Process

AI Adoption: TechGreen adopted an AI-based recruitment platform, focused on sustainable sourcing, and operated through three key areas:

- Digitized Recruitment Platforms: The company digitalized the recruitment processes; all processes had been made entirely online. This includes applicant access to the portal, online submission of application documents, and conducting interviews. Hence, no paper documents were necessarily needed, and most of the environmental impacts relating to printing and then mailing the documents were saved.

- Video Interviewing across a distance using AI-empowered video Interviewing Technology: No longer was it necessary to visit for an in-person interview. The platform carried out both interviewing and assessment with AI-powered algorithms used for studying responses and body language studies generating fewer carbon emissions. In addition, AI tools enabled the automation of candidate selection through accelerated and more productive decision-making processes.
- Talent Needs Predictive Analytics: The use of AI tools to predict future staffing needs was used by forecasting past data and future requirements of business operations. Given new hires' number predictions and the correct skill sets required, TechGreen was in a better position to organize streamlined recruitment efforts and to prevent over-recruiting and waste in hiring (D. Rosenberg, 2019).

Outcome and Impact

- Carbon Footprint: The use of AI in the remote recruitment process helped to reduce carbon footprints because less travel was necessary, and thus there was less carbon emission.
- Operational Efficiency: AI helped to optimize candidate selection through hiring thus saving time and resources with respect to recruitment.
- Easy Accessibility and Inclusivity: Through remote interview options and online media, TechGreen has the ability to hire talent from virtually any background or geography. Such accessibility leads to increased diversity and inclusion in the workplace.

Case Study Example 2

GreenEarth Ltd: AI-Based Employee Engagement in Green Practices

AI Implementation the employee engagement tool developed on AI technology-enabled GreenEarth to motivate the internal sustainability behavior of the company:

- Artificial Intelligence-Based Gamification: The company employed the use of AI to gamify sustainable activities. In this context, acts that were environment-friendly were considered “rewards.” Through the platform, employees could compare results. The company also used AI algorithms to monitor employee activities and provide them with personalized recommendations that could better assist them in enhancing their sustainability scores.

- Immediate Feedback and Employee Acknowledgement: Instant feedback concerning the employees' sustainability actions from the system, as an example, a point for recognizing a particular employee who lowered the amount of energy they use in the office or participates in a green volunteer program. Employees received badges, points, and recognition for their efforts with immediate AI-enabled instant feedback that would serve to reinforce proper behavior.
- AI-based Customized Recommendations: AI recommended options to employees of things they could do better and reduce further their carbon usage. For example, office employees would get AI-driven suggestions for cutting down on using more paper, adjusting the temperature settings, or saving more energy. The carbon footprint was also tracked and what measures could be taken to lower it.

Results and Impact

- Higher Employee Engagement: Through the proper AI-driven gamification and real-time feedback, higher employee engagement of the employees in green initiatives ensued. The platform ensured instant recognition and rewards to spur employees into engagement in sustainability practices.
- Behavior Change: GreenEarth successfully measured a decrease in the consumption of energy, waste, and transportation emissions because of the newly formed habits of sustainable living by the employees.
- Organizational Culture Change: The AI system enabled sustainability to find its way into the organizational culture. From there, sustainability would penetrate all aspects of business practice and educate employees on friendly behavior towards the environment at work and in private life.

Case Study Example 3

EcoLogics: Monitoring and Providing Real-Time Feedback on Sustainability

EcoLogics implemented its development of an integrated AI-based tool that would continuously monitor actual levels of sustainability metrics at all organizational levels. The three major areas for tracking AI tools related to the real-time aspect of monitoring as listed below:

- Resource Consumption and Carbon Footprint: The IoT sensors and AI analytics on company office buildings tracked energy consumption, water usage, and waste generated. HR and facilities management teams could track real-time data related to sustainability performance in order to gain an understanding of usage patterns and lay outlines for improvement.

- Employee AI-Driven Sustainability Feedback: The employees received real-time sustainability feedback based on an application they could carry in their mobile that communicated with the company's AI. This mobile application presented rich information about their consumption and waste generation, sustainability metrics, and recommendations that were aligned with their behavior. For example, when an employee turns off and forgets to switch on the lights of a meeting room, it will prompt the employee to switch the lights off.
- Automated Reporting and Insights: The AI system prepared completely independent sustainability reports, which provided a very holistic view of the usage of resources in an organization and the carbon emissions due to them. These were shared with the employees for the cognizance of their individual and organizational implications. The AI presented suggestions for the improvement of sustainability at an individual as well as at the organizational level.

Outcome and Impact

- Optimization of Resources: With real-time monitoring, it was easy for EcoLogics to identify wasted resources and, therefore, measure reductions in energy and water consumption.
- More Aware Employees and Part of the Action: Immediate feedback plus customized recommendations have triggered employees to act right and make sustainable decisions in real-time. Employees were empowered in creating action towards direct contributions toward the company's sustainability goals.
- Better Sustainability Performance: The data-based approach helped EcoLogics monitor its sustainability progress as well as informed it to make decisions based on the improvement of its practices. The company reduced the amount of its products that were being disposed of as waste and emitted carbon amounts in significant quantities over the course of one year.

Case Study Example 4

Unilever: AI-Based Recruitment and Alignment with Sustainability

- Application: Unilever, the giant consumer goods firm, is employing AI to drive recruitment, tying it to sustainability goals. Its AI-based recruitment platform, HireVue, leverages predictive analytics for screening and selection based on a broader spectrum of skills, reducing bias, and increasing diversity.

- Sustainability Focus: In automating recruitment, Unilever has reduced the resources traditionally consumed in the recruitment process (paper and interview travel expenses, etc.). The system also values fairness and inclusion in hiring, aligning with the social sustainability aspect of the company's goals.
- Outcome: This has enhanced Unilever's efficiency in recruitment, reduced environmental impact, and improved diversity of talent pool, thereby further aligning the company's efforts with the United Nations Sustainable Development Goals (SDGs), specifically SDG 5 (Gender Equality) and SDG 10 (Reduced Inequalities).

Case Study Example 5

Siemens: AI-Based Employee Learning and Development

- Application: Siemens, a global technology conglomerate, has utilized AI-based learning platforms to align employee development with sustainability. Its AI-based Learning Management System (LMS) tailors learning content based on individual learning habits, skills gaps, and career aspirations.
- Sustainability Focus: The AI platform helps Siemens build sustainable training initiatives by delivering content online, minimizing the use of physical materials and travel (reducing carbon footprint). The system also helps ensure employees' skills are still aligned with evolving sustainability needs of the company.
- Outcome: The system not only enhances employee engagement but also enhances organizational agility, fostering a culture of continuous learning and development. It also contributes to SDG 4 (Quality Education) and SDG 8 (Decent Work and Economic Growth) by empowering employees to enhance their skill sets in a sustainable way.

Case Study Example 6

IBM: AI for Diversity and Green HR Practices

- Application: IBM integrates AI with its HR practices, particularly with talent management and diversity programs. Its AI-powered tool, Watson Career Coach, offers personalized career advice to employees and helps HR teams with workforce planning, ensuring that the company's recruitment, training, and development plans are aligned with their sustainability initiatives.

- Sustainability Focus: IBM employs AI-based analytics to track diversity programs and ensure that its recruitment plans are diverse, with an emphasis on social sustainability as well as creating a more sustainable work culture. IBM also employs AI to optimize its resource usage in HR activities, reducing unnecessary overhead expenses and resource usage in recruitment and training activities.
- Outcome: IBM's use of AI in HR ensures the preservation of a diverse and capable workforce, while their approach to sustainable resource management aids SDG 5 (Gender Equality) and SDG 12 (Responsible Consumption and Production).

Case Study Example 7

Google: AI-Driven Employee Well-being and Sustainability

- Application: Google has developed a set of AI tools that manage employee well-being and automate HR processes. Google's AI-based employee wellness programs, such as personalized work-life balance advice and stress management programs, are designed to enhance employees' mental and physical well-being.
- Sustainability Focus: These AI solutions contribute to the social sustainability area by ensuring employees' needs are addressed, leading to greater engagement, productivity, and retention. Google's HR also works to reduce environmental impact through the use of AI to track energy consumption in office spaces and suggest greener alternatives.
- Outcome: Google was able to effectively improve employees' well-being, and AI-based analysis of resource use helps achieve their environmental sustainability objectives. This also supports SDG 3 (Good Health and Well-being) and SDG 7 (Affordable and Clean Energy).

Case Study Example 8

Nestlé: AI for Ethical Talent Management and Sustainability

- Application: Nestlé uses AI to improve sustainability in their human resource management, particularly in keeping their talent management ethical. With AI solutions, Nestlé tracks employees' engagement, performance, and well-being and ensures the recruitment and management process of the company is sustainability-oriented.

- Sustainability Focus: The use of AI allows Nestlé to instill a culture of sustainability, where employ
- Employes are engaged and committed to the company's mission and values. By streamlining their recruitment process, Nestlé ensures the hiring of diverse talents that are aligned with the company's sustainability objectives, improving their long-term environmental and social sustainability.
- Outcome: Nestlé improved its talent management, recruitment process, and employee engagement, which supports SDG 12 (Responsible Consumption and Production) and SDG 8 (Decent Work and Economic Growth).

Conclusion: That is with the AI-driven real-time sustainability monitoring and feedback system, companies has been able to highly enhance resource efficiency as well as engage employees in achieving sustainability. It was thus possible for the company to cut a lot on energy and resource consumption by being able to track and provide immediate feedback on sustainability metrics, one example of how AI can enforce sustainability at the organizational level (Vrontis, 2021).

10. SUSTAINABLE HRM FRAMEWORK WITH THE USE OF AI

The integration of AI into HRM for sustainability needs an approach that would help in effectively leveraging AI to achieve environmental and social goals. An approach that integrates strategic recommendations for incorporating AI, sets robust sustainability metrics and facilitates and promotes continuous improvement might turn out to be critically important in aligning the HRM practices of an organization with the overall objectives of sustainability. We outline below a holistic way of building such an approach.

10.1 Strategic Suggestions for AI Implementation

Let us strategize so that AI is implemented in HRM in a way that will be an enabler of the sustainability goals of any organization. Let technology remain subservient to human elements in the practice of HRM. Some strategic recommendations for the effective implementation of AI include:

1. **Define Clear Sustainability Objectives:** Sustainability goals must be clearly defined and aligned with the general business strategy before AI integration can be undertaken in HRM. Therefore, organizations need to define clear environmental and social outcomes that they would like to achieve, such as lower carbon

emissions, diversity and inclusion, or increased engagement by employees in green practices.

2. **Gradual Implementation:** AI acceptance should be gradual, considering the requirements of the organizations. Companies cannot overnight change their HR practices. Companies can start by implementing AI in a specific area of their organization, such as in recruitment or employee training, and then proceed incrementally. In a gradual approach, changes can also be made subject to employee feedback and the efficiency of the AI tools.
3. **Engage Employees in the Process**: Besides getting the employees bought into it, successful AI integration in HRM requires engaging the employees through the process. First, organizations should explain how AI aligns with sustainability goals and improves HRM practices without invading the privacy or autonomy of an individual. Educating the employees on the understanding of AI, among other issues, will further comfort the employees with the technology and its applications in sustainability issues.
4. **Ethical AI Practice:** One should ensure that the infusion of AI in HRM is aligned with ethical concerns, like transparency, fairness, and data privacy. This extends to validating non-biased algorithms, protection of employee data, and explainability of AI decisions. As such, a clear ethical guideline for AI use will facilitate its congruence with both sustainability, as well as social equity goals (Russell, 2020).

10.2 Sustainability Metrics and Goals

A more naturalistic step after integrating AI tools into HRM is the establishment of metrics and goals that are measurable and sustainable to monitor the progress of various improvements. Thus, the metrics act as a benchmark for indicating the success or failure of AI initiatives in terms of their outcomes at the organizational and environmental levels. The following are the steps involved in establishing metrics and goals.

1. **Align the metrics with the UN's SDGs**: Organizations need to connect their unique definition of sustainability metrics by using the framework provided by the UN's SDGs. For example, SDG 8 on Decent Work and Economic Growth can guide employee engagement and fair treatment of employees in the workplace, while SDG 12 on Responsible Consumption and Production can guide metrics around resource usage and waste reduction.
2. **Measure key performance indicators:** Typical KPIs of sustainable HRM via AI include

   a. **Carbon footprint:** Carbon footprint lowered by AI-enabled methods like working from home and also by using AI-based energy-saving gadgets
   b. **Employee engagement:** For example, the level at which employees are engaging with sustainability efforts is measured by AI systems.
   c. **Diversity and inclusion:** Improvement in aspects of diversity in the recruiting process is tracked by AI tools.
   d. **Resource utilization:** Waste, paper usage, energy consumption, and its reduction using AI analytics.
3. **Application of AI in real-time data gathering and reporting:** Data can be followed and analyzed from continuous, AI-driven observations about the sustainability goals of an organization. It may then be used in real-time reporting; with this kind of reporting, organizations track the effectiveness of their sustainability strategies and can make actual data-driven decisions.
4. **SMART Goals:** Sustainability goals need to be Specific, Measurable, Achievable, Relevant, and Time-bound, which are all part of SMART. An example of such a goal could be a business that aims to reduce carbon emissions by 20% in the next two years through AI-facilitated energy-saving measures.

10.3 Continuous Improvement and Feedback from Employees

This therefore requires continuous improvement practices in place as well as continuous employee feedback that will help ensure the sustainability framework improves with time. In this manner, organizations can adapt their AI-driven HRM practices to evolving needs and improve the effectiveness of their sustainability activities (Guerci, 2016).

1. **Continuous Monitoring and refinements:** AI-based applications must be monitored over time for the maximum desirable output. The real-time gathering of data which also shall include an information feedback loop can help in the real-time detection of issues so that changes could be made in AI models or HRM practices. For example, if AI systems operating in the recruitment sector are found holding bias or not delivering the desired diversity, they can further be retrained over more diversified data to reduce such biases.
2. **Employee Feedback Mechanisms:** Another indispensable element when it relates to the alignment of AI-based HRM systems with the needs and values of employees is the collection of feedback. This can be achieved through surveys, focus groups, and/or individual meetings with employees to bring to the fore any concerns and improvements required regarding the implementation of AI systems. The input of the employees should be very significant and considered when aligning AI systems.

3. **Sustainable Culture:** To ensure that these improvements are continuing, the organization should embrace sustainability in work culture. This would involve incentives for active contributors toward reaching the goals of sustainability, recognition of their successes, and support for personal and professional development related to environmental and social issues.

The framework for sustainable HRM with AI will stimulate continuous improvement toward an objective of sustainability for organizations. This will create a responsible and ethical working environment.

Example Content: Clear, quantifiable sustainability goals associated with SDGs must be launched first by the organizations as a foundation for building an AI-powered sustainable HRM framework. Formulation of the metrics for the energy used, resources saved, and employee engagement forms a starting point for continuous improvement. Additionally, employees must be consulted on getting feedback so that it is ensured that AI is as ethical as it is effective and aligned with organizational values toward sustainability and inclusion.

11. CONCLUSION AND FUTURE DIRECTIONS

Through this chapter, we have searched for the growing intersection of sustainable HRM and AI to create the greenest and most socially responsible workplace. Key themes explored covered the alignment of HRM to sustainability goals, using AI for recruitment and employee engagement, and sustainability monitoring (Guerci, 2016). With AI, immense possibilities lie in totally transforming HRM practices that contribute towards environmental conservation, social equity, and employee satisfaction. However, the embracing of AI is to be cautiously done considering ethical values and long-term objectives.

11.1 Summary of Key Findings

1. **AI-Driven Efficiency and Sustainability:** AI can transform HRM practices in significant ways through efficiency savings that bring in less environmental influence like reduced carbon emissions, paper waste, and savings in energy consumption. Furthermore, AI practice facilitates and fosters sustainable work environments through practices such as remote work and personalized employee engagements.

2. **AI enhances the HRM practice** while raising important ethical issues such as data privacy, algorithmic bias, and the need for transparency in the use of AI for decision-making. Clear policies and ethics in AI need to be settled to ensure that HR would responsibly use this kind of technology (B. Sypniewska, 2023).
3. **Employee Engagement and Continuous Improvement:** Continuous monitoring and employee feedback are critical to refining AI systems and aligning them toward sustainability goals. Through AI, employees can be engaged in sustainability practices, including gamification and real-time feedback systems, to be more sustainable (Bersin, 2021).
4. **SDGs Strategic Alignment:** Organizations need to align their AI-enabled HRM practices with the United Nations' Sustainable Development Goals so that the outcome of the organizations will be in alignment with global sustainability targets formulated by the SDGs. AI-enabled HRM will solve various SDGs ranging from responsible consumption and decent work to economic growth (Nations).

11.2 Limitations and Challenges

While AI has tremendous potential for sustainable HRM, there are a few limitations and challenges that organizations must consider:

1. **Technological Limitations:** The AI system is still evolving; it may pose technical limitations that prevent AI from fully potentializing in HRM, such as the accuracy of analytics that can predict and the ability to nullify bias in the algorithms used by AI.
2. **Organizational Resistance:** For some employees, AI tools could represent an alteration because AI tools would substitute them, or remove the personal factor in an HR process. These resistance issues can be overcome by effective communication and education, and by getting employees involved in the adoption of AI tools in the organization (Cascio, 2016).
3. **Data Privacy and Security:** Mass collection and processing of employee data for AI technologies raise issues related to data privacy and security. Organizations need to ensure that they are at least compliant with privacy legislation and have strong cybersecurity measures to protect the sensitive data of employees.
4. **Integration Costs:** The cost of integrating AI into an HRM system is very high, at least during the initial investment. The justifications for such investments can prove difficult, especially for smaller organizations due to resource constraints (Deloitte, 2021a).

11.3 Potential Areas for Future Research

Some areas require further study on AI and sustainable HRM that may explicate more of the areas to be explored in greater detail to enhance the role of AI in enhancing sustainability efforts:

1. **The Impact of AI on Organizational Culture:** There is a potential study that may explain how the absorption of AI in HRM affects the organizations' culture and employee satisfaction-whether it enhances or weakens employee trust and engagement with sustainability practice.
2. **AI and Diversity in HRM:** Further exploration will be needed into the way AI may feed into the scope of diversity, equity, and inclusion within HRM practice. The research focus would most probably remain on how AI may be an effective tool in reducing biases when hiring or doing performance or career advancement evaluations.
3. **Environmental and Social Impact for a Longer Period:** There is a need for longitudinal studies to determine the long-term environmental and social impact of AI-enabled HRM practices. Research can be designed to look at the sustainability outcomes of AI-driven recruitment, employee engagement, and resource management in various industries (Jabbour, 2008).
4. **AI and Ethical Frameworks:** There is also a need to carry out research in developing ethical frameworks that may be applicable in the implementation of AI in HRM. This shall include guidelines that are fair, transparent, and accountable with respect to both sustainability and ethical standards for AI.

Example Content: This chapter explains how Sustainable HRM with AI has much to contribute towards achieving the SDGs. In terms of optimizing resource use, containing bias, and engaging employees on issues of sustainability, AI is a means by which HRM can push forward with environmental goals. Challenges persist, however namely the bias that already exists must be overcome to preserve data privacy. For instance, further research would be needed into whether AI produces long-term sustainability impacts, whether it offers sectoral applications, and perhaps more controversially, whether could AI contribute to new concepts of sustainability frameworks.

Theoretical Contribution

The theoretical contribution of this study is to span two necessary but heretofore siloed fields: Artificial Intelligence (AI) and Sustainable Human Resource Management (HRM). Though each field has been studied in isolation, their intersection has been the subject of very little scholarship in the literature, particularly in terms of how AI can initiate sustainability efforts within HRM systems. This study provides a new theoretical perspective by thinking of AI not just as a tool for operations efficiency but as a strategic enabler of sustainable action in HRM (B. Sypniewska, 2023). In bridging AI technologies and the tenets of Sustainable HRM, the study provides a new paradigm that spans human capital development with environmental and ethical goals, thus providing an integrated model for organizations to pursue sustainability.

The essence of this theoretical contribution is the conceptualization of an integrated model that rethinks HRM in the light of AI and sustainability. The model includes not just traditional HRM functions such as recruitment, training, and performance management but also environmental sustainability initiatives, such as reducing the environmental footprint of HR operations and promoting green programs (D. Rosenberg, 2019). In demonstrating the possibility of AI in automating and streamlining HR processes in a way that promotes environmental sustainability, the study opens the door for subsequent studies on "green" HRM practices, which have been mostly underdeveloped in HRM literature.

Apart from this, this study offers a theoretical analysis of the HR triad—HR professionals, line managers, and employees within an AI-based, sustainable HRM system. By exploring the role and responsibility redrafted by AI among these players, this study offers an expanded understanding of evolving dynamics at work as businesses begin to adopt AI in their sustainability initiatives (Guerci, 2016). Such an offering is timely as it not only analyzes the technical side of AI in HRM but also the human side of how AI influences decision-making, behavior, and collaboration across an organization's hierarchy.

Furthermore, the theoretical model offered here goes beyond existing knowledge on AI's application in HRM by considering the interdependence between technological progress and sustainable development goals (SDGs) (Nations). The study offers a new understanding of how AI makes such goals easier to achieve through aligning organizational and employee development initiatives with wider environmental and social responsibility goals. It calls for organizations to embrace an integrated system where AI not only makes business easier but also induces a more sustainable and socially responsible workforce.

Overall, this study's theoretical contribution offers rich insights into the intersection of AI and Sustainable HRM, offering a holistic, integrated model that repositions both fields. By calling attention to AI's potential to induce sustainable behaviors in HRM and reimagining our understanding of the HR triad's role in accomplishing such an end, this study offers a platform for future studies and practical applications to enable organizational prosperity and a cleaner, more responsible future.

Practical Implication

- Strategic AI Integration in HR Practices: Organizations may employ AI not only for optimizing processes but also as a strategic enabler of sustainability. By integrating AI with Sustainable HRM, organizations may automate and streamline HR processes recruitment, training, and performance management to meet environmental sustainability goals as well. This can eliminate wastefulness, streamline processes, and minimize the carbon footprint of HR processes. For instance, AI-based platforms may optimize virtual recruitment processes, hence minimizing the carbon cost of travel and paper-intensive processes.
- Environmental Sustainability of HRM: AI may be a great enabler of green HRM practices. The current study believes that AI tools may assist in measuring and minimizing the environmental impact of workplace practices, such as energy usage, resource usage, and waste disposal. HR departments may employ AI to monitor and track sustainability projects and ensure the organization is making progress toward environmental goals. For instance, AI-based analytics can spot areas of energy efficiency or optimize remote work solutions to lower carbon emissions.
- Better Employee Well-being and Engagement: The study identifies that AI may enhance employee development and, in the process, contribute to sustainability goals. AI-powered personalized learning experiences may ensure employees are suitably equipped with the skills and knowledge needed to enable sustainable practices. Moreover, AI may enable HR departments to develop strategies to enhance employee engagement, well-being, and job satisfaction by creating personalized experiences under individual needs, hence aligning with the long-term goals of Sustainable HRM.
- Redefining the Role of HR Professionals: HR professionals need to learn to function in the new AI-based HRM world. As per this study, HR leaders need to not only learn to implement AI technologies but also influence employees and managers to adopt sustainable practices. As AI reshapes HR functions, HR professionals will play a pivotal role in managing the social and ethical dimensions of AI integration, ensuring organizational values and sustainability goals are aligned with the technology.

Policy Implications

- Regulation and Ethics of AI in HRM: Policymakers must address the ethical implications of AI in HRM, including issues of data privacy, bias, and transparency. Since AI technologies are embedded in HR practices, it will be important to have regulations ensuring equitable, fair, and responsible use of AI. This includes protecting against potential biases in AI algorithms and ensuring AI-based decision-making is carried out transparently and ethically.

- Promoting Sustainability Standards in HRM: Governments and international organizations could encourage the inclusion of sustainability standards in HRM by offering incentives to companies that align HR practices with environmental and social sustainability goals. By encouraging initiatives that enable green HRM, policymakers can enable broader organizational change towards more sustainable work practices.
- Workforce Development for AI and Sustainability: To achieve the full potential of AI in Sustainable HRM, policies must encourage education and training programs that enable HR professionals to leverage AI effectively and ethically. Governments and educational institutions could also collaborate to create specialized curricula that address the intersection of AI, HRM, and sustainability, enabling a future-ready workforce.

REFERENCES

Armstrong, M., & Taylor, S. (2020). *Armstrong's Handbook of Human Resource Management Practice*. Kogan Page.

Bersin, J. (2021). *AI in HR: The Future of Work*. Deloitte University Press.

Bessen, J. E. (2019). AI and Jobs: The role of demand. *NBER Working Paper Series*, 24235.

Bostrom, N. (2014). *Superintelligence: Paths, Dangers, Strategies*. Oxford University Press.

Brynjolfsson, E., & McAfee, A. (2014). *The Second Machine Age: Work, Progress, and Prosperity in a Time of Brilliant Technologies*. W. W. Norton & Company.

Cascio, W. F., & Aguinis, H. (2020). *Research in Personnel and Human Resources Management: AI and the Future of Work*. Emerald Publishing Limited.

Cascio, W. F., & Montealegre, R. (2016). How Technology Is Changing Work and Organizations. *Annual Review of Organizational Psychology and Organizational Behavior, 3*(1), 349–375. DOI: 10.1146/annurev-orgpsych-041015-062352

Choudhury, P., Foroughi, C., & Larson, B. (2020). Work-from-anywhere: The productivity effects of geographic flexibility. *Strategic Management Journal, 41*(9), 1748–1770.

Deloitte. (2021a). *Global Human Capital Trends: The social enterprise at work*. Deloitte Insights.

Deloitte. (2021b). *Global Human Capital Trends: The social enterprise at work*. Deloitte Insights.

Dessler, G. (2019). *Human Resource Management*. Pearson.

Dhar, V. (2020). Digital Transformation and Human Resource Management. *Journal of World Business, 55*(1), 1–13.

Emerald Insight. The artificial intelligence-enabled customer experience in tourism: a systematic literature review.

Forum, W. E. (2018). The Future of Jobs Report 2018. *World Economic Forum*.

Freeman, R. E. (2010). *Strategic Management: A Stakeholder Approach*. Cambridge University Press. DOI: 10.1017/CBO9781139192675

Gartner. (2022). Top Trends in Human Resource Management and Artificial Intelligence. *Gartner Research.*

Garvin, D. A. (1993). Building a learning organization. *Harvard Business Review*, *71*(4), 78–91. PMID: 10127041

Goodfellow, I., Bengio, Y., & Courville, A. (2016). *Deep Learning*. MIT Press.

Guerci, M., & Carollo, L. (2016). A paradox view on green human resource management: Insights from the Italian context. *International Journal of Human Resource Management*, *27*(2), 212–238. DOI: 10.1080/09585192.2015.1033641

Hammer, M. (2004). Deep Change: How Operational Innovation Can Transform Your Company. *Harvard Business Review*, *82*(4), 84–93. DOI: 10.1109/EMR.2004.25106 PMID: 15077369

Jabbour, C. J. C. (2008). Environmental training and environmental management maturity in companies. *The case of Brazil.Journal of Cleaner Production*, *16*(3), 216–226.

Jackson, S. E., & Renwick, D. W., & Jabbour, C. J. C. (2011). Sustainable human resource management: A comprehensive review and research agenda. *Business &amp. Society*, *50*(2), 299–323.

Jatobá, A. M. B., Pimentel, M. D. G., & Alvarenga, M. A. (2019). Applications of artificial intelligence in human resource management: A systematic review. *Brazilian Journal of Operations & Production Management*, *16*(3), 509–524.

Kahn, W. A. (1990). Psychological conditions of personal engagement and disengagement at work. *Academy of Management Journal*, *33*(4), 692–724. DOI: 10.2307/256287

Kaplan, A., & Haenlein, M. (2019). Siri, Siri, in my hand: Who's the fairest in the land? On the interpretations, illustrations, and implications of artificial intelligence. *Business Horizons*, *62*(1), 15–25. DOI: 10.1016/j.bushor.2018.08.004

Kaushal, N., Kaurav, R. P. S., Sivathanu, B., & Kaushik, N. (2023). Artificial intelligence and HRM: Identifying future research Agenda using systematic literature review and bibliometric analysis. *Manag Rev Q*, *73*(2), 455–493. DOI: 10.1007/s11301-021-00249-2

Lawler, E. E., & Worley, C. G. (2006). *Built to Change: How to Achieve Sustained Organizational Effectiveness*. John Wiley & Sons.

Nations, U. Goal 8: Decent Work and Economic Growth. .

Noe, R. A., Hollenbeck, J. R., Gerhart, B., & Wright, P. M. (2021). *Fundamentals of Human Resource Management*. McGraw-Hill Education.

Porter, M. E., & Kramer, M. R. (2011). Creating shared value. *Harvard Business Review*, *89*(1/2), 62–77.

Raisch, S., & Krakowski, S. (2021). Artificial intelligence and management: The automation-augmentation paradox. *Academy of Management Review*, *46*(1), 192–210. DOI: 10.5465/amr.2018.0072

Rosenberg, D. (2019). AI and HR: New Partners in Talent Management. *Harvard Business Review*, *97*(2), 42–50.

Rosenberg, D. (2019). AI and HR: New Partners in Talent Management. *Harvard Business Review*, *97*(2), 42–50.

Russell, S., & Norvig, P. (2020). *Artificial Intelligence: A Modern Approach*. Prentice Hall.

Stone, D. L., Deadrick, D. L., Lukaszewski, K. M., & Johnson, R. (2015). The influence of technology on the future of human resource management. *Human Resource Management Review*, *25*(2), 216–231. DOI: 10.1016/j.hrmr.2015.01.002

Sypniewska, B., Baran, M., & Kłos, M. (2023). Work engagement and employee satisfaction in the practice of sustainable human resource management–based on the study of Polish employees. *The International Entrepreneurship and Management Journal*, *19*(3), 1069–1100.

Vrontis, D., Christofi, M., & Makrides, A. (2021). Artificial intelligence, robotics, advanced technologies and human resource management: A systematic review. *International Journal of Human Resource Management*, *32*(21), 4599–4627.

Wamba, S. F., Akter, S., Edwards, A., Chopin, G., & Gnanzou, D. (2017). How 'big data' can make big impact: Findings from a systematic review and a longitudinal case study. *International Journal of Production Economics*, *165*, 234–246. DOI: 10.1016/j.ijpe.2014.12.031

Chapter 9
Sustainable Consumption:
Barriers and Catalysts for Green Purchase

Sylvia Marshal
Christ University, India

Deeksha Gupta Ganguly
https://orcid.org/0000-0002-5502-8276
Christ University, India

ABSTRACT

This chapter examines factors influencing green purchasing in both developed and developing nations, addressing catalysts and barriers. Key obstacles include price sensitivity, socioeconomic disparities, lack of standardized eco-labeling laws, and ingrained cultural beliefs. It highlights the roles of government regulations, corporate social responsibility (CSR), environmental awareness, and sustainable packaging in shaping consumer behavior. The study explores knowledge gaps, financial limitations, and cultural influences that affect sustainable choices. Strategies such as eco-labeling, tiered pricing, and financial incentives are analyzed, alongside CSR's impact on consumer trust. Emerging trends like post-COVID green consumption, blockchain for eco-label transparency, AI-driven marketing, and circular economy practices are also discussed. Ultimately, the chapter proposes an integrated framework combining technological, cultural, and economic perspectives to enhance sustainability.

DOI: 10.4018/979-8-3373-0139-6.ch009

INTRODUCTION

Sustainable consumption is a very focal point in international environmental discourse Since the realisation of consumer behaviour impacts the ecological outcome considerably (Young et al., 2009). In response to the environmental implications of consumption, Sustainable consumption calls for decisions that serve social justice and sustainability within the economy (Peattie, 2010). Despite the increasing desire for sustainable consumption, many barriers Still hinder customers from opting for green products, while some catalysts Facilitate these behaviours. First, let us define barriers and catalysts.

What are barriers and catalysts?

Especially In terms of sustainable consumption, to critical drivers of customer behaviour are catalysts and barriers. Catalysts are drivers that motivate and facilitate sustainable consumption, whereas barriers are drivers that prevent people from adopting green behaviours. Politicians and firms promoting green purchasing need to be aware of how a range of factors, such as logistical, social, informational and economic ones affect both.

Due to various barriers to adoption of sustainable consumption, green product is less accessible or appealing to average consumer. The most significant barrier is price sensitivity. Sometimes consumers wonder if the benefit to the environment by green products are enough to outweigh their higher price as compared to conventional alternatives (Glem et al., 2013). Price sensitivity could be particularly challenging for low-income customer to manage because they are likely to place short-term financial concern over long term environmental goals (Joshi & Rahman, 2019). As such many consumers opt for less expensive, unsustainable choices that are within their Budget better, especially in developing countries (Diamantopoulos et al 2003).

Other challenge is represented by socioeconomic features including the, geographical location, income levels and educational level. People in lower income levels might not care more about sustainability due to pressing financial limit. Geographical location is also significant since, as (Young et al., 2009, P. xx) points out, a customer in an urban setting will easily find green product while the one in rural setting will have fewer options.

Another effect is that these sustainable consumption habits also depend upon the cultural and social norms that one follows in daily life. It has been studied that green items are less opted for consumers of some civilzations who do not share environmental thoughts or greenish way of purchase behavior as widespread practices in daily lives (Chekima et al., 2015). Since these have become a customary pattern, so customers under societal force, unless someone practices green purchasing activities, even customers will have lesser tendency.

Another major barrier is the lack of credible and consistent eco-label and certifications. Although many products are labelled as eco-friendly, consumers get confused and skeptical because there is no standard, reliable certifications. Many consumers are unable to differentiate between sustainable and "greenwashed" items, and the credibility of these labels is often doubted (Cerri et al., 2018). The inconsistent and unclear labelling makes the customers' decision-making process much more complicated.

Even though there are a lot of hurdles to sustainable consumption, many factors are encouraging consumer behavior towards environmental awareness. As consumers become increasingly aware of the negative impacts caused by traditional products and unsustainable methods, they become more likely to opt for solutions that reduce ecological damage (Zhang & Dong 2020). Environmental considerations and an increasingly developed awareness of the necessity to adopt sustainable processes are compelling consumers to opt for greener options (Chekima et al., 2016).

Willingness to pay for sustainable products also represents another incentive. According to (Mahmoud et al., 2022), An important percentage of consumers are ready to pay additional amounts for perceived environmentally friendly products. According to numerous studies, consumers' interest in protecting the environment and future generations is greater than in the short-term monetary benefits of using green products (Prakash et al., 2019).

Government policies are increasingly becoming crucial in determining sustainable consumption. Powerful change agents include laws that promote environmentally friendly businesses, provide financial incentives for sustainable goods, and advance environmental education. For instance, the OECD stipulates that strict environmental loss and incentive for sustainable consumption in member nation are important (Arthur et al., 2023).

CSR (corporate social responsibility) Initiatives are the third motivating force. The majority of the eco-sensitive customers seeks such companies, who reflect sustainability by having moral sourcing and responsible production patterns. Those purchasing behaviour which focus more on social value preferences and green lifestyle favour businesses who take responsible measures toward social responsibility and environmentally sensitive behaviour (Duarte et al., 2024). CSR Initiatives with a focus on sustainability also encourage companies to develop new, more reasonably priced sustainable products and services.

The sustainable and minimalist packaging also plays a significant role in the green purchasing decision of customers. Customer's opinion about their environmental impacts of a product is highly affected by its packaging. Consumer's choices are being influenced more and more by the sustainable packaging materials, which include recyclable, biodegradable or reusable packaging and simple aesthetics. As companies continue to move towards greener packaging alternative, customers are more likely to favour companies that focus on environmental issues which it comes to making packaging choices (Piracci et al., 2023).

Since the environmental impact of consumer behaviour threatens ecological balance ever more, sustainable consumption has become a global priority issue. The United Nations' Sustainable Development Goals (SDGs), particularly Goal 12 (Responsible Consumption and Production) recognize that sustainable consumption is fundamental to achieving world sustainability. By advanced circular economy concepts, reducing waste, and closed-loop utilisation of resources, principles of industrial ecology also sustainable consumption.

SDG Goal 12 promotes resource efficiency, waste minimization, and sustainable production practices to ensure sustainable consumption and production patterns. Hence, this chapter promotes Goal 12.

The chapter highlights the need for businesses to minimise pollution and waste, as encouraged by Goal 12, through insisting on the functions of Eco labelling, CSR and Sustainable Packaging.

To guarantee that individual is equipped with knowledge and awareness to adopt sustainable life styles target 12.8 (promote universal understanding of sustainable lifestyles) is buttressed by the address of environmental literacy, willingness to pay, and behaviour change strategies.

Target 12.6 (Encouraging companies to adopt sustainable practises and sustainability reporting) promotes companies to implement sustainable practises and include sustainability reporting, and the examination of government policies, corporate sustainability initiatives, and fiscal incentives align with this objective.

Gold 12.5 (Substantially reduced waste generation), whose theme is the prevention, reduction, recycling, and reuse of waste, aligns with this chapter's focus on the importance of recyclable and minimalist packaging.

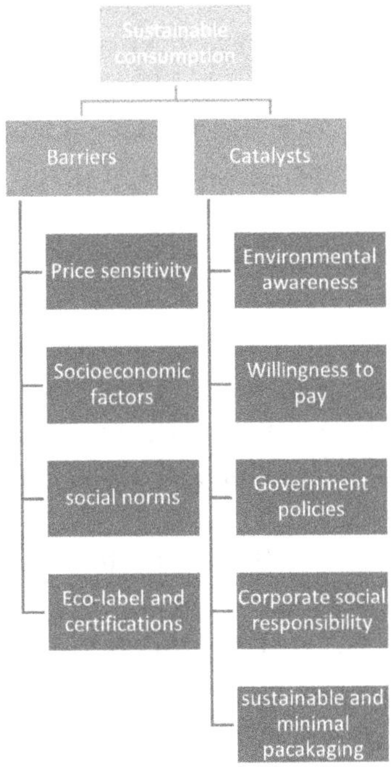

BARRIERS TO SUSTAINABLE CONSUMPTION

1. Price Sensitivity

Several factors hinder the implementation of sustainable consumption practises. Top given that environmentally friendly products are usually sold at a higher price than their traditional counterparts, price sensitivity remains one of the biggest hurdles (Chekima et al., 2016). Several studies have indicated that socioeconomic traits and green buying behaviour are positively related while demographic factors such as income, education level and social class play a considerable rule in the sustainable buying behaviour (Diamantopoulos et al., 2003). When price is the most critical attribute of our product, shoppers may select cheaper alternatives with minimum regard for others whose quality or ethical superiority is more superior. Price sensitivity normally leads to our vicious paradox in the market for green products customers covet green goods yet they are usually unwilling to pay the premiums such items command (Gleim et al., 2013; Chekima et al., 2016; Diamantopoulos et al., 2003). Therefore, customers perceptions, income levels, and options of alternative products influence the price due to several factors in addition to the price of a product. For example, according to research Chekima et al., (2015), by in regards to sustainable product options customers being more or less sensitive to price is influenced by demographic characteristics and culturally particular beliefs. This complexity shows up in consumer behaviour in the various markets wherein the issue of perceived value to cost varies drastically (Young et al., 2009; Martinho et al., 2015; Chakraborty & Mukherjee 2013).

Sustainable consumption promotes the use of goods and services that benefit environmental conservation, although the goods are usually costlier because they are sourcing and production methods are pricey (Joshi & Rahman,2015; Ghosh, 2016; Sheth et al., 2010). Not even the environment-conscious consumer will buy a sustainable solution because they think green products are expensive states (Cerri et al., 2018).

A study by (Joshi & Rahman, 2015) states that price sensitivity is one of the major hurdles towards the implementation of green buying behaviour. Although environmentally concerned consumers are willing to pay more for sustainable goods, their decision-making process usually gets affected by budgetary restriction. Because sustainable consumption is perceived more often as a luxury that commands a huge price, this enhances the effect of price sensitivities of environmental products among mid-level and lower customers (Young et al., 2009). Price sensitivities not only impact individual's decision to consume certain products, but also bound broader societal incorporation of environmentally conscious items (Martinho et al., 2015; Diamantopoulos et al., 2003; Chakraborty & Mukherjee, 2013). According to research people become less price sensitive and more willing to pay spend money on green products when they understand these long-term benefits (Martinho et al., 2015; Prakash et al., 2019). The argument has been supported by evidence of Nichifor et al. (2025), revealing economic constraints to be the most powerful disincentive to adoption practises of sustainable consumption in food industry. This is our challenge especially in developing countries where consumers with limited disposable income prefer affordability over sustainability (Arthur et al., 2023).

2. Socioeconomic Factors

Affordability perhaps one of the major barriers to buying green products often more expensive is still being rationed. This is because price is a critical barrier to green purchase is has most customers from low to middle income groups have expressed that they view the elevated prices of green items to be too high. Socioeconomic factors largely influence consumer behaviour, particularly concerning sustainable consumption. Diamantopoulos et al., 2003; Gleim et al., 2013; Chekima et al.,2016 reveals that the economic gap will only make such environmentally friendly products accessible to weldy consumers primarily in developed nations. Evidence by Arias (2024) in his study shows the role of education programmes and influencing consumer behaviour towards sustainability. The adoption of sustainable consumption in these industrialised countries is not possible due to financial constraints in the low-income population groups. Low-income customers have no spare money hence financial needs supersede environmental needs and prefer the same old cheap regular products rather than environmentally friendly products (Duarte et al., 2024; Joshi & Rahman, 2015; Martinho et al., 2015).

Geographical and socioeconomic factor aggravate the access barrier to green products and therefore enhance the affordability gap. Green products and services are highly accessible in urban and upscale sectors but lack access in rural or low-income sectors making there a separation in accessing green consumption (Young et al., 2009; Chekima et al., 2015; Nyborg et al., 2016). As for underdeveloped regions they suffer from not only financial but also physical lack of accessibility to sustainable alternatives which again enlarges the socio-economic gap in green buying behaviour.

Socio economic groups also vary drastically in education and awareness, which in turn impacts green purchasing behaviour. Due to less exposure to educational and informative tools and resources lower income population are often left uninformed about environmental problems thus making them less aware of sustainable practises and their long-term benefits (Chekima et al., 2015, Cerri et al., 2018; Duarte et al., 2024).

Consequently, these groups are probably not very aware of environment-friendly products and how their purchasing decisions affect the environment. It is well documented that highly educated consumers understand and please value on green practice, (Diamantopoulos et al., 2003). Therefore, socioeconomic barriers associated with educational disparities from an information gap that affects green consumer behaviour across various income sections (Gleim et al., 2013; Young et al., 2009; Nyborg et al., 2016).

In addition, opinions regarding green products are also subject to the social values that are linked to various socio-economic classes. The upper socioeconomic classes are likely to regard sustainable consumption as a lifestyle behaviour, and by extension eco-friendly products are regarded as morally good and a measure of social status (Chekima et al., 2015; Prakash et al., 2019; Joshi & Rahman, 2019). The lower socioeconomic strata tend to focus on low cost and easy use, in which green goods are considered frills rather than necessities. For instance, finding by (Gleim et al., 2013) and (Nyborg et al., 2016) states that this value gap in the social dimension determines purchasing priorities. This makes the promotion of sustainable consumption more difficult among a vast cross section of consumers.

Socioeconomic factors thus present significant barriers to sustainable consumption which may assume many forms including affordability excess education and social norms even with the growing environmental awareness the price premium of the eco-friendly products is the measure deterrent for the low-income consumer (Duarte et al., 2024; Gleim et al., 2013; Cerri et al., 2018). These challenges further call for integral policies that make such sustainable purchasing reachable to a significant population to assure that green consumerism is a privilege not practised only by high income earners but also be reached by common people (Arthur et al., 2023; Chakraborty & Mukherjee, 2013).

3. Social Norms

Social norms play an important role in determining people's attitudes toward sustainable consumption and also have a significant influence on how consumer behave. What people decide, such as purchasing or not buying environmentally friendly products are determined by social norms, which are the implicit rules or expectations that govern behaviour within a particular social group or community. According to the theory of plant behaviour (TPB), social norms influence behaviour intentions through the pressure of choosing behaviours that are considered desirable or acceptable by peers and society in general (Ajzen, 1991; Han et al., 2009). The incorporation of Weder et al. (2025) contributes to the argument of cultural and social effects towards the implementation of sustainability. People have a social responsibility to behave in an environmentally responsible way when sustainable buying is consistent with these expectations, which makes them more likely to make green purchases (Young et al., 2009). Moreover, a sustainable behaviour is observed in social networks people are more likely to engage in it because they want to be as environmentally responsible as their friends (Goldstein et al., 2008).

Social norms have a cascading effect where the pro social behaviour by one individual encourages others in the social network to develop similar behaviours (White et al., 2019). For example, studies in social psychology shows that identifiable environmental behaviours such as recycling or reusable bags, are often indicative of norm following and encouraging others to follow suit as well (Schultz et al., 2018; Th⊘gersen, 2009). This imitation effect is particularly prominent when role models within a culture adopt sustainable behaviours, as people tend to mimic the behaviour of people they respect (Goldstein et al., 2008). Social acceptance is an important factor that contributes to development of green consumption because individuals often feel a social pressure to act in line with social norms despite their own attitudes toward sustainability (Nyborg et al., 2016).

However, the current consumption pattern in most society emphasises fast fashion, rapid technological innovations, or purchasing luxury goods all of which discourage green consumption. According to White et al. (2019), even in consumers have personal environmental concern, social conventions may causeway friction, and thus it is difficult for them to switch to greener options. This is a serious challenge because many buyers avoid eco-friendly products due to criticism or social exclusion (Goldstein et al., 2008).

Green practises being adopted as minimalist or non-status driven, have lower chances of being embraced in societies where material wealth symbolises success or status. Peer might perceive sustainable choice as being unrealistic or unnecessarily utopian. As such, (Gleim et al., 2013) found green consumers to suffer subtle stigmatisation. Those are the social beliefs that make citizen shun developing countries sustainable consumption as they are predisposed to behaviour considered inappropriate or from standards (Han et al. 2009).

However, societal norms can be a constraint and an encouragement for sustainable consumption depending on the prevalent attitudes present in a particular group or community (Th⊘gersen, 2009). In countries with not so developed pro environmental values there may be a feeling to take decision related to sustainability which could then result in lesser sustainable consumption level (White et al., 2019). On the contrary in highly enviro-conscious societies, social norms are a strong river and support the large-scale adaptation of sustainable commodities by those even who have nothing to do with the environment (Han et al., 2009).

In the absence of various initiatives, social norms will likely remain a barrier since people will be coerced to comply with cultural norms that give preference to convenience, price, and prestige over environmental responsibility (Peattie 2010).

4. Eco-Labels and Certification

Eco-label and certifications are often promoted as powerful tools of encouraging sustainable consumption that can provide information for consumers in clear terms regarding the impact of products on the environment. As declared by (White et al., 2019), Eco labelling is 1 form of communicating information regarding the sustainable feature of a product. However, when such issues regarding implementation and understanding arise these labels can also serve as potential barriers to sustainable purchases. This may result in customer confusion or mistrust over the truthfulness of the claims. There are numerous eco labels each having its certification requirements. Due to the lack of a universal standard for eco-labels, customers may not know the difference between products that are actually sustainable and those that are just labelled as such, as goods with similar sustainability attributes may bear different labels (Peattie, 2010, p. xx; Gleim et al., 2013). There could be inconsistencies in the messages delivered to the consumers regarding the environmental qualities of a product or service because certain certifications may discuss only one aspect, like waste management or energy savings, while the others may include a more encompassing assessment. Research has shown that if consumers are faced with too many eco labels choices they may suffer from the element ailment known as "label fatigue" where a consumer develops a disinterest in or skepticism regarding the truthfulness of the environmental claim (Martinho et al., 2015). If eco-labels emphasise a few environmental characteristics, like carbon neutrality or zero waste, without any context or comparison, then consumers may become confused about what the actual environmental benefits of the products are buying.

The other challenge that eco labels encounter is the likelihood of green washing. Green washing, in simple terms, is when an enterprise misrepresents its products as environmentally friendly. Such behaviour has tarnished the customer's trust in certifications (Mahmoud et al., 2022). This erosion of trust is worsened by the fact that the process of certification lacks transparency. Therefore, consumers have a doubt regarding whether the products with the equilibrium indeed comply with the stringent standards of sustainability (Chekima et al., 2016). This contradiction reduces the prospects that eco-labels may develop sustainable confidence overtime while simultaneously misleading consumers (Van den Bergh, 2008).

Let's further discuss Greenwashing and Eco-labelling standardization:

Green washing is among the most significant obstacles to sustainable consumption, having a considerable impact on consumer trust and involvement in environmentally friendly products. Green washing happens when businesses provide false information regarding the environmental advantages of their products, leading to confusion and mistrust among consumers (Zhang et al., 2023).

One of the main issues with green washing is the absence of uniform regulations and standards for eco-labelling. Different organisations and sector have varying labelling standards, and it becomes challenging for consumers to confirm the legitimacy of claims of sustainability (Gao et al., 2023).

Government and regulatory agencies across the globe have started enforcing tighter policies to address greenwashing and harmonize eco-labelling procedures. For example, the European Union has put in place the Green Claims Directive, which compels businesses to present scientific evidence for their green claims (Udeagha & Muchapondwa, 2022). Likewise, the Federal Trade Commission (FTC) of the United States imposes rules on environmental marketing claims to avoid misleading advertisement (Faccia et al., 2023).

The psychological impact that sustainability choices make is also focused upon since green labels may provoke a sense of guilt or accountability, which leads consumers tuition shun rather than indulge (Chekima et al., 2015).

Eco-labels certification might be too pricey, considering small and medium enterprises, if these forms were willing to apply eco label on their products. Some of these costs might range from the payment for certification to audit and to meet the environmental applicable laws in operation. Due to the constraint on supply in this market some consumers will lose a number of options as several smaller businesses and a financially unstable unable to gain the same certification (Peattie et al., 2010; Gleim et al.,2013). It becomes difficult for customers to find eco-friendly solutions since there is less availability of products that are eco-labeled, especially in marketplaces where SMEs dominate, which further deters them from purchasing such products.

Furthermore, the persuasive force of eco-labels is also undermined because they generally do not provide enough information to consumers about exactly what environmental benefit certified products generate (Prakash et al., 2019). Consumers see detailed explanations for the environmental gains of eco labelled products whereas most certification hardly provide this. Instead, a majority of them settle for Vague phrases (Van den Bergh, 2008).

CATALYSTS FOR SUSTAINABLE CONSUMPTION

1. Environmental Awareness

The most dominant catalyst is environmental awareness. As the awareness of the public regarding the environmental crisis, climate change, increases, many consumers are going for a re-evaluation of their purchases. The role of awareness in influencing sustainable consumer behaviour is also justified by Arias (2024). When people get increasingly aware of the disastrous effect of traditional products and unsustainable practices, they will be more receptive to picking the products that promote less harm to the environment (Zhang & Dong, 2020).

Environmental awareness refers to the comprehension and awareness of issues in the environment, including natural ecosystems, and the need to ensure sustainability. In most cases environmental awareness influences people to back legislative reforms and embrace eco-friendly habits. The emotional attachment towards nature and cognitive knowledge of environmental challenges are integrated into environmental awareness (Cerri, Testa, & Rizzi, 2018).

According to (Peattie, 2010) environmental awareness is the knowledge and understanding of environmental issues and how they influence human behaviour. On the other hand, sustainable consumption is the use of resources in a manner that supports ecological balance and reduces environmental effects (Chekima et al., 2015). The two concepts are related in that consumer are opting for greener product and minimising their personal waste due to increased awareness of environmental damage. For example, a rare individual of how single use plastics impact the environment are more likely to avoid them and opt for reusable options (Chakraborty & Mukherjee, 2013). Awareness breeds a sense of responsibility that changes the way people perceive consumption and encourages changes towards behaviour that support long term environmental well-being (White & Simpson, 2013). The theory of planed behaviour (TPB) that sustainable behaviours are influenced by social norms, attitudes and perceived behaviour control (Ajzen, 1991; Joshi & Rahman, 2015).

Environmental awareness is the bedrock of this transition to sustainable consumerism, as it both encourages eco-friendly behaviour and arms consumers with a sense of authority to call the companies and the government accountable for their environmental policies (White & Simpson, 2013).

Consumer's behaviour is directly influenced by environmental concerns as a factor that makes consumers choose sustainable purchasing habits. Environmentally concerned consumers are more likely to contemplate about product life cycle, responsible sourcing and the environmental footprint of their product (Chen et al., 2024). Study findings reveal that environmentally concerned consumers often select regionally produced or procured products select brands that demonstrate environmental leadership and make selection with environmentally friendly labels (Cerri et al., 2018).

2. Willingness to Pay

A key concept when understanding sustainable consumption is the concept of willingness to pay, which reveals how much consumers will spend on such goods that come with a lowered environmental awareness, more people are ready to spend for sustainable commodities, like efficient energy appliances and organic food stuffs, and eco-friendly way of transport. Environmentally conscious consumers also value sustainability more and are willing to pay extra for goods that align with their moral values, as stated by (Gleim et al., 2013). Research suggests that consumers who align their values with environmental conservation are more likely to pay more for sustainable goods (Bisschoff & Fullerton, 2025). The rising demand for sustainable products, which is being stimulated by awareness in ethical consideration, support the idea that WTP acts as in a potent catalyst for encouraging ecologically friendly purchasing patterns as stated by (Chekima et al., 2015).

The first factor to Influence WTP for sustainable goods is environmental awareness. With increasing consciousness over the ecological harm caused by regular consumption patterns customers are deciding to promote those items that lower environmental degradation. As a result of this increased understanding customers view sustainable products as an investment in the future of the planet rather than merely and purchasing decision, thereby increasing their WTP (Peattie, 2010).

There has been a direct relation relationship between increased WTP and sustainable products, where the latter was seen as the consequence of ethical consumerism. According to Chekima et al. (2015), ethical consumerism motivates individuals to spend extra for products deemed socially responsible. Moreover, even if electric or hybrid automobiles are more expensive during the initial purchase as opposed to the traditional gasoline-powered cars, customers are willing to pay extra for them (Gleim et al., 2013). This indicates that ethical consumer behaviour is encouraging the trend towards a more ethical and sustainable market, apart from increasing WTP for environmental-friendly products (Joshi & Rahman, 2015).

Yet another thing that brings an element of WTP in respect to sustainable items is the value customers assigned to these items as something desirable for sustainable purchase. (Peattie, 2010) holds that it is the psychological satisfaction a consumer derives from making purchases that align with his value in life that compels him to pay more money. Symbolic value is much more pronounced when status and expression of self-interest dominate the psyche of the individual consumer in particular industries such as fashion and food (Chekima et al., 2015; Gleim et al., 2013).

The WPT is also highly influenced by the availability of transparent information regarding a goods environmental impact. According to Cerri et al., (2018) "Eco-labels and certification, such as the energy star of air trade labels, provide consumers with credible information that empowers them to make informed choices about a product sustainability".

In an attempt to lower price sensitivity and increase WTP for sustainable products, government policies and initiatives might be crucial. Many governments offer subsidies, tax credits, and refunds for green technologies like solar panels, electric cars, and energy-efficient appliances, promoting their adoption. The rules tend to make environmentally conscious goods cheaper in comparison to conventional goods hence easier for the consumers to purchase (Chekima et al., 2015). For instance, tax incentives for electric automobiles have been shown to increase customer demand for these products by lowering their effective price.

All things considered, willingness to pay in the context of eco-friendly goods and activities place a pivotal role in enhancing their adoption. As a significant problem for higher adaptation, price sensitivity still continues and persists even today. Aligned to consumer preference through reduced cost associated with green purchases, WTP can eventually become a powerful driver for environmental change.

3. Government Policies

Consumer behaviour business operation, and market dynamics all impact on how a government intact in acts heat policies toward ensuring sustainability, and for that reason the catalyst to the adoption of sustainable consumption behaviours in economies cannot miss this point. Government in different countries all around the world embracing the adoption of sustainability policies reveals their importance towards environmental goals and environmental long-run equilibrium. Policies can promote sustainable consumption by creating incentives, support system, and policies aimed at reducing environmental effect and promoting environmentally friendly products and service. For example, carbon pricing and green technology subsidies are policies where a financial framework is created, which encourages businesses and consumers to favour sustainability over other options. Mandates, financial incentives, and public awareness campaign through government policies help shape supply and demand in the direction of these sustainable substitutes (Arthur et al., 2023; Stevens, 2010). The addition of Weder et al. (2025) makes the argument stronger and consular and social influence towards adopting sustainability. This policy interventions make environmentally conscious choices accessible while exerting influence in shifting societal norms to adopt such choices by setting them as a standard (Stevens, 2010; Chakraborty & Mukherjee, 2013).

Most countries complement regulation with market-based tools such as cap-and-trade systems. The system will cap emissions but then let the forms trade the permits. Such an approach compels companies to produce green technology given a strong financial reason for reducing their emissions. The OECD has created a global model of policy-driven sustainability by encouraging these market-based solutions among its partner nations, which other countries have begun to emulate (Arthur et al., 2023). Moreover, taxation and subsidies to firms making use of eco-friendly practices are some financial incentives that help bear the cost of high initial green technologies together with encouraging more firms to adopt the sustainable models (Chekima et al., 2016; Duarte et al., 2024).

The impact of subsidised R&D has been most effective in energy-intensive industries, where switching to sustainable methods is not cost-effective (Kumar & Chandra, 2010). Govt grants hand funding for R&D in environmental-friendly technologies promote the innovation of green technologies, supporting sustainable consumption. Government fund research into renewable energy, biodegradable materials, and sustainable production procedures that are not only advancing technology but also making it more efficient and cheaper for businesses to supply sustainable products. Take renewable energy; for example, India's national solar mission set ambitious targets aimed at increasing solar power and subsequently reducing dependence on fossil fuels (Dey, 2010).

According to (Raina & Sinha, 2019), the government of India made solar energy affordable to a more extensive segment of the population with the provision of subsidies and tax benefits for installations of solar panels. Unlike some international examples where lack of uniform federal incentives has led to less effective adoption rates and dependence on state policies in developing renewable energies, these incentives therefore overcome the monitory barriers to adoption while demonstrating that, through government policy sustainable options are not only accessible but also attractive from an economic standpoint for consumers (Jordan, 2014).

In addition, through certification standards and labelling laws, government direct customers to choose greener options. As pointed out by (Martinho et al., 2015), labels like "organic"," Energy efficient", or "recyclable" can decrease the information imbalance between producers and customers and are reliable makers of a product's sustainability.

4. Corporate Social Responsibility (CSR)

Corporate social responsibility what is it?

The CSR where firms voluntarily look for ways through which they would be able to operate in order to go about things that are environmental and social with more than merely profit-driven approaches, is of course are dominant reason for finding sustainable consumerism (Basil & Erlandson, 2008). This association leads to sustainable consumption because consumers increasingly make choices that reflect their environmental values and this enhances the circular economy, which reduces the wastage of resources (Arthur et al., 2023). Corporate social responsibility (CSR) is an important catalyst of sustainability because studies show that when companies openly show their dedication to environmental sustainability, it encourages customers to adopt sustainable consumption practices (Luo & Bhattacharya, 2006).

For example, (Becker-Olsen et al., 2006) shows that CSR programs enhance customer trust, which becomes a strong motivation for customers to choose businesses that have the same values, and as indicated by (Chekima et al., 2015), customers who are knowledgeable about sustainability issues are more likely to favor companies that adopt green methods.

According to a study by (Martinho et al., 2015), consumer prefer products with greener packaging as it depicts the commitment of a company to the environment. CSR provides companies with an opportunity to incorporate environmentally friendly practices that are not only limited to the design, sourcing, and production of a product but also in its distribution.

CSR not only reduces barriers to sustainable consumption but also lines business operations with social standards, which can make customers embrace sustainable practice even more. Social norms can encourage customers to adopt environmental behaviour, particularly if they feel that greater community supports their decision (Nyborg et al., 2016). CSR ads, which point out the social and environmental advantages of sustainable product take advantage of these dynamics by promoting desirable behaviour and creating sustainable consumption as the right thing to do, this helps attract environmentally conscious customers.

As consumer interest in fair trade and labour practices grows, ethical manufacturing CSR programs are also integral to building a sustainable consumption process. Consumers are getting increasingly interested in fair labour and ethical sourcing and are more likely to favour firms that demonstrate an interest in corporate social responsibility within the supply chains. According to Prakash et al. (2019), consumer willingness to buy environmentally friendly product is influenced positively by consumers' perception regarding those companies that have fair and ethical production process. Also, according to (Becker-Olsen et al., 2006), firms which use socially responsible sourcing are more likely to win the market because it is also indicating the responsibility of taking care of the community's well-being and human rights.

The policies of the government that force encourage CSR activities, such as waste reduction targets or renewable energy requirements, foster a condenser conducive environment for sustainability and the collaboration of CSR with the government policy helps increase sustainable consumption as policy systems are often in incentives to firms to embrace responsible practices (Chakraborty & Mukherjee, 2013). Business enterprises that associate with their CSR with policy objectives get both regulatory assistance and better corporate reputation.

The CSR programs address the growing concern of consumers about the safety of the products and impacts that they leave on the environment. Large sectors like food and cosmetics respond to the aforementioned concerns by consumers through CSR practices. Consumers aware of the risk pose to their health and the environment by certain products have shifted preference and chosen companies that are focused on these issues. For example, companies implementing CSR programmes focussing on non-toxic materials and chemical-free production procedures attract environmentally and health safe-conscious customers. Customers are likely to support a company that is transparent and honest about the safety of its products because it helps them make choice that are congruent with their beliefs according to claim of (Cerri et al., 2018). It ensures sustainable consumption as a result of effective decision-making government policies in collaboration with CSR encourage a sustainable consumption because the framework of laws that govern our society often present the opportunity for enterprises to embrace positive responsible business conduct by offering inducements. It creates an atmosphere that favors the sustainability culture, as regulations or requirements that create a mandate to engage in CSR activities such as waste reduction or renewable energy obligation. According to (Chakraborty & Mukherjee, 2013) companies that associate their CSR activities with policy goals receive regulatory support in addition to enhancing the business reputation. Businesses can boost their reputation and encourage sustainable consumption among customers who like a responsible corporate sector and regulation adherence by simply advocating for a collaborative approach between CSR and government policy.

5. Sustainable and Minimal Packaging

Recently sustainable development is being carried out through sustainable packaging. According to environmental demands, sustainable strategies in packaging appears to be ever more relevant now. Analysing the new trend toward "green" friendly packaging is imperative to understanding this change by describing how packaging practises affect consumer response. According to (Duarte et al., 2024) the behaviour of purchasing sustainable packaging is influenced by several relevant factors associated with customers believe system and awareness of environmental issues, as well as the perceived benefit from using eco-friendly packaging.

(Ghosh, 2016) States that packaging is the most critical variable that affects consumer purchasing behaviour in industries like food and privileges beverages the analysis of mother dairy products in Kolkata reveals that consumers prefer product using eco-friendly packaging over those that utilise traditional packaging. The perception is what influence consumers purchasing decision, has most willing consumer are willing to pay a premium for environmentally friendly packaging (Duarte et al., 2024). The study indicates that aside from the security it offers the merchandise packaging forms and essential communication tool to project the corporate thoughts and value statement of the organisation since people nowadays are highly concerned with their decision at the checkout line since they impact their environmental legacy, this factor becomes much beyond just in sales point for consumers customers today (Ghosh, 2016).

In addition, (Martinho et al., 2015) looked at the factors influencing consumers decisions regarding sustainable packaging when purchasing and recycling products. Establish that attitude towards recycling and ease of disposal of sustainable packaging are influential factor in the choice of consumers. Even more significant as firms turn to' green' and more convenient packaging that will ease customer access to environmentally conscious consumption (Martinho et al., 2015).

Altruistic and egoistic personal value are also very important in determining how consumers feel and act when it comes to sustainable packaging. According to (Prakash et al., 2019) consumer perceptions of environmentally friendly goods are affected by both altruistic (such as caring for environment) and egoistic (such as the demand for safer or healthier products) values. environmental awareness and pro social values are intertwined since consumer are motivated to reduce their environmental footprint. Under the motivation of choosing ego friendly package product this double impulse leads to a more complex choice that depends both on the environment and self-consideration of customers (Prakash et al., 2019).

Many consumers have been reported to pay extra for goods packaged in more environmentally friendly materials, and these consumers are said to be highly influenced by sustainability of packaging to make purchasing decision according to a variety of research that demonstrates sustainable packaging impact the consumer's behaviour (Gleim et al., 2013).

In addition, minimalist packaging has gained traction as an environmentally friendly alternative to traditional packaging solution. The aim of minimalist packaging is to limit the packing materials to only the essential while also reducing the resultant waste and its associated environmental cost. The practise is therefore consistent with the tenants of the circular economy that encourages recycling, reusing, and reducing the material (Arthur et al., 2023). A study revealed that minimal packaging is not only beneficial to the environment but also profitable for businesses companies can reduce production cost and product weight which subsequently reduces transportation charges, by reducing packaging sizes and material usage (Mahmoud et al., 2022).

There are also factors concerning the research study on green marketing and consumer behaviour that drive this trend toward greener packaging. For instance, studies show that consumer view on CSR as well as the tendency to participate in green campaigns are positively linked (Becker-Olsen et al., 2006).

HOW INDIA'S NATIONAL SOLAR MISSION IS A PRIME EXAMPLE OF SUSTAINABLE CONSUMPTION AND DEVELOPMENT?

Established in 2010 as a part of the National Action Plan on climate change India's national solar mission has evolved into a historic initiative that supports sustainable consumerism worldwide and paves the way for the large-scale deployment of solar energy. Earlier known as Jawaharlal Nehru National Solar Mission, or JNNSM, the National solar mission, or NSM seeks to make India the world leader in solar power generation by providing essential energy supply sustainably since the country is heavily dependent on fossil fuel especially, coal this machine has been seeking to reduce the emission of greenhouse gases, strength and energy security and create an energy framework that is both economic and environmentally sustainable (Dey, 2010).

The ambitiousness of the NSM was high at the outset. The target of 20 GW of solar power by 2022 was revised to 100 GW primarily due to positive solar conditions in India, economic feasibility, and high enthusiasm among foreign investors (Raina & Sinha, 2019). This increased objective comparised 40 GW from rooftop solar systems and 60 GW from utility-scale projects to position India as a global leader in renewable energy. India's aims under the Paris Agreement, essentially a more integrated objective to meet 40 per cent of non-fossil sources, and by 2030 reduce carbon-emission intensity against 2005 levels by between 33 - 35% were in support of these attempts (Yenneti, 2016).

The mission's scope includes not only technology innovation but also policy reform and social transformation. There are, therefore, complex interactions between the impediments and catalysts affecting implementation and success (Dey, 2010, p. xx; Yenneti, 2016; Raina & Sinha, 2019).

Various legislative and regulatory steps were taken by the government in order to facilitate solar investments towards the solar mission. These included Renewable Purchase Obligations (RPOs), which mandatorily directed utilities to obtain a certain percentage of their electricity from renewable sources. Financial incentives such as feed-in-tariffs, tax exemptions, and capital subsidies were also introduced to help ease the burden on early investors (Raina & Sinha, 2019).

In India the Ministry of new and renewable energy also formed the solar energy cooperation of India also known as SECI which oversees the tendering process of solar projects so that bidding would be transparent and price competitive.

However, many challenges have accompanied the practice of putting the NSM into action. The process of acquiring lands is among the key difficulties. Generally, in any solar project, large pieces of land are required, but these have proved to be very hard to come by, especially in the populated areas, which have their high density because their land is almost wholly occupied with residential and agricultural activities. Many projects have been delayed by land ownership issues and opposition from the surrounding communities, who fear eviction or unfavorable environmental impacts (Dey, 2010). As a result, the government asked the solar companies to place their projects in dry and semi-arid regions where land prices are cheaper and locals will be more willing to use otherwise unusable land for energy production (Yenneti, 2016).

Cost sensitivity is one among other substantial barriers which is deemed as one of the chief reasons for this non-implementation of India's solar mission. Economics remain largely uneven in this country and, therefore, huge initial capital for solar installments bars its comprehensive acceptance. Even after government initiatives, research indicates that the cost premium of solar technology is still a major factor in consumer choices, especially for lower-income households (Chen et al., 2024; Gleim et al., 2013). The cost-benefit analysis hampers solar adoption across different socio-economic strata and frequently favors traditional energy sources in the short term.

Now how does the Government of India tackle this problem?

To overcome the substantial challenge of price sensitivity, India adopted a holistic approach. The government introduced innovative funding schemes and subsidy programs to facilitate easy adoption of solar technology by different levels of economic society. The introduction of RECs and solar-specific subsidies proved effective in reducing the upfront costs for the customers (Chakraborty & Mukherjee, 2013; Stevens, 2010; Arthur et al., 2023). To offset the above costs, the government gave subsidies and incentives. The price of solar energy has decreased in the past from an approximate INR 15 per kwh in 2010 to less than INR 3 per kwh due to improvements in technology and economies of scale (Raina & Sinha, 2019,).

Other structural requirements were in the form of specialized funding, and co-operation was required with the foreign funding organization. The project involved guiding investment and competitive bids that reduced solar power costs substantially; SECI proved to be highly instrumental in doing so. As a consequence of such endeavors and technological advancement, the tariff of solar power has decreased from almost INR 17 per unit in 2010 to below INR 3 per unit in recent times (Dey, 2010; Kumar & Chandra, 2010; Yenneti, 2016).

Social norms and cultural aspects further complicate the successful execution of India's solar mission. Dominant social norms and cultural traditions are known to play a significant role in acceptance at the community level for new energy technologies (Nyborg et al., 2016; Schultz et al., 2018; Goldstein et al., 2008). The problem goes beyond the mere adoption of new technology but rather has to do with changing the underlying behavioral patterns and the social norms regarding energy use.

The Indian Government undertook efforts of awareness campaign and community engagement in order to resolve this situation. The Indian Government collaborated with the local communities, schools, and NGOs with a view of enlightening the mass people and increase its usage to generate solar power (Schultz et al., 2018; White et al., 2019; Goldstein et al., 2008).

Developing model solar villages and carrying out widespread outreach programs in rural regions were two other elements of the implementation approach. The demonstrations eliminated the distrust of the technology from traditional societies, making them believe in it further. Educational programs held in schools and colleges also helped bring about a new generation of solar energy enthusiasts (Nagarajan et al., 2022; Th⊘gersen, 2009; Young et al., 2009).

Socioeconomic factors also make the adoption of solar energy in India complex. For example, it has been proven that access to and understanding of solar technology depend on several economic levels, educational backgrounds, and urban-rural divides (Diamantopoulos et al., 2003; Joshi & Rahman, 2015; Young et al., 2009).

Solar solutions have extensively been incorporated into rural electrification programs; the Schemes like Deen Dayal Upadhyaya Gramme Jyoti Yojana targeting agricultural consumers for stable power supply. The government also encouraged the implementation of solar technology in middle-class families by providing special provisions for rooftop solar systems in urban areas (Gleim et al., 2013; Han et al., 2009).

This becomes more complex for the solar mission in India due to complicated eco-labels and certification procedures. It is found from the studies that various certification schemes have been misinterpreted and the consumers lack confidence due to which they are confused in the marketplace, and so the mistrust continues (Horne, 2009; Cerri et al., 2018; Tanner & Wölfing Kast, 2003).

India has a proper system of standardization to address concerns over certification and eco-labels. For solar installations and equipment, the Ministry of New and Renewable Energy (MNRE) defined clear criteria and quality standards. These include procedures for certification and harmonized monitoring systems for compliance, respectively (Horne, 2009; Cerri et al., 2018; Tanner & Wölfing Kast, 2003).

The initiatives streamlined the certification procedure for installers and manufacturers, increased consumer confidence, and guaranteed the calibre of solar installations (Prakash et al., 2019; Sheth et al., 2010; White & Simpson, 2013).

ROLE OF DIGITAL PLATFORMS AND E-COMMERCE IN PROMOTING SUSTAINABLE CONSUMPTION

The increasing power of digital platforms and ecommerce is driving sustainable consumption is a critical field of research. The chapter examines extensively how online platforms, search algorithms, and AI recommendation shape consumer behavior towards sustainable options.

Digitalization has transformed consumer access and purchasing of sustainable goods. Online stores, including Amazon, Alibaba, and specialized sustainable marketplaces, have incorporated sustainability filters to make it easier for consumers to discover environmentally friendly products (Faccia et al., 2023). These online stores make it possible for small and medium sized businesses (SMEs) with sustainable goods to sell to a worldwide market without physical stores saving carbon footprints on logistics and property.

AI-based recommendations are an integral part of individualized marketing of sustainable consumption. Algorithm scans for consumer behaviour and recommended green products based on what they have purchased in the past, what they have searched for, and social interactions. Research shows that digital marketing campaigns supplemented with ai can boost the adoption of green goods considerably by presenting their advantages in our personalised context (Gao et al., 2023).

Social media has also been instrumental in the embrace of sustainable consumption behaviour. Influencers marketing, in which those with a big following promote green brands, and has been especially useful in determining what people want. It has been proven that sustainability influencer has more influence on consumers' purchasing behaviour compared to regular corporate promotions, as they are viewed as being more genuine and reliable (Roszko- Wójtowicz et al., 2024).

Further, the contribution of digital platform to encouraging sustainable consumption is towards consumer education. Sustainability blogs and e-learning platform offer essential information for embracing green lifestyles. For example, Google another search engines rank sustainability-related content high in search results, making awareness and intelligent decision-making by consumers more likely (Oláh et al., 2023).

Corporate accountability is also changing online. Several of the largest e-commerce players have begun to include sustainability target in their logistic and operation. Amazon's climate pledge friendly initiative, for instance, badges eco-friendly products and rewards sellers who achieve environmental certification. Other the major e-Commerce player has similar scheme that encourage businesses to be greener (Jawadi et al., 2024).

POST-COVID SHIFTS IN GREEN PURCHASING

The COVID-19 Pandemic saw drastic changes in consumer behaviour, especially with sustainable practices. During the pandemic, the breakdown of supply chain, enhance concern for health and hygiene, and economic uncertainty shape the way consumer engaged with sustainability-led options (Bowen et al., 2023). After the pandemic, the changes have been lasting, creating growing demand for locally sourced, ethically made, and sustainable products. Research shows that customers today value brands will definite sustainability promises, especially those with an emphasis on carbon footprints reduction and building resilience against future disruption (Faccia et al., 2023).

Green consumption trades in post-pandemic recovery have been driven by efforts in economic recovery. Governments have implemented stimulus packages and green recovery strategies that will foster sustainable forms and promote eco-friendly consumer behaviour (Gao et al., 2023). For example, the green recovery plan launched by the European Union focuses on renewable energy sources, sustainable agriculture, and circular economy systems, thus reinforcing longer-term sustainability buying patterns among consumers.

The pandemic also spurred digitalization, which had a bearing on green consumerism. Online marketplaces experienced an unprecedented surge in screen shopping habits as consumers turned to online platforms sustainable goods (Oláh et al., 2023). With greater dependency or digital business, companies now use AI and big data to improve transparency of sustainability assertions and offer consumer credible information regarding the environmental footprint of their purchases.

Corporate sustainability efforts have gained traction post-COVID many multiple corporations have integrated sustainable business strategies into their recovery plans. This includes adopting carbon-neutral operations, investing in sustainable supply chains, and implementing transparent reporting on environmental and social governance (ESG) matrics (Roszko-Wójtowicz et al., 2024).

EMERGING INNOVATIONS

Circular Economy and Waste Reduction

The shift to circular economy has accelerated with industries and policymakers acknowledging that waste needs to be reduced and products lifespan need to be lengthened (Nath et al., 2023). Circular business models focus on sustainability through the elimination of waste maintenance in use of materials, and the regeneration of nature.

Government across the globe are putting in place regulatory policy that support circular economy approaches. For instance, the European Green Deal features policies that foster product durability, reparability, and recyclability, thereby minimising waste generation. Extended producer the responsibility (EPR) legislation requires manufacturers to assume responsibility for managing post-consumer waste, motivating them to develop product with end-of-life recyclability (Gao et al., 2023).

Companies are increasingly embracing circulatory led-innovations. Large companies have launched take-back schemes, refurbishments facilities, and closed-loop manufacturing system minimising virgin material dependence (Bowen et al., 2023). For example, the fashion sector is experiencing a boom and rental service, resale websites, and upcycling programs designed to reduce textile waste and encourage sustainable consumption.

Technological innovation is a central factor in waste minimization. Artificial intelligence-based waste sorting technologies, blockchain-based traceability, and biotech applications in materials science enable better circularity across packaging, electronics, and food manufacturing industries (Oláh et al., 2023).

Blockchain for Eco-Labeling Transparency

The application of blockchain technology to sustainability measures has transformed the transparency of eco-labelling, with consumer able to cheque the credibility of green assertions (Zhang et al., 2023). with mounting green washing concerns, blockchain-powered verification system present and unalterable ledger to guarantee supply chain traceability and empower consumer to make educated purchase decisions.

Blockchain increases transparency by documenting every step of the production and distribution chain, from raw material procurement to final product delivery. Example of companies using blockchain for sustainability are food companies certifying organic product, fashion companies checking for ethical label practises, and consumer electronics companies ensuring ethical mineral sourcing (Faccia et al., 2023).

Regulatory institutions and Industry Associations are appreciating the capability of blockchain in reinforcing eco-label validity. North American and European Union governments are seeking legislative systems for requiring blockchain-enabled reporting of sustainability assertions. Through that adoption of this technology into compliance framework, regulatory authorities are able to successfully track and validate corporate sustainability promises (Udeagha & Muchapondwa, 2022).

Consumer Confidence in green certification is necessary for mass application of blockchain technologies. Surveys shows that consumers are more open to using environmentally friendly brands with blockchain-confirmed information on carbon emissions, ethically sourced ingredients, and responsible production methods (Roszko-Wójtowicz et al., 2024).

Future progress in blockchain technology will additionally advance verification of sustainability. Smart contract development will allow compliance checking to be automated to ensure that companies stick to their sustainability commitments without involvement of humans (Jawadi et al., 2024).

Blockchain implementation in supply chain management and eco-labelling is a step in the right direction towards curbing misinformation and enhancing accountability of corporations in sustainability efforts. As more sector embrace the technology, environmentally aware consumer will receive verifiable, temper-free information, allowing them to make more sustainable choices.

Through the mitigation of these rising trends and innovation, this chapter offers an overview of how circular economy practices, and blockchain transparency are all leading to sustainable consumption.

CURRENT STUDY LIMITATIONS, IMPLICATIONS AND FUTURE RESEARCH

Current Study Limitations

Lack of Empirical Data from Developing Countries

One of the key limitations of existing research on sustainable consumption is the dominance of consumer behaviour in developed nations at the expense of socio-economic realities in developing economies. The chapter identifies price sensitivity and economic limitation as hurdles to green buying but draws significantly on studies from high income countries where disposable income and government incentives enhance the affordability of sustainable products. The lack of empirical information from low-income areas leaves a knowledge void since consumers here are confronted with special challenges like lack of access to green options, law enforcement of environmental policies, and lower levels of awareness. The chapter would be enhanced by case studies of consumer attitudes in emerging economies, especially where affordability and infrastructural challenges mean sustainability as a concern is less feasible.

Inconsistencies in Eco-Labeling and Greenwashing Research

Eco-labelling plays a critical role in directing consumers towards environmentally sustainable decision, but variability in certification standards represents a major limitation in existing literature. The chapter addresses the necessity for a universally acceptable eco-labelling framework and the menace of greenwashing, where business es misleads customers by falsely labelling their products as green (Cerri et al., 2022). Yet, it does not contain in-depth case studies or regulation comparisons that assess how various countries address eco-certification. While Blockchain an artificial intelligence is hinted at as remedies to increase transparency, little actual implementation is discussed. Most of the studies analyse consumer attitudes toward eco-labels instead of labelling systems efficiencies in inhabiting deceptive claims of sustainability. In the absence of an internationally standardised format, consumers cannot tell the real green products from those that are greenwash for sales.

Unclear Policy Effectiveness

Government policy plays a central role in determining sustainable consumption but existing research tends to not to evaluate the long-term effectiveness of policies. The chapter sites policies like India's National Solar Mission as a successful story of government intervention (Raina & Sinha, 2023), but there is limited debate regarding comparative policy effectiveness between regions. Although economic incentives, subsidies, and regulation can stimulate short-term changes towards sustainability, their long-term effect is uncertain. Furthermore, policy success is context-specific, depending on enforcement tools, societal acceptance, and economic environments. There are limited longitudinal studies that review how policies affect consumer purchasing behaviours overtime, constraining the capacity to ascertain whether these initiatives are indeed transformative or require ongoing reinforcement.

Cultural And Social Norms Influence

Sustainable consumption is greatly shaped by social norms and cultural values, but most existing research tends to ignore regional differences that determine consumer behaviour. The chapter recognises that social expectations can support or deter green buying (White & Simpson, 2021), but it does not explore deeply how sustainability is viewed in various societies, material possessions and consumerism are more highly valued, so there will be less acceptance of green consumption. Social norms also determine what is prestigious or desirable in local places. For example, in a few high-consumption societies, convenience and luxury are more important than sustainability, so green behaviour will be more challenging to adopt.

Implications

Price sensitivity, social and socioeconomic constraint, cultural practices, and mixed interpretation of ecological labelling are some of the factors encouraging and discouraging sustainable buy. So, strategies in markets should be calibrated and adjusted quickly. For instance, price sensitivity suggests that, while there is an affordability gap of sustainable good buys, there may be price bundling suggested as a counter-solution for affordability concern of customers; Chekima et al. (2016) & Gleim et al. (2013). Further, White, Habib, and Hardisty (2019) indicate that social norms are involved in the influence of sustainability on purchasing and help campaigns in marketing by creating an appeal to socially desirable green products.

Problems in eco-labeling and certification have consequences for companies and lawmakers. As pointed out by (Peattie, 2010), there is no uniformity, hence the necessity for certifications that are globally recognized. As observed by (Martinho et al., 2015; Gleim et al., 2013), complexity in the landscape of eco-labels would most likely increase the confidence of consumers and reduce consumers' reluctance toward greenwashing.

Economical and regulatory policy can alter market incentives for the sustainable products offered. For example, (Arthur et al., 2023) and (Stevens, 2010) gave detailed examples that subsidies and tax incentive could close this price gap as well as create an opportunity to encourage the industry to take greener practices. As noted by (Dutta & Hwang, 2021), it has been suggested that policies as such have a strong impact on green automobile adoption with the mere presence of financial incentives.

Furthermore, studies on social norms and the efficacy of eco-labels (Nyborg et al., 2016; White et al., 2019) indicate that the government should take the initiative to support environmental initiatives and create standard certification requirements so that sustainable practices become the rule.

Future Research

Expanding Socio-Economic Analysis of Green Consumers

Socioeconomic determinants such as location, income, and education have a high influence on judgments toward sustainable purchasing (Young et al., 2009; Diamantopoulos et al., 2003). Lower-income customers are often likely to believe that green products are scarce because they have a higher price; therefore, more research needs to be done on which of the socioeconomic profiles are more prone to engage in green shopping and the problems that they face in this regard (Chekima et al., 2016, Joshi & Rahman, 2015). Such studies could concentrate on particular constraints and incentives in each of the groups. These streams of research can help direct marketing efforts as well as tailor policy interventions towards reducing the access gap for green products across these different socioeconomic groups (Arthur et al., 2023; Gleim et al., 2013).

Further, future research may look into the socioeconomic determinants of green consumption in developed and developing countries to better understand how social structures and inequality in wealth influence sustainable purchasing practices across the world (Chekima et al., 2015). Such research might be useful, for instance in the design of tiered pricing or government subsidies that makes the sustainable product cheaper, thereby appealing to a population of a lower income-class, or even for wider intervention in market-incentive mechanisms (Martinho et al., 2015; Chakraborty & Mukherjee, 2013).

Deepening Insights on Social Norms and Cultural Influences

In a study by Nyborg et al. (2016), the consumer's social expectations and the cultural pressure can drive their behaviors toward sustainable products. Acceptance and visibility in the social environment about the cultural context are what influence one's desire to have sustainable products. Most people will buy products in a society in accordance with social expectations, as opposed to having a personal say in choosing such products (White et al., 2019; Chekima et al., 2015). Further research may delve into the dynamics that take place across societies in which reasons for being environmentally conscious are ascribed to peer influence, media influence, or public figures advocating sustainable practices in order to understand how social norms change with time.

Future research can also focus on the concept of "norm spillover," whereby green behavior in one area of life carries over into another sustainable practice. This can include ascertaining if recycling or saving energy impacts a person's desire to buy more sustainably and vice versa (White & Simpson, 2013). Moreover, knowledge concerning cultural influence could explain why a certain customer may believe that an environmentally friendly practice reflects corporate action following culturally relevant values.

Examining the Impact of Eco-Labeling Standardization

Undoubtedly, one of the most helpful means of customer awareness about environmental benefit from the use of the product is by an eco-label. Nevertheless, since certification requirements are different depending on the various schemes, that would decrease an effectiveness of such an eco-label and may mislead and deceive the consumer about it (Horne, 2009; Peattie, 2010). This research might explore how different types of eco-labels influence the capability of consumers to make environmentally aware decisions, that is, carbon neutrality, recyclable packages, or fair labor conditions (Gleim et al., 2013).

Yet another interesting topic that can be followed up in more depth is the possibility of "label fatigue", that is, when consumers become bombarded by too many labels and just give up making sustainable choices altogether (Martinho et al., 2015). Study may identify how to best communicate sustainability metrics to consumers in ways that are informative without being overpowering through appropriate presentation techniques. This may involve the development of holistic labels that offer multi-attribute assessments or a more simplified labeling hierarchy rank that would rate items based on their impacts on the environment, as proposed by research into environmental marketing (Cerri et al., 2018).

Investigating the Longitudinal Impact of Government Interventions

These include tax credits, subsidies, and green certifications by governments. Stevens, 2010; Chakraborty & Mukherjee, 2013. Still, there is a need to do further studies in order to measure the expected long-term effects of these strategies on consumers' decisions and market conditions. For instance, experiments may consider whether short-run incentives, including tax breaks for electric cars, influence consumer behaviour over a more extended period of time or the effect tends to evaporate as soon as the incentive is withdrawn (Dutta & Hwang, 2021).

Further, the best policies between countries can be compared to determine the most effective ones. The studies can compare the effectiveness of direct subsidies and grants against market-based mechanisms like carbon taxes or cap-and-trade schemes to determine which approaches will result in more lasting changes in green purchasing (Arthur et al., 2023). This research may look at how the policies play out in developed and developing countries and how government's programs could close socioeconomic disparities which have allowed limited accessibility to green products (Jordan, 2014; Joshi & Rahman, 2019).

Analyzing Willingness to Pay and Emotional Factors in Sustainable Consumption

According to Peattie, 2010; Prakash et al., 2019, WTP for green items depends on both rational and emotional aspects, such as brand trust, ethical concern, and environmental consciousness. Future research might focus on the psychological underpinnings of WTP through examination of whether consumers' affective ties towards sustainability and personal values influence the amount they pay for green goods, as claimed by Gleim et al., 2013. For example, researches might examine whether it is the strength of either egoistic or altruistic motivations underlying WTP because it is coming more from the environmentalistic motivation or through the benefits of consumer safety about the product, Chekima et al., 2015; Prakash et al., 2019.

It has been established that consumer trust and WTP for sustainable products increase when transparent information is given about the environmental impact of a product, for example, an eco-label showing less carbon footprint (White & Simpson, 2013; Cerri et al., 2018).

CONCLUSION

It is much more than a shift in consumer preference, as this chapter so eloquently illustrates. It involves a new paradigm: that of orienting economic, social, and individual values to ecologically conscious action. Increasingly pressing global concerns--shortages of resources, climate change, environmental degradation--cannot be met without more sustainable consumption. However, there are a few reasons that do not allow consumers to choose it, such as price sensitivity, socioeconomic barriers, cultural practices, and the complexity of eco-labels. This is one of the major barriers seen over time. Consumers with lower incomes and rural backgrounds face great challenges when the barriers, such as high costs and unavailability of environment-friendly products, have more influence. Social norms also bring out the challenge of changing the consumption patterns, as they sometimes resist the change toward sustainable behavior. Social and cultural norms for green practices have the power to advance sustainable consumption into the mainstream or impede its advancement.

However, there are very strong catalysts that can get over these hurdles. Environment awareness is perhaps the most potent motivation since it enables consumers to relate to how their purchasing choices affect the environment. A greater understanding of the importance of long-term environmental well-being over immediate financial savings frequently results in a willingness to invest more in sustainable choices.

One of the key advances in sustainable consumption is that of digital platforms and e-commerce. Digitalisation has facilitated green products' easier availability, and consumers can now easily compare and check for credibility of sustainability claims. Online e-commerce sites also have separate green product sections now, and recommendation engines driven by artificial intelligence align sustainable options according to consumers' preferences.

Another such innovative imperative is blockchain for transparency in eco-labeling, which has become a game-saver against greenwashing. Blockchain technology offers a tamper-proof and verifiable history of a product's life cycle, from sourcing to end sale, making claims of sustainability transparent.

The circular economy philosophy and waste minimization are strongly advocates of sustainable consumption. Governments and companies are embracing circular business models where companies aim to stop waste, retain materials in a loop, and regenerate natural systems. Recycling and upcycling are becoming ever more popular with companies introducing take-back schemes and extended producer responsibility (EPR) programs for promoting the reusing and reconsecrating of material. In addition, technologies such as AI-driven waste segregation and biodegradable packaging are complementing these circular economy initiatives to minimize landfill waste and pollution significantly.

Policies of the government also transform through providing right support in terms of incentives, subsidies, and laws that boost the availability and demand for such products. Such regulations not only change the approach of the consumers but also align the businesses toward higher ethical norms. CSR and Eco-labeling policies also promote accountability and trust which enable the consumer to make an appropriate choice towards his ethical as well as environmental values.

Sustainable packaging is a push factor for the green purchase decision. Packaging has often been found to be one of the primary impressions of the product, therefore the trend for eco-friendly and minimalist packaging encourages consumers to waste-free decisions. Minimalist packaging uses less material, making the solution affordable but still satisfies consumer demands for being accountable in businesses they support.

This example, India's National Solar Mission, is one of the compelling ways that a huge, consistent change can be made with governmental support. It further shows that it is possible to adopt renewable energy through such examples and demonstrates how the removal of social and economic barriers can initiate sustainable behavior nationwide.

Even with the progresses that have already been made through many businesses and geographical areas, there is always much to do. The challenges identified in the chapter indicate a constant need to innovate and seek ways to upgrade sustainable consumption activities. Consumers, businesses, and decision-makers all have to contribute to an environment in which everyone can have sustainable choices, find them desirable, and consider them feasible. Future research may delve deeper into the socioeconomic, cultural, and psychological factors that may influence green consumption behavior and incorporate efforts to promote appealing sustainable consumption behavior for all.

The post-COVID trends in green shopping completely reshaped consumer behaviors towards placing greater values on sustainability. The pandemic revealed weakness in global supply chains and had the impact of reinforcing the merits of purchasing locally, ethically, and sustainably (Bowen et al., 2023).

This chapter is of immense value to policy makers and academicians through a comprehensive examination of driver's sustainable consumption. For policymakers, the chapter offers pragmatic strategies, which can be used in the formulation of effective interventions so that sustainability interventions are effective and inclusive. For academicians, this chapter contributes to the literature through establishment of key gaps in green consumer behaviour, policy, and market reaction. Through the bridging of the gaps between practice and theory, this chapter is a platform for evidence-based decision-making in the academic world and among policymakers.

In a word, sustainable consumption must be stimulated holistically while employing the correct levers fairly and eliminating blockers. The governments, businesses, and consumers together need to cooperate constantly in crafting a future for themselves where practices of sustainability shall become the norms rather than deviations from the common norm. This could then become eco-friendly consumerism, which can be transformed to the betterment of the well-being of the planet and creation of the most just society. However, this would be achievable only if these green products constantly become available, affordable, and appealing to all sections of society.

Key Findings

The key findings of this book chapter on sustainable consumerism are that how the complex barriers and catalysts interact to influence consumer behavior toward greener buying. Although more awareness about sustainability has been generated across the globe, several barriers exist which are stopping its wide acceptance. The largest obstacle is price sensitivity: middle-class and lower-class customers will pay more for a product that is seen as less crucial to savings or reducing spending. Since economic disparities tend to occur with greater probability within developing nations, the assertion that sustainable goods constitute a luxury exacerbates the dilemma.

Socioeconomic factors, such as geographic region, financial status, and educational attainment, have a significant impact on consumers' perceptions and behavior regarding environmentally friendly consumption. The environmental consumer is more likely to opt for green products than the rural consumer, who is restricted by educational and logistical barriers because they have more choices and more purchasing power. As against it, in the progressive culture, cultural norms and societal pressures allow for more pro-environment behavior, whereas in the traditional society, they are trying to fortify the conventionally consumption manners. The present serious problem because customers are incapable of understanding or believing the environment claims made by companies is inconsistent with universal requirements for eco-labels.

Since such catalysts would be brought into use that could encourage sustainable consumption even though the same seems directly opposite to the basic concept of that, therefore, the reality of the given situation is presented by this chapter. Environmental consciousness increases because of the efforts the people make while raising awareness on different media avenues regarding the problem of resource degradation and climate change in this world. Government policies and regulatory frameworks are the enabling agents of the green options. They provide monetary incentives, subsidies, and infrastructure assistance and thus make green options accessible and attractive to consumers. CSR programs are also an important driver of this trend. The CSR programs make business practices more acceptable to customers, who believe that the interactions a company has with them must be morally and ecologically responsible. For instance, proactive green practices of such businesses, such as the use of eco-friendly packaging and clear eco-labeling, are more likely to gain customer trust and loyalty. They are indispensable for fortifying critical market trends toward sustainability. In general, it means eco-friendly and minimalist options have emerged as highly decisive factors to customers, which reveal how mere presentation changes are having a wide influence on greening practices of products.

REFERENCES

Arias, C. (2024). Understanding the drivers and barriers towards sustainable consumption: An approach by linking a pedagogical strategy of active learning with adopting sustainable behaviors by students. *Journal of Sustainability Perspectives, 4*, 270–293. DOI: 10.14710/jsp.2024.25041

Arthur, E. E., Gyamfi, S., Gerstlberger, W., Stejskal, J., & Prokop, V. (2023). Towards circular economy: Unveiling heterogeneous effects of government policy stringency, environmentally related innovation, and human capital within OECD countries. *Sustainability (Basel), 15*(6), 4959. DOI: 10.3390/su15064959

Basil, D. Z., & Erlandson, J. (2008). Corporate social responsibility website representations: A longitudinal study of internal and external self-presentations. *Journal of Marketing Communications, 14*(2), 125–137. DOI: 10.1080/13527260701858497

Becker-Olsen, K. L., Cudmore, B. A., & Hill, R. P. (2006). The impact of perceived corporate social responsibility on consumer behavior. *Journal of Business Research, 59*(1), 46–53. DOI: 10.1016/j.jbusres.2005.01.001

Bisschoff, C., & Fullerton, S. (2025). Comparing green customer citizenship attitudes and behaviours in South Africa and South Korea. *Acta Commercii, 25*(2). Advance online publication. DOI: 10.4102/ac.v25i2.1336

Bowen, R., Dowell, D., & Morris, W. (2023). Hospitality SMEs and the circular economy: Strategies and practice post-COVID. *British Food Journal, 126*(1), 80–97. DOI: 10.1108/BFJ-10-2022-0932

Cerri, J., Testa, F., & Rizzi, F. (2018). The more I care, the less I will listen to you: How information, environmental concern and ethical production influence consumers' attitudes and the purchasing of sustainable products. *Journal of Cleaner Production, 175*, 343–353. DOI: 10.1016/j.jclepro.2017.12.054

Chakraborty, D., & Mukherjee, S. (2013). Fiscal subsidies and environmental sustainability: What does the cross-country empirical estimates suggest? *Foreign Trade Review, 48*(3), 383–397. DOI: 10.1177/0015732513496619

Chekima, B., Chekima, S., Syed Khalid Wafa, S. A., Igau, O., & Sondoh, S. L. (2015). undefined. *International Journal of Sustainable Development and World Ecology, 23*(2), 210–220. https://doi.org/ DOI: 10.1080/13504509.2015.1114043

Chekima, B., Syed Khalid Wafa, S. A., Igau, O. A., Chekima, S., & Sondoh, S. L. (2016). undefined. *Journal of Cleaner Production, 112*, 3436–3450. DOI: 10.1016/j.jclepro.2015.09.102

Chen, C., Zhang, D., Zhu, L., & Zhang, F. (2024). undefined. *Resources, Conservation and Recycling, 207*, 107682. DOI: 10.1016/j.resconrec.2024.107682

Dey, D. (2010). The political economy of India's national solar energy mission. SSRN *Electronic Journal*. https://doi.org/DOI: 10.2139/ssrn.1633911

Diamantopoulos, A., Schlegelmilch, B. B., Sinkovics, R. R., & Bohlen, G. M. (2003). Can socio-demographics still play a role in profiling green consumers? A review of the evidence and an empirical investigation. *Journal of Business Research, 56*(6), 465–480. DOI: 10.1016/S0148-2963(01)00241-7

Duarte, P., Silva, S. C., Roza, A. S., & Dias, J. C. (2024). Enhancing consumer purchase intentions for sustainable packaging products: An in-depth analysis of key determinants and strategic insights. *Sustainable Futures : An Applied Journal of Technology, Environment and Society, 7*, 100193. DOI: 10.1016/j.sftr.2024.100193

Dutta, B., & Hwang, H. (2021). Consumers purchase intentions of green electric vehicles: The influence of consumers technological and environmental considerations. *Sustainability (Basel), 13*(21), 12025. DOI: 10.3390/su132112025

Faccia, A., Le Roux, C. L., & Pandey, V. (2023). Innovation and e-Commerce models, the technology catalysts for sustainable development: The emirate of Dubai case study. *Sustainability (Basel), 15*(4), 3419. DOI: 10.3390/su15043419

Gao, J., Siddik, A. B., Khawar Abbas, S., Hamayun, M., Masukujjaman, M., & Alam, S. S. (2023). undefined. *Sustainability, 15*(2), 1594. DOI: 10.3390/su15021594

Ghosh, B. (2016). Impact of packaging on consumers' buying Behaviour:A case study of mother dairy, Kolkata. *Parikalpana: KIIT Journal of Management, 12*(1), 63. DOI: 10.23862/kiit-parikalpana/2016/v12/i1/133042

Gleim, M. R., Smith, J. S., Andrews, D., & Cronin, J. J.Jr. (2013). Against the green: A multi-method examination of the barriers to green consumption. *Journal of Retailing, 89*(1), 44–61. DOI: 10.1016/j.jretai.2012.10.001

Goldstein, N. J., Cialdini, R. B., & Griskevicius, V. (2008). A room with a viewpoint: Using social norms to motivate environmental conservation in hotels. *The Journal of Consumer Research, 35*(3), 472–482. DOI: 10.1086/586910

Han, H., Hsu, L., & Lee, J.-S. (2009). Empirical investigation of the roles of attitudes toward green behaviors, overall image, gender, and age in hotel customers' eco-friendly decision-making process. *International Journal of Hospitality Management, 28*(4), 519–528. DOI: 10.1016/j.ijhm.2009.02.004

Horne, R. E. (2009). Limits to labels: The role of eco-labels in the assessment of product sustainability and routes to sustainable consumption. *International Journal of Consumer Studies*, *33*(2), 175–182. DOI: 10.1111/j.1470-6431.2009.00752.x

Jawadi, F., Pondie, T., & Cheffou, A. (2024). New challenges for green finance and sustainable industrialization in developing countries: A panel data analysis. https://doi.org/DOI: 10.2139/ssrn.4916541

Jordan, P. G. (2014). Federal and state energy policies. *Solar Energy Markets*, 65-109. https://doi.org/DOI: 10.1016/B978-0-12-397174-6.00006-4

Joshi, Y., & Rahman, Z. (2015). Factors affecting green purchase behaviour and future research directions. *International Strategic Management Review*, *3*(1-2), 128–143. DOI: 10.1016/j.ism.2015.04.001

Joshi, Y., & Rahman, Z. (2019). undefined. *Ecological Economics*, *159*, 235–243. DOI: 10.1016/j.ecolecon.2019.01.025

Kumar, S., & Chandra, N. (2010). Sustainability concerns in Indian agriculture: An analysis of subsidies and investment. SSRN *Electronic Journal*. https://doi.org/DOI: 10.2139/ssrn.1709898

Luo, X., & Bhattacharya, C. (2006). Corporate social responsibility, customer satisfaction, and market value. *Journal of Marketing*, *70*(4), 1–18. DOI: 10.1509/jmkg.70.4.001

Mahmoud, M. A., Tsetse, E. K., Tulasi, E. E., & Muddey, D. K. (2022). Green packaging, environmental awareness, willingness to pay and consumers' purchase decisions. *Sustainability (Basel)*, *14*(23), 16091. DOI: 10.3390/su142316091

Martinho, G., Pires, A., Portela, G., & Fonseca, M. (2015). Factors affecting consumers' choices concerning sustainable packaging during product purchase and recycling. *Resources, Conservation and Recycling*, *103*, 58–68. DOI: 10.1016/j.resconrec.2015.07.012

Nagarajan, M., Saha, R., Kumar, R., & Sathasivam, D. (2022). Impact of peer influence and environmental knowledge on green consumption. *International Journal of Social Ecology and Sustainable Development*, *13*(6), 1–16. DOI: 10.4018/IJSESD.292039

Nath, O., Kumar, S., Sharma, V., Lal Meena, M., & Jain, R. (2023). Design of strategic framework for green sustainable practices in e-Commerce companies. *Materials Today: Proceedings*. Advance online publication. DOI: 10.1016/j.matpr.2023.03.032

Nichifor, B., Zai , L., & Timiras, L. (2025). Drivers, barriers, and innovations in sustainable food consumption: A systematic literature review. https://doi.org/DOI: 10.20944/preprints202502.0273.v1

Nyborg, K., Anderies, J. M., Dannenberg, A., Lindahl, T., Schill, C., Schlüter, M., Adger, W. N., Arrow, K. J., Barrett, S., Carpenter, S., Chapin, F. S.III, Crépin, A., Daily, G., Ehrlich, P., Folke, C., Jager, W., Kautsky, N., Levin, S. A., Madsen, O. J., & De Zeeuw, A. (2016). Social norms as solutions. *Science*, *354*(6308), 42–43. DOI: 10.1126/science.aaf8317 PMID: 27846488

Oláh, J., Popp, J., Khan, M. A., & Kitukutha, N. (2023). Sustainable e-Commerce and environmental impact on sustainability. *Economics & Sociology (Ternopil)*, *16*(1), 85–105. DOI: 10.14254/2071-789X.2023/16-1/6

Peattie, K. (2010). Green consumption: Behavior and norms. *Annual Review of Environment and Resources*, *35*(1), 195–228. DOI: 10.1146/annurev-environ-032609-094328

Piracci, G., Boncinelli, F., & Casini, L. (2023). Investigating consumer preferences for sustainable packaging through a different behavioural approach: A random regret minimization application. *Environmental and Resource Economics*, *86*(1-2), 1–27. DOI: 10.1007/s10640-023-00785-3

Prakash, G., Choudhary, S., Kumar, A., Garza-Reyes, J. A., Khan, S. A., & Panda, T. K. (2019). Do altruistic and egoistic values influence consumers' attitudes and purchase intentions towards eco-friendly packaged products? An empirical investigation. *Journal of Retailing and Consumer Services*, *50*, 163–169. DOI: 10.1016/j.jretconser.2019.05.011

Raina, G., & Sinha, S. (2019). Outlook on the Indian scenario of solar energy strategies: Policies and challenges. *Energy Strategy Reviews*, *24*, 331–341. DOI: 10.1016/j.esr.2019.04.005

Roszko-Wójtowicz, E., Deep Sharma, G., Dańska-Borsiak, B., & Grzelak, M. M. (2024). Innovation-driven e-Commerce growth in the EU: An empirical study of the propensity for online purchases and sustainable consumption. *Sustainability (Basel)*, *16*(4), 1563. DOI: 10.3390/su16041563

Schultz, P. W., Nolan, J. M., Cialdini, R. B., Goldstein, N. J., & Griskevicius, V. (2018). The constructive, destructive, and reconstructive power of social norms: Reprise. *Perspectives on Psychological Science*, *13*(2), 249–254. DOI: 10.1177/1745691617693325 PMID: 29592653

Sheth, J. N., Sethia, N. K., & Srinivas, S. (2010). Mindful consumption: A customer-centric approach to sustainability. *Journal of the Academy of Marketing Science, 39*(1), 21–39. DOI: 10.1007/s11747-010-0216-3

Stevens, C. (2010). Linking sustainable consumption and production: The government role. *Natural Resources Forum, 34*(1), 16–23. DOI: 10.1111/j.1477-8947.2010.01273.x

Tanner, C., & Wölfing Kast, S. (2003). Promoting sustainable consumption: Determinants of green purchases by Swiss consumers. *Psychology and Marketing, 20*(10), 883–902. DOI: 10.1002/mar.10101

Th⊘gersen, J. (2009). The motivational roots of norms for environmentally responsible behavior. *Basic and Applied Social Psychology, 31*(4), 348-362. https://doi.org/DOI: 10.1080/01973530903317144

Udeagha, M. C., & Muchapondwa, E. (2022). Environmental sustainability in South Africa: Understanding the criticality of economic policy uncertainty, fiscal decentralization, and green innovation. *Sustainable Development (Bradford), 31*(3), 1638–1651. DOI: 10.1002/sd.2473

Van den Bergh, J. C. (2008). Environmental regulation of households: An empirical review of economic and psychological factors. *Ecological Economics, 66*(4), 559–574. DOI: 10.1016/j.ecolecon.2008.04.007

Weder, F., Golob, U., & Podnar, K. (2025). Sustainable consumption in context: A cross-cultural study of social representations. *Sustainability (Basel), 17*(4), 1531. DOI: 10.3390/su17041531

White, K., Habib, R., & Hardisty, D. J. (2019). How to SHIFT consumer behaviors to be more sustainable: A literature review and guiding framework. *Journal of Marketing, 83*(3), 22–49. DOI: 10.1177/0022242919825649

White, K., & Simpson, B. (2013). When do (and don't) normative appeals influence sustainable consumer behaviors? *Journal of Marketing, 77*(2), 78–95. DOI: 10.1509/jm.11.0278

Yenneti, K. (2016). The grid-connected solar energy in India: Structures and challenges. *Energy Strategy Reviews, 11-12*, 41–51. DOI: 10.1016/j.esr.2016.06.002

Young, W., Hwang, K., McDonald, S., & Oates, C. J. (2009). Sustainable consumption: Green consumer behaviour when purchasing products. *Sustainable Development (Bradford), 18*(1), 20–31. DOI: 10.1002/sd.394

Zhang, S., Ding, Q., & Ding, J. (2023). Return strategy of e-Commerce platform based on green and sustainable development. *Sustainability (Basel)*, *15*(14), 11188. DOI: 10.3390/su151411188

Zhang, X., & Dong, F. (2020). Why do consumers make green purchase decisions? Insights from a systematic review. *International Journal of Environmental Research and Public Health*, *17*(18), 6607. DOI: 10.3390/ijerph17186607 PMID: 32932797

Chapter 10
Water and Energy Nexus in the Circular Economy Context

Erkin Cihangir Karataş
https://orcid.org/0000-0002-4906-9696
Independent Researcher, Turkey

ABSTRACT

The relationship between water and energy is strong enough to create a nexus. It is clear that water and most of the resources we obtain energy from today are finite. Although it is possible to benefit from infinite energy sources, the water side of the nexus is dependent on finite resources. Therefore, it is important to make the nexus a circular structure for sustainability. In this study, the water side of the nexus seems prone to circular sustainability. The perspective from the energy side is more pessimistic. However, the available data mainly covers developed economies. Therefore, it has become clear that all stakeholders must take responsibility in order to create a global circular economy. It was concluded that more funds are needed from developed economies to strengthen the water-energy nexus in developing economies, and developing economies need to reduce their own population pressures.

INTRODUCTION

The beginning of life on Earth took place in water. Since this beginning, water has been the basic component of life on Earth. All life forms have continued their existence in places where water was available according to their needs. Humans are also one of the creatures that continue their lives dependent on water. For this reason, people continued their lives in places where there was access to water. When

DOI: 10.4018/979-8-3373-0139-6.ch010

they transitioned to an agricultural society, they chose regions where access to water was easy, as in the hunter-gatherer period. Since agricultural societies were more dependent on water, they clustered on the coasts and started civilization in these regions. In essence, it would not be wrong to associate the entire history of humanity with the economy. Because people have had to use relatively scarce resources efficiently during every period of their lives. While hunter-gatherers needed water for their most basic needs, farmers also needed it to continue their production. The situation has not changed since the industrial revolution. Water has remained a natural part of human life and a basic input in the economy. As humanity and civilization have flourished, economic relations have become more complex. Accordingly, the relationship between water and economy has become increasingly stronger. Today, water is a basic production input in every field, from the manufacturing industry to agriculture, from the energy sector to urban use. Without water, it is not possible to continue production activities in any sector.

In today's world, the inputs of life and production activities have changed compared to the past. This change will continue to change as civilization progresses. For example, the food items that humans need to sustain life have become more diverse and complex compared to the past. Similarly, production has become more complex and diverse with increasing and changing human needs and technology. In direct proportion to this increase, the demand for water, which continues to be an input for every production, has also increased. At this point, another input that is as necessary for production as water comes to mind. This input is energy. As the production process, which used to be based on muscle power, began to be replaced by machines over time, the demand for energy has also increased. While the energy required at the beginning of the mechanization process was provided by coal, this was eventually replaced by oil and then by fossil-based fuels such as natural gas. Although electricity was later included, the partial dependence on fossil fuels for electricity production could not be eliminated.

However, there is a dependency on water for all the processes that are in effect to obtain energy. This can be seen in all sectors where the necessary inputs are provided to produce energy. From mining to the metalworking industry, from oil refining to distribution, water is a production input that cannot be ignored in every field. For example, it is a requirement for water cooling systems in facilities that operate with nuclear and thermal power to generate electricity. It is used for washing in solar systems. Hydroelectric systems are already directly dependent on water. Even wind energy is not free from a water footprint. While water is used in processes such as waste storage and cooling in mining, it is used for process and cooling purposes in oil refineries. In the simplest terms, all the equipment required to produce energy has a certain water footprint. This shows that the nature of energy production is directly and/or indirectly dependent on water.

It is obvious that there is a dependency on water to obtain energy. However, this dependency cannot be said to be a one-way dependency. When it comes to water, the need for energy to enable integrated and efficient water management is undeniable. Without energy, the purification required to protect water resources in terms of quality is not possible. Energy is absolutely needed for water abstraction from resources and for the distribution and fair sharing of water. In addition, energy stands before us as a necessity again to measure and evaluate the quantity and quality status of all water resources. In short, just as there is a clear dependency on water to obtain energy, the dependency on energy for access to water and management of water is equally clear.

The question that comes to mind here is, from an economic perspective, is the dependence on water and energy, which are basic inputs for all sectors, in a sustainable cycle? This will be the main research question of this study. To do this, both qualitative and quantitative methods can be followed. Both methods have their own strengths and weaknesses. It is clear that the quantitative method will give clearer results if the data series are reliable. However, the qualitative method seems more efficient in terms of evaluating the effects caused by human behavior and being able to better analyze negative externalities. When it comes to water resources, water management involves intense economic and environmental externalities due to being an input in every sector of the economy and a basic component of life. For this reason, a qualitative method was preferred in this study.

The first step to be taken while pursuing a possible answer to the research question of the study is to address the factors that transform the water and energy relationship into a real nexus. Secondly, the current demand for water and energy and the predictions about how the demand for this duo will change in the future will be evaluated qualitatively using numerical data. Then, in the third step, the mutual interaction of the duo and the pressures they create on each other will be examined in the context of sustainability. In the last step, an economic relationship framework will be tried to be drawn by considering the duo as an interconnected whole.

The purpose of writing this section is to examine whether the global water-energy nexus is circular, to examine the progress and to contribute to the creation of this cycle in the future. If the water-energy nexus can be made fully circular, this means sustainable water-energy management. In other words, the need for additional resources to obtain water and energy will no longer be required. Bringing the nexus into a sustainable cycle is a big step towards achieving the development goals (SDGs) "Goal 6. Ensure availability and sustainable management of water and sanitation for all", "Goal 7. Ensure access to affordable, reliable, sustainable and modern energy for all" and "Goal 12. Ensure sustainable consumption and production patterns". The 2024 SDGs progress report stated that each stage of production offers an opportunity to reduce resource and fossil fuel use, save energy, and for a circular economy approach (UN Progress towards the SDGs, 2024). From this point of view, it is clear that the circularization of the water-energy nexus will make a fundamental contribution to achieving these goals. It can also ensure that the funds in this area, which are constantly increasing due to decreasing costs, are transferred to projects that will increase total welfare on a global scale.

1. ON THE RELATIONSHIP BETWEEN WATER AND ENERGY

When trying to make sense of the relationship between water and energy, the first thing to do is to define the dependency of each on the other. Then, it would be a good approach to draw a picture of how the two are in a circular connection through this description. First of all, it is useful to remember that water is a production input in almost every sector. Water is also a basic production input in the energy sector. Especially in developed countries, the industry branch where water is used the most is the energy sector. With a general definition, the energy sector can be examined in two main wings. The first can be determined as primary resource production and the second as power production. Primary resources are divided into three as bioenergy, fossil fuels and hydrogen. A three-part classification can be made for the resources required for power production. These resources can be listed as nuclear, fossil fuels and renewable(IEA, 2024). Water is needed to obtain energy from all of these energy sources.

Energy obtained directly or indirectly from biomass is called bioenergy. Bio-energy resources are used in many areas from heating to transportation(Pelkmans, 2020). In addition to traditional resources such as wood and crop/animal residues, modern resources such as acid oil, vegetable and animal oils, used cooking oils, and vegetable ethanol can be given as examples of bioenergy resources(Bioenergy, 2024). In order to obtain these resources, raw materials that are generally related to agriculture and/or soil are used. It is clear that water and irrigation are needed to produce these raw materials. At this point, it is useful to remember that irrigation in agricultural activities constitutes the largest portion of water drawn in the world today. In addition, water is required for the production process required to obtain the final product, such as waste disposal, cooling and washing.

Fossil fuels are a type of resource used not only as a primary source but also as fuel for power generation. In both cases, it is the most relied upon resource. The transportation sector in the world is almost completely dependent on fossil fuels with a rate of over 90% (IEA, 2024). The dependence on fossil fuels in heating systems is still at a large rate of 75% (IRENA, 2024). Fossil fuels are still dominant in electricity generation with a share of over 60% (Statista, 2024). First of all, the mining process is needed to obtain fossil fuels. In this process, water is used for processes such as transportation, separation, dust suppression, purification and washing (Güney, 2022). In the refining process required for the conversion of the ore into the final fossil fuel, water is needed for washing, evaporation, cooling, purification and separation processes (Febregas, 2023). In facilities where fossil fuels are used as a source for electricity generation, there is a water-dependent process, 90% of which is for cooling purposes. In short, to obtain energy from fossil fuels, processes that use water as a basic input are needed two or three times.

In hydrogen production facilities, which is one of the primary energy sources, the amount of water used varies depending on the type of production. In these facilities, water is separated by various methods to obtain hydrogen, so it is a raw material. In addition, it is also used for cooling purposes. However, since every type of hydrogen production facility requires energy, it also uses water indirectly (IRENA and Bluerisk, 2023). Mining activity is required to provide the fuel required to produce nuclear power, another energy option. The need for water at this stage is stated above. Although the carbon footprint of nuclear power is quite low, it requires a large amount of water for production and cooling in nuclear power facilities where finite resources are used. The turbines where power is transferred during the production phase operate with steam and this mechanism is used for all reactors. In some reactors, water is also used as a neutron moderator. Water is also needed for nuclear waste management (Ma et al., 2023).

Renewable methods that use solar energy, wind and water movement to generate power require water. Water usage intensity varies from method to method. In hydroelectric plants, the power obtained from the movement of water turns the turbines. Therefore, water is the basic input here. While generating power from solar energy, water is not required for the direct production phase, but water is used to wash the PV panels. When generating electricity from wind energy, water is not directly required during the production process, but water is still an input during the maintenance and management process. In this respect, wind energy power generation is the least water-intensive method (Mörk, 2024). All power generation methods are dependent on water for the maintenance and management process. In addition, the equipment to be used in power generation facilities also has a certain water footprint. It should not be forgotten that technology-intensive equipment in particular has a higher water footprint.

Figure 1. Water energy nexus

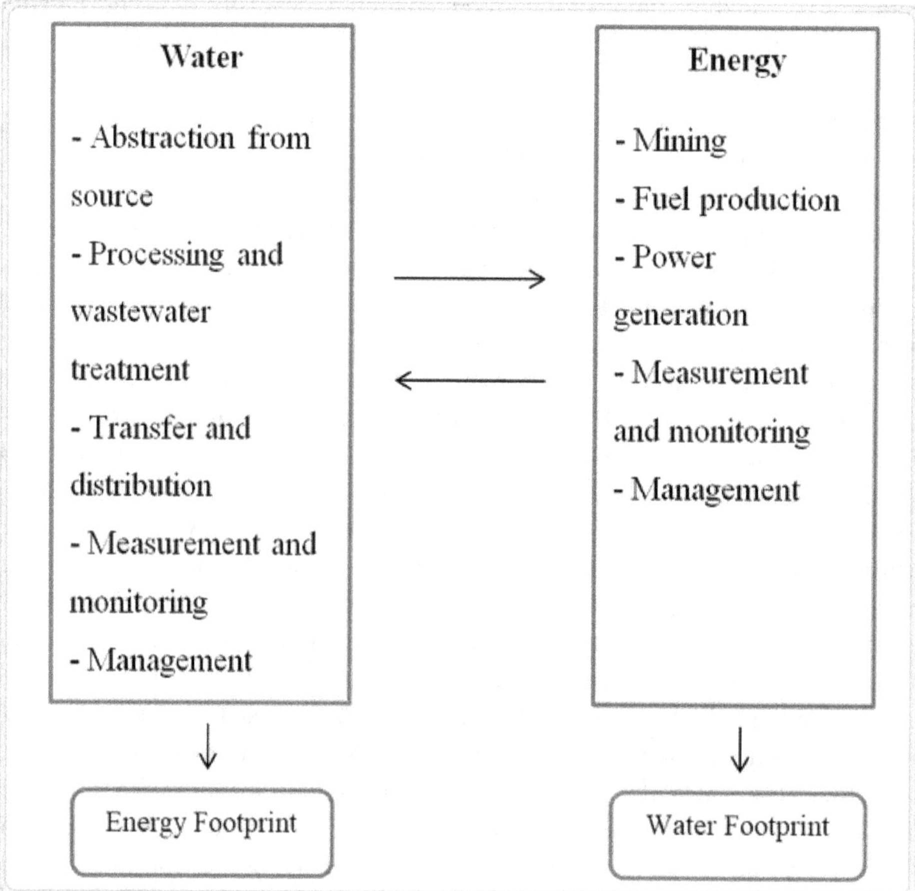

Source: *Author's own design*

Just as energy is dependent on water, effective water management is also dependent on energy. Since water is the basic input of life and economy, it is exposed to very intense negative externalities. Therefore, an integrated water management is needed for sustainable use of water. This management effectively coordinates all stakeholders and makes them parts of an integrated whole. The systems required to withdraw water from the source, load it into the distribution network and distribute it to users require energy. Energy is also required for the fair distribution and measurement of the withdrawn water. The process of collecting and purifying used water is the most energy-consuming process. This process is inevitable since discharging used water into natural resources without purification will harm the environment and human health. Discharging water classified as black and brown, especially into natural resources even after purification, puts pressure on their resources in terms of quality. Therefore, studies are ongoing to develop more advanced purification systems.

Preventing water resources from coming under quality and volume pressure and ensuring that water is shared in an accessible and fair manner for everyone are components of integrated water management. For this purpose, the methods used to observe and measure, classify and store the obtained data are also energy dependent. Even the equipment used throughout this entire process required for water management has an energy footprint. In addition, energy is used in activities to raise awareness of users and coordinate stakeholders. All of these make water management completely dependent on energy. At this point, it becomes clear that the relationship between energy and water is not one-sided, and that both are dependent on each other, and this situation creates a nexus (Figure 1). Similar studies that have been conducted before revealing this nexus are available in the literature (Helerea et al., 2023; Kheirinejad et al., 2022; Fayiah et al., 2020; Dai et al., 2018).

2. MUTUAL USE OF WATER AND ENERGY

Looking at the use of water and energy from a reciprocal relationship window is perhaps the most important indicator in revealing the water-energy nexus. When it comes to water use, water consumption and withdrawal are two basic concepts encountered. They are often used in the same direction but do not mean the same thing. Before starting to evaluate the data in a qualitative sense, it is useful to explain these two concepts in order to understand the difference between them. First of all, it is necessary to say that when water withdrawal is mentioned, all of the water taken from the sources is meant. If the withdrawn water needs processing, it is loaded into the network after processing or delivered directly to the end user. The obtained water is discharged to the source after being used according to the sectoral need. Before this discharge process, used water is collected and since this water undergoes quality deterioration at various rates, it is subjected to different levels of purification process. The purification process often cannot bring the water to its former quality. For this reason, wastewater discharge causes pressure on water resources in terms of quality. Used water may not have experienced any quality loss depending on the purpose of use. However, in every case, there is a volumetric loss of water from the withdrawal phase to the discharge phase. This loss may have occurred due to different reasons such as evaporation and being trapped in the plant structure. This volumetric loss is called consumed water. Both consumed and withdrawn water put resources under volumetric pressure. This is also called quantity pressure. This pressure is an undesirable situation because it disrupts the natural cycle of water. As the quantity and quality pressure on resources increases, the safe usage rate decreases. In other words, there is a contraction in the available part of water. The use that will be made without increasing this contraction is called sustainable water use.

Figure 2. Water usage volume in the energy sector (bcm)

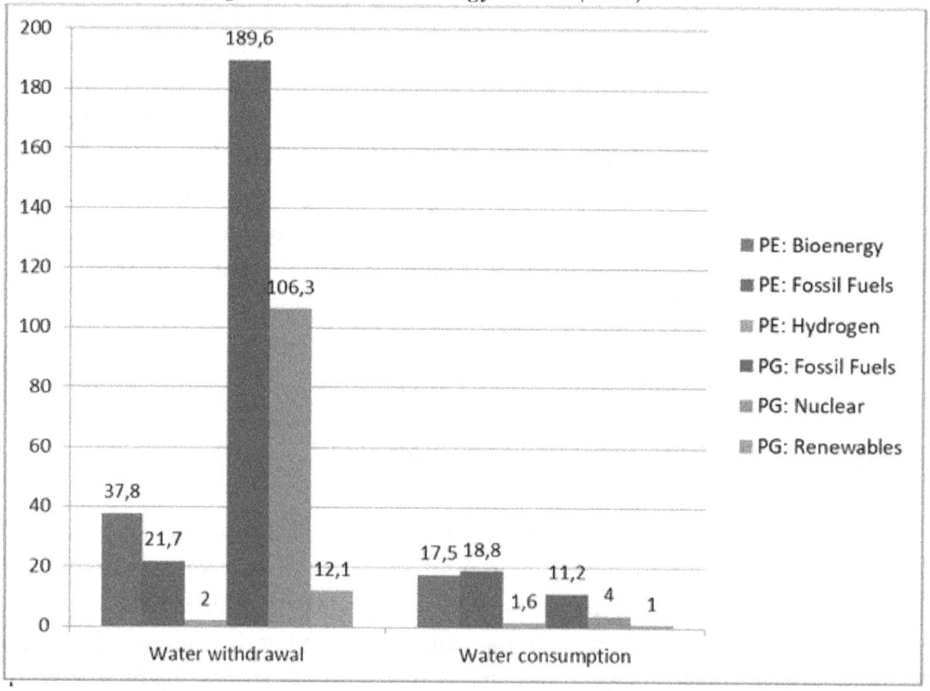

Source: *Bredariol, T. O., (March 22, 2023). Clean energy can help to ease the water crisis, 2021 and 2030.*

Today, the largest portion of the water used by the industrial sector in the world belongs to energy production (Hossain, 2019). The volumes of water withdrawn and consumed in 2021 to obtain energy are shown in Chart 1. According to the information here, the global energy sector has withdrawn a total of 369.5 bcm of water from resources annually. Approximately more than 14.5% of this volume, i.e. 54.1 bcm, is the amount consumed for energy. While the highest water withdrawal for primary energy (PE) is made for bioenergy sources, the use of fossil fuels for power generation (PG) is where the highest water withdrawal is made. In terms of water consumption, the highest consumption is again fossil fuels as the primary source. In power generation, the highest water consumption belongs to fossil fuels. The current picture shows that energy production from fossil fuels is the area that uses the highest amount of water in the energy sector. In addition, it is noticeable that bioenergy sources are not as frugal as thought in terms of water use. Although nuclear power generation consumes a small amount of water, it draws a very high amount of water from the source. This means that it creates a quantitative pressure on water resources. Since it is a relatively new technology, the share of the hydrogen production process in energy production is quite small. Keeping this in mind, we see that the water withdrawal it makes is low. However, the fact that it has consumed 80% of the water it draws raises questions about this new source. It is somewhat comforting that the water used in this method is drawn from seawater. Because what is really scarce in the world is not seawater, but available freshwater. In other energy sources, seawater is usually the next choice due to its various negative characteristics such as causing high wear.

At this point, how water usage in the energy sector causes quality pressure on water resources is a separate issue. For example, although nuclear power generation facilities consume a low amount of water, it is extremely important that the discharge they make is not contaminated. Discharges made without using the right methods will put water resources under quality pressure. Therefore, the presence of nuclear pollution in the discharged water can make water resources completely unusable. The production of fossil fuels as primary energy creates high quality pressure on water resources. However, since the water discharged after power generation from these sources is generally used for cooling purposes, if it does not contain high heat, it does not cause a high quality pressure. At this point, renewable energy is impressive in terms of water consumption. However, the ratio of the amount of energy obtained from renewable energy to the water it consumes is an issue that should be kept in mind. At this point, it makes sense to look at the water use efficiency of the energy type.

Graph 2 is taken from the article prepared by Cleveland and Louf (2024) based on the study of Jin et al. (2019). The graph shows the average liter of water consumed per TWh according to the source type in power generation. The water consumption efficiency of power generation can also be seen from here. Accordingly, the source with the lowest unit of water consumption per unit of energy produced is wind energy. Natural gas is more economical than others in this sense among fossil fuels. However, it is surprisingly seen that hydropower systems and biological fuels are actually quite inefficient in terms of water consumption. Another interesting fact is that infinite source systems such as solar energy are not always more efficient in terms of water consumption than finite sources such as natural gas. However, it should not be forgotten that the sun has a high potential to provide high efficiency by increasing the techniques used as a fertile energy source.

Figure 3. Water consumption in power generation

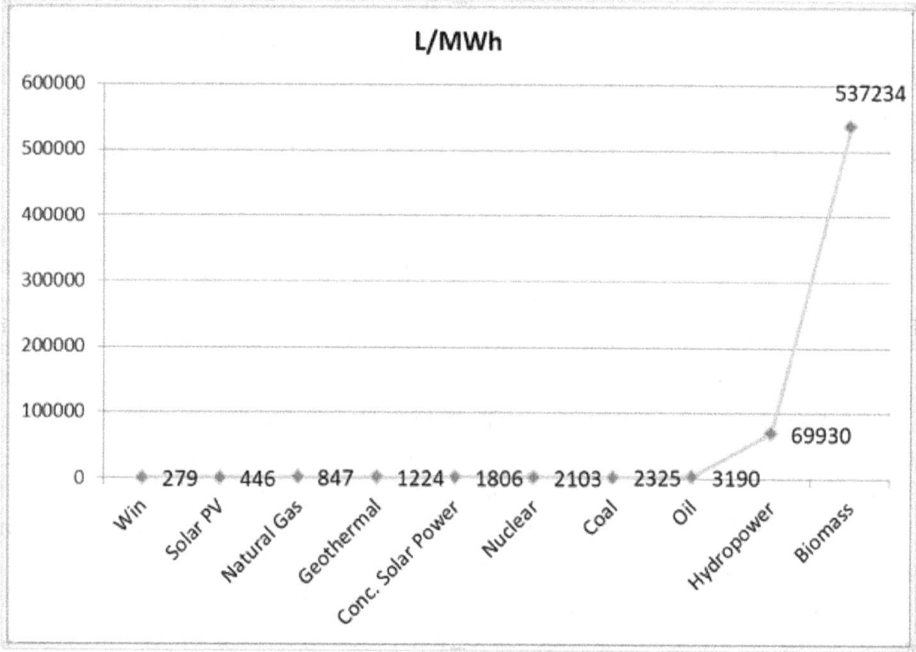

Source: *Cleveland, C. & Louf, N. (February 24, 2024). What methods of electricity generation use the most water?*

The world is still highly dependent on coal (35%) and other fossil fuels (25%) for power generation (Energy Institute, 2024). Although power generation methods using fossil fuels are below average in terms of water consumption efficiency, they have the highest carbon footprint. This has a very high negative impact on global climate change. Therefore, generating energy from a finite and high carbon emission source such as coal has an indirect but very large negative impact on global climate (Wieczorek-Kosmala, 2020) and water resources.

Up to this point, a compilation as up-to-date as possible has been made on the water required for primary energy and power generation by moving from energy to water in the water-energy nexus. At this point, the same process will be carried out from the other direction. In other words, the energy required for the water sector to continue its activities will be evaluated in the context of consumption. Graph 3 includes an overview of how the water sector consumes the energy it needs. The UN Water Report, published in 2024, divided the electricity consumed by the global water sector in 2021 according to processes. The total energy consumed by the sector is 978 TWh, which constitutes approximately 4% of the global electricity demand. The supply phase is the initial processing of water by drawing it from underground and surface resources (UN Water Report, 2024). The electricity required for this is the step in which the highest electricity is consumed in the sector. The transfer phase refers to the processes of transporting the drawn water to where it is needed and loading it onto the distribution network. This process, although not as large as distribution, requires a lot of energy. Water distribution is the third most energy-consuming process in the sector. It is not surprising that a lot of energy is needed to provide access to water for as many users as possible. The global energy consumption for this process is approximately 195 TWh.

Figure 4. Electricity consumption in the water sector

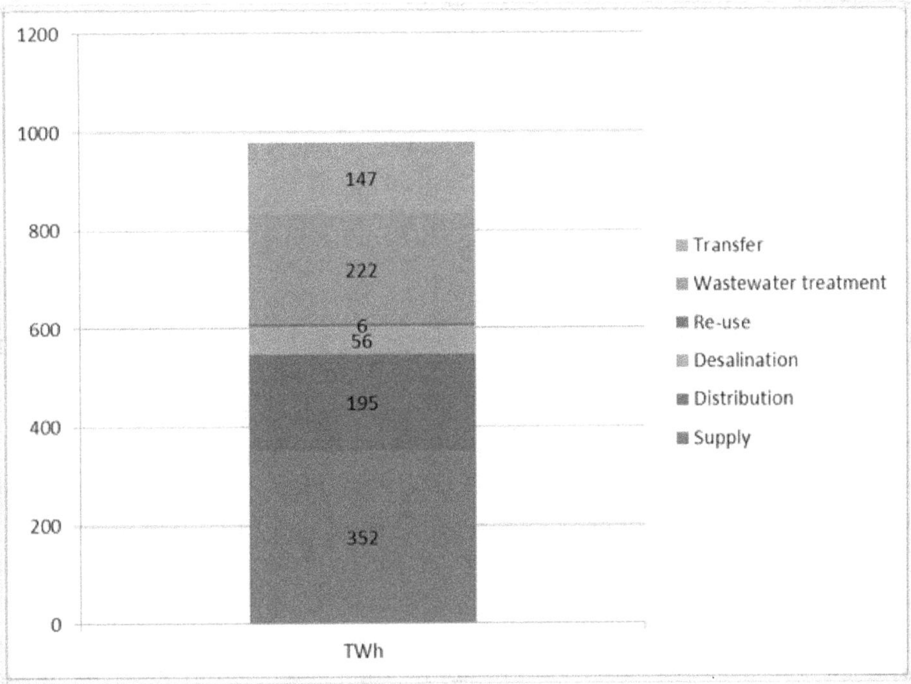

Source: *UN Water Report, 2024, p.68.*

The wastewater treatment stage is the most important stage required to protect water resources from quality pressure. Since water is a basic input in every sector, it contains many known and unknown externalities. For this reason, it is necessary to discharge wastewater as close to its natural state as possible. The treatment processes required to do this are usually energy intensive. This makes the wastewater treatment stage the second most energy-consuming stage in the sector.

Desalination of salt water and making it usable seems like a good alternative source for the limited fresh water resources in the world. This situation makes it a stage that is still being intensively worked on. However, this method has serious problems that have not yet been overcome. One of these is that current desalination methods consume a lot of energy. This makes it a very expensive and energy-dependent method. The other problem is the disposal of the solution containing high salt. High-salt waste product that is not managed properly causes serious environmental damage. For this reason, the desalination method is generally preferred in regions with low water needs and rich financial and energy resources due to the existing concerns. Gulf countries generally use this method, but they cause a high carbon footprint due to their dependence on fossil fuels as an energy source. This causes an indirect increase in pressure on water resources through global warming. When we look at the energy distribution used in the water sector, we see that the desalination method with high environmental costs has found its place.

3. FUTURE PROJECTIONS FOR WATER AND ENERGY DEMAND AND CIRCULAR SUSTAINABILITY

There are various predictions for the future within the water and energy nexus. For this, different predictions have been made that vary according to various scenarios such as zero emissions (Bredariol, 2023). These scenarios have been produced by taking into account resource efficiency and environmental degradation. However, in order for them to be valid, current production processes must have reached the desired level. Since this is not yet possible, it seems more accurate to analyze the current situation scenarios. Therefore, in this study, predictions made in the light of the current situation, just like UN Water did (2024), were used. The third section will start with the water use of the energy sector, as in the previous section. As a first step, it is necessary to look at the prediction of the amount of water used in the energy sector in Graph 4. Here, increases are seen in the water use of the energy sector for 2030 in terms of both water withdrawal and water consumption. The water withdrawal to be made on a global scale for the entire sector is 393.2 bcm. In other words, it is approximately 23.7 bcm above the last measurement made for the current situation. It is predicted that 15.4% of this volume, 61.2 bcm, will be the consumed part. This consumed part is 7.1 bcm above the last measurement.

Figure 5. Water usage volume projection in the energy sector (2030)

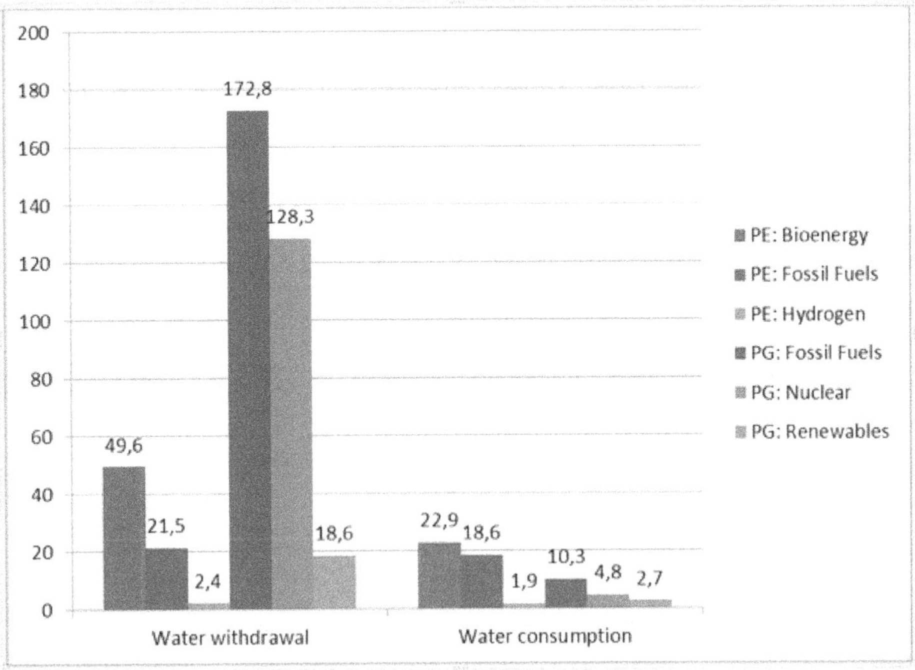

Source: *UN Water Report, 2024, p.64.*

In the near future, there is an expectation of an increase in water withdrawals from primary sources for bioenergy and hydrogen in the energy sector. It is certain that this general increase will create volume pressure on freshwater resources. In addition, it is highly likely that it will increase quality pressure on all water systems. The increase in water withdrawals for biomass in particular is significant. There is an expectation of a slight decrease in water withdrawals for fossil fuels as a primary source. Although there is a slight decrease in water withdrawals for power generation for fossil fuels, it is anticipated that a higher volume of water will be withdrawn in the future for the power generation process using nuclear fuels. There will also be an increase in power generation from renewable energy sources. In terms of water consumption, an increase in consumption is expected in almost all energy sectors except fossil fuels. While almost no change is expected in the water consumption of fossil fuels in primary energy generation, a decrease of approximately 1 bcm is expected in power generation.

When we look at the future change pattern of water consumption in energy, it is seen that there will be a large proportional increase in renewable energy. There is an expectation of a decrease in water withdrawal and consumption in energy production from fossil fuels. This may be due to the preference moving away from these areas or the development of technical capacity in the production process. There is an expectation of an increase in the use of biomass (Graph 3), which is in the worst situation in terms of water/energy efficiency. Since there will be no regression in the current technology, this increase can only be due to the increased use of biomass for energy.

The predictions about the future situation of the water-energy nexus in terms of energy consumption, just like the water usage, are on an increasing trend. Graph 5 shows that the total energy consumed in the sector will be 1252 TWh in the prediction made for 2030. It is particularly important to note that the energy consumed for water supply in 2030 is almost constant. In addition, a very high rate of increase is observed in the energy consumed for the desalination process in the near future. Considering that most of the work on the desalination process is aimed at reducing energy consumption or using renewable resources (Mahmoudi et al., 2023), the widespread use of the desalination process will be behind this striking increase.

Figure 6. Electricity consumption in the water sector (2030)

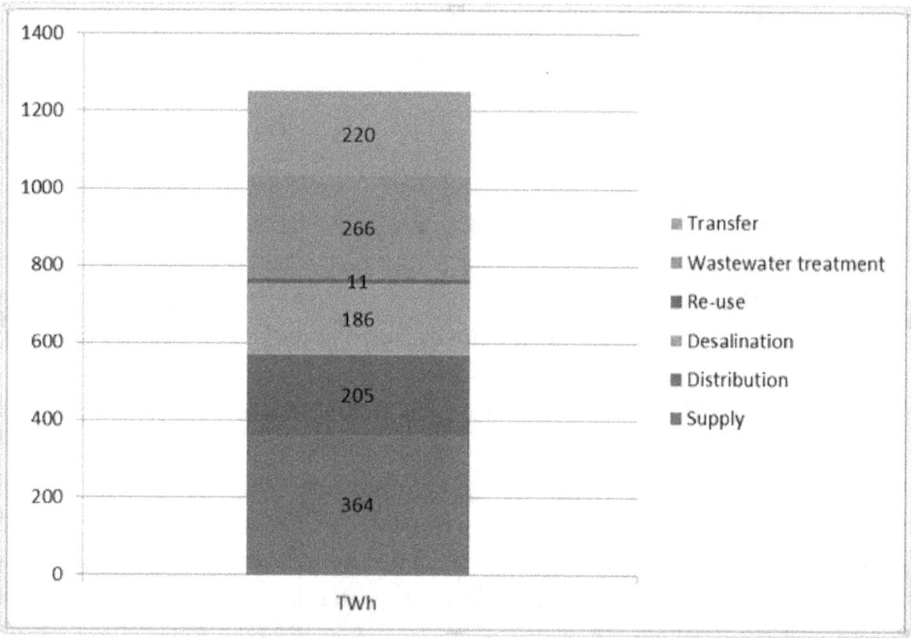

Source: *IEA (March 3, 2020, Last updated 20 Mar 2023). Introduction to the water-energy nexus.*

The second largest increase expectation for the energy to be used in the water sector is for the purpose of transfer. Based on this information, it can be said that the problems of access to water resources will become apparent in the future. In addition, the increase in the energy consumed for the distribution of water in the near future will also be considerable. It would definitely not be wrong to say that the energy to be consumed for distribution will increase significantly in the medium and distant future. Because the distribution of water is not only a technical but also a social issue. If we keep in mind that access to water is a fundamental right, it is clear that access to water for everyone will be one of the factors necessary to ensure social peace at national and international levels. For example, the UN Water Report was published in 2024 under the title "Water for prosperity and peace". Therefore, all stakeholders have a responsibility regarding the fair distribution of water. At this point, although the distribution process in the water sector is the third process that consumes the highest energy, billions of people in the world still have difficulties in accessing safe drinking water (Soliman, 2024). Therefore, it would be prudent to expect that the energy demand for equitable distribution will tend to increase.

There is also an expectation of an increase in the energy to be used for wastewater treatment. At the current point, it is not possible to completely return used water to its natural state. This means that the treatment process will be carried out more intensively in the future. It can be said that energy demand will also tend to increase due to the possibility of providing water to masses that do not have access to water in this way. Even if the amount of energy to be spent for the reuse of water is high as a percentage, it is noteworthy that it will not constitute a high amount in the future as it is now. In fact, reuse is an important step in the transition to a circular water economy. All stakeholders must show will to expand R&D and initiatives in this area.

The possible change expectation in the mutual relationship of the water-energy nexus is visualized in Figure 2. According to the available data, it is expected that there will be a decrease of 0.063 and 0.01 bcm in water drawn and consumed for each TWh of power used in the water sector, respectively. In other words, by 2030, there will be a decrease in the volume pressure of the water-energy nexus on the water side. Although the decrease seems low, the situation of remaining stable or decreasing in the face of increasing energy consumption (Cozzi et al., 2024) is positive. Since the possibility of new production methods causing higher damage to water resources in terms of quality compared to the methods used today is low, the nexus appears to be in a sustainable cycle on the water side.

Figure 7. Possible changes in the water-energy nexus in a circular context

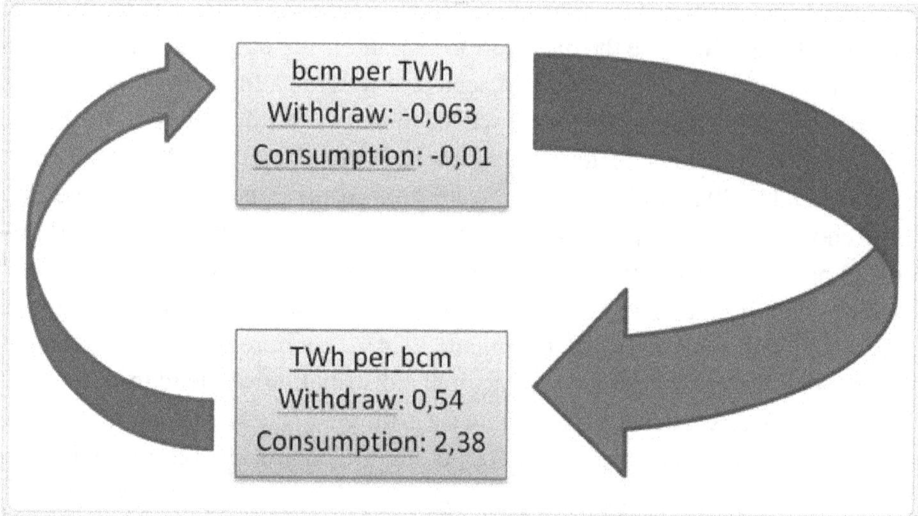

Source: *Prepared using data from Chapters 2 and 3.*

Currently, the situation on the energy side of the nexus is not as encouraging as it is on the water side. In the water energy nexus, there will be an increase of 0.54 TWh in the energy consumed for each bcm of water withdrawal. The situation is worse in the energy consumed for each bcm of water consumed, because there is a forecast of an increase of 2.38 TWh. If we also take into account the forecast that the entire increase in energy consumed for both consumed and withdrawn water cannot be obtained from renewable resources, there is no sustainable situation in the use of the energy side of the nexus. In other words, the circular process of the nexus is under pressure in the energy side.

The future estimates made for the energy use of the water sector and the water use of the energy sector have been made into the current scenarios we use in the study and the more optimistic scenarios not used here. However, when examining the water-energy nexus, numerical values should not be allowed to create a misconception. When the series of concerns listed above are taken into account after the estimates regarding the interdependence of the nexus, a different picture emerges. The concerns, especially regarding water, mentioned indicate that inferences should be made that the demand for water will or should increase. As Soliman (2024) states, if 4.4 billion people, more than half of the world's 8.2 billion population, cannot access safe drinking water (2024), this indicates that the demand for water will increase significantly in the future. At this point, the possibility of the water-biased cycle of the nexus, which is sustainable according to current estimates, remaining in this state is significantly reduced. Moreover, when it comes to drinking water, an intensive treatment process will be added to the supply, transfer and distribution processes. This means that the nexus will further strain the already unsustainable energy cycle.

4. POLITICAL ECONOMY OF THE WATER-ENERGY NEXUS: A BRIEF ASSESSMENT

Looking at the water and energy duo from an economic perspective can be one of the tools that can be used to get rid of the limited perspective by reducing the water-energy nexus to numerical values. In order to do this, a brief assessment will be made based on the available data and estimates about the present and future of the nexus from an economic perspective. The distribution of available freshwater resources in the world is independent of political borders. Generally speaking, South Asian and MENA countries are in a problematic situation in terms of water resources. While most of them draw more than their available water resources, the situation is generally good in Europe and America in this regard (UN Water, 2024). In terms of safe access to drinking water, the situation is quite good in North America, Europe, South Korea and Japan, while the situation is quite bad in Central Africa. While it is better in MENA countries, access to safe drinking water is problematic in South Asian countries (EPI, 2024).

In terms of energy use, the global distribution is generally concentrated in regions where water resources are relatively scarce. The intensity of energy use per capita appears high in industrialized areas where water resources are not scarce (EEA, 2024). Total energy use can also be high in areas where water resources are scarce due to high population pressure and proximity to regions where primary energy sources are obtained (Ritchie et al., 2020b). For example, in South Asia, energy use is intense due to population pressure. However, access to water is still low in this region. In the Middle East, high oil and gas reserves increase energy use due to price advantage.

When it comes to economic significance, both sides of the nexus are open to economic association. If we look at this significance from the water side of the nexus, there are studies in the literature that water affects economic indicators in terms of both quality and volume (Hernandez et al., 2020; Kong et al., 2021; Ferasso et al., 2021; Russ, et al., 2022; Guo et al., 2022). Likewise, there are many studies in the literature, including causality testing, that there is a relationship between energy and economic variables (Fotis et al., 2017; Mahmood & Ahmad, 2018; Carfora et al., 2019; Li & Leung, 2021; Xiong & Xu, 2021; Ayhan et al., 2023; Kumari et al., 2024; Ben Saad et al., 2024). In fact, there is no need to give separate scientific examples to show that there is a relationship between water, energy and economic indicators. The water and energy connection shows that one-sided evidence has the quality of two-sided evidence.

Although total energy use in today's world varies depending on population factors, there is a significant geographical overlap in terms of GDP per capita and energy use per capita. These parts of the world generally fall into the category of economically developed countries (IMF, 2024) and their water withdrawals from their resources are generally at a level that will not cause water scarcity. With a few exceptions, access to safe drinking water is also parallel to the distribution of GDP per capita. At this point, even if sub-Saharan Africa's water resources are not depressed, it has quite a problem in terms of access to safe water and energy use. The same region is also far behind in economic terms. This is not surprising when you consider that access to water is largely dependent on energy. In short, economic wealth, energy use and freshwater resources are distributed parallel to each other on a per capita scale in the world.

There seems to be a relationship between the water-energy nexus and economic variables on a global scale. Especially when viewed from the GDP per capita window, although there are some exceptions, developed countries present a much more balanced structure in terms of the water-energy nexus. However, the future predictions made on the water-energy nexus in the third section indicate that there will be an increase in production/consumption on both sides of the nexus. Although the water side of the nexus seems sustainable, an increase is seen as inevitable for 2030. So what will the increase/decrease situation be in terms of economic output on a global scale? At this point, the expectation of an increase in total GDP in the predictions made until 2030 has not changed over the years (ESPAS, 2019; IMF, 2024). There is also an increase prediction in the context of GDP per capita, but in this context, enrichment will be seen in developed economies (Luxton, 2016; Boretti&Rosa, 2019; IMF, 2024). According to Luxton, a writer for the World Economic Forum, the gap between developed countries and others in GDP per capita will remain constant, while according to the IMF, a more up-to-date source, it will grow. Based on this data, it can be said that the world's economic enrichment will be concentrated towards developed countries and this concentration will continue to increase.

Global energy demand is expected to increase in general terms by 2030. The majority of this increase in developed economies will be in the form of an increase in power generation from renewable sources. On the other hand, the increase in developing economies is expected to be in the use of fossil fuels for primary energy sources and power generation. Although the energy demand of developed economies will continue to be met from more sustainable sources, the demand for finite energy sources such as fossil fuels will not disappear completely (Cozzi, et al., 2023). This means that in developing economies, where access to water is already problematic and resources are under pressure, the pressure on the water side of the nexus will increase. The mutual increase in the energy and water nexus on a global scale is due to the developing technology and diversified human needs in developed economies. In developing countries, the pressure of a regularly increasing population is also added to this situation (INED, 2024). The accumulation in the increase in individual economic wealth will also be towards developed economies and there will be no individual economic enrichment in countries experiencing problems in terms of water. Energy consumption per capita will increase in countries where individual wealth is higher. In other words, the rich will consume energy, but the aquatic pressure required to produce this energy will be shared by the entire globe. This sharing will increase water pressure to produce primary energy in geographies rich in energy resources. Since energy production also produces the most carbon emissions globally (Ritchie et al., 2020a; Blok et al., 2020), geographies that are fragile in terms of water will be hit a second time through global warming. The processes required to obtain alternative water sources, such as desalination, also mean high energy consumption and high costs. Therefore, considering the tendency of low-income countries not to become rich, it seems that they do not have the chance to implement such alternative technologies. When all these are considered, it becomes clear that access to safe water and putting the water-energy nexus into a circular structure does not yet seem possible globally.

5. CONCLUSION

There is empirical evidence that there is a connection between water and energy. Many academic studies have been published to support this. The study has been able to cite only a small part of the literature in this regard. It is seen that the use of water in the energy sector, both primarily and for power generation, is inevitable. The same dependency is also revealed in the opposite direction. Today, it is seen that the majority of the energy produced in the world is obtained from fossil fuels. As a result of this, producing energy from fossil fuel sources today uses the highest volume of water compared to production from other sources. However, it is seen that it is less efficient than "renewable" sources such as biomass in terms of direct water use efficiency in power generation. In this context, sources such as solar PV and wind are renewable sources with much higher efficiency. In the water sector, it is seen that the areas with the highest energy needs are supply and wastewater treatment areas.

The predictions made are that energy consumption in the water sector will increase. Although a major change is not foreseen in the distribution of water consumed according to the energy source, there is an expectation of an increase in favor of renewable resources. However, there is no expectation of a decrease in the energy consumed. It has been observed that the energy required for supply and distribution has not increased sufficiently, but there will be an increase in energy consumption in stages such as transfer, wastewater treatment, and desalination. There is an expectation of increased efficiency in water use in the energy sector. It would not be wrong to say that the increase in the use of high water efficiency methods such as solar and wind resources will be effective in this. In the current situation, while there is an expectation of increased efficiency on the water side of the water-energy nexus, there is an expectation of a decrease on the energy side. Under the assumption that there must be a mutual balance for the nexus to have a circular structure, it can be seen that the cycle is broken on the energy side of the nexus.

However, using only water and energy data is insufficient to make a circular economy interpretation. If the economic aspect of the connection is examined, the situation is not so clear. Considering the expectations in the global economy, it is seen that the accumulation in economic enrichment will continue to increase towards developed economies. In other words, the gap between developed economies and others will increase. In places where water resources cannot be used sustainably or where access to safe water is low, the population will increase, and accordingly there will be no individual enrichment or a very limited increase. The increase in energy use per capita will generally occur in developed economies. At this point, it is useful to remember that almost 2/3 of global emissions belong to the energy sector. It is clear that the entire globe will bear the cost of consumption due to the integrated structure of water management and the global economy. This means that the rich will consume more energy, and the vulnerable will bear the burden of the total aquatic pressure created in the globe.

The predictions made for the change of the link also seem to support this result. The increase in the supply side of energy consumption in the water sector is low. On the other hand, the growth in treatment and transfer is remarkable. This situation can be interpreted as the supply will not increase in regions where access to water is limited, but to provide more and better quality water to developed economies. In the context of the water-energy link, there are things that water and energy stakeholders in developed and developing economies need to do to transition to a circular economy. Here, the increasing population pressure of developing economies is becoming overwhelming for the entire globe. Innovations are far from being sufficient for sustainable resource use. Under these conditions, it is essential to keep the increasing population pressure under control in these regions. On the other hand, even if developed economies have turned to relatively sustainable resource management, their consumption and production processes are pressuring the resources of the entire globe. Therefore, in order to establish global peace and sustainable resource use, there is a need for educational measures for all stakeholders, especially from developed economies to developing ones, sustainable water and energy use tools, ensuring fair sharing and allocating more funds for regional development. In conclusion, although it does not seem possible to create a sustainable water-energy nexus within a circular economy in the near future, it is clear that taking such measures will save time in the transition to a circular economy.

5.1. Practical Implications and Theoretical Contribution

The practical implications and theoretical contributions of this section are mentioned in the conclusion. Their itemized list is as follows.

Practical Implications:

- Energy sources considered as renewable such as biomass and hydropower in power generation are quite inefficient in terms of water consumption.
- The water-energy nexus has not yet settled into a circular context.
- In the projections made for 2030, the water-energy nexus will not exhibit a fully circular structure.
- When the change in the water-energy nexus is evaluated per unit for 2030, the water-biased pressure tends to decrease but the energy-biased pressure tends to increase.
- In the future, economic prosperity will continue to accumulate in developed economies, while the population will increase in developing economies.

Theoretical contribution: Bringing the water-energy nexus to a circular structure is vital for achieving the SDGs. However, the increasing population in developing economies and the unbalanced economic pattern that grows in favor of developed economies are problematic for creating a circular economy in the context of the water-energy nexus. At this point, the demand growth caused by the population in developing economies should be balanced. In addition, developed economies have responsibilities that they must fulfill both economically and morally to create a circular economy on a global scale.

5.2. Current Study Limitations

The biggest factor limiting the studies in the current situation is the limited data sets and their reliability problems. In developing economies, measuring and recording data is problematic for many reasons. These include economic inadequacies, lack of technical capacity, administrative corruption, social corruption and reluctance to share data. In addition, there are very few institutions and organizations operating on a global scale to process the existing data. Since there is no global authority to impose sanctions to achieve the SDGs, the measures taken and academic studies remain advisory. In addition, it is extremely difficult to predict the uncertainties created by new technologies such as artificial intelligence (AI). For example, there are serious concerns about the water and energy consumption of AI models. The increasing scope of activity and number of users of AI means that water and energy demands are constantly increasing. In addition, as the LLMs used for more advanced AI grow, water-energy consumption also grows (Ren, S. et al., 2023). All of this means more difficulties in designing and creating a circular water-energy nexus. In addition, it is not possible to take into account all aspects of the externalities that will be created by multifaceted inputs such as water and energy.

5.3. Future Research Work

It can be said that AI will take the most important place in this regard for future research studies. Many studies need to be done to provide AI models with a structure that consumes less energy and therefore less water. The reason for this is the increasing area of use, the increasing number of users and larger LLMs. This situation is important enough to affect the global financial market. The most recent example of this is the financial leap made by DeepSeek recently. One of the reasons for this leap is the emergence of the speculation/fact that the DeepSeek model consumes less energy (Küçük G., 2025). Future studies on the resource efficiency of AI will contribute to the creation of a circular water-energy nexus. Another contribution that AI will make to the circularity of the nexus comes from the revolutionary impact it will create in the management process. The impact of AI use in processes such as production, measurement, monitoring and R&D for water and energy management should be studied.

One of the fields of study that will contribute to the circularity of the nexus is human development in the future, as it was in the past. Studies on the theme of human development are important in the transition to SDGs and therefore to a circular economy by contributing to the main consumer, the human being, becoming a more responsible consumer. Another classical field of study that will maintain its importance in the future is the balanced economic distribution problem. One of the practical implications of this section is the expectation that the economic distribution gap between developed and developing economies will indirectly increase the unbalanced distribution in individual welfare. Therefore, an unbalanced economic development structure poses a serious problem for the SDGs and the transition to a circular economy.

REFERENCES

Ayhan, F., Kartal, M. T., Kılıç Depren, S., & Depren, Ö. (2023). Asymmetric effect of economic policy uncertainty, political stability, energy consumption, and economic growth on CO2 emissions: Evidence from G-7 countries. *Environmental Science and Pollution Research International*, *30*(16), 47422–47437. DOI: 10.1007/s11356-023-25665-7 PMID: 36737567

Ben Saad, M., Ben Saad, R., Kertous, M., & Samir, B. M. (2024). Asymmetric impacts of renewable energy and economic development in North Africa countries: Evidence from the NARDL model. *Environmental Economics and Policy Studies*. Advance online publication. DOI: 10.1007/s10018-024-00425-0

Bioenergy (2024). Products. Retrieved October 24, 2024, from https://bioenergy.com.tr/products/

Blok, K., Afanador, A., van der Hoorn, I., Berg, T., Edelenbosch, O. Y., & van Vuuren, D. P. (2020). Assessment of Sectoral Greenhouse Gas Emission Reduction Potentials for 2030. *Energies*, *13*(4), 943. DOI: 10.3390/en13040943

Boretti, A., & Rosa, L. (2019). Reassessing the projections of the World Water Development Report. *NPJ Clean Water*, *2*(1), 15. DOI: 10.1038/s41545-019-0039-9

Bredariol, T. O. (March 22, 2023). *Clean energy can help to ease the water crisis, 2021 and 2030*. Retrieved October 24, 2024, from https://www.iea.org/commentaries/clean-energy-can-help-to-ease-the-water-crisis

Carfora, A., Pansini, R. V., & Scandurra, G. (2019). The causal relationship between energy consumption, energy prices and economic growth in Asian developing countries: A replication. *Energy Strategy Reviews*, *23*, 81–85. DOI: 10.1016/j.esr.2018.12.004

Cleveland, C., & Louf, N. (February 24, 2024). *What methods of electricity generation use the most water?* Retrieved October 25, 2024, from https://visualizingenergy.org/what-methods-of-electricity-generation-use-the-most-water/?

Cozzi, L.. (2023). *World Energy Outlook 2023*. IEA.

Cozzi, L.. (2024). *World Energy Outlook 2024*. IEA.

Dai, J., Wu, S., Han, G., Weinberg, J., Xie, X., Wu, X., Song, X., Jia, B., Xue, W., & Yang, Q. (2018). Water-energy nexus: A review of methods and tools for macro-assessment. *Applied Energy*, *210*, 393–408. DOI: 10.1016/j.apenergy.2017.08.243

EEA. (Published Feburary 28, 2015; Modified September 11, 2024) *Correlation of energy consumption and GDP per person*. Retrieved November 17, 2024, from https://www.eea.europa.eu/en/analysis/maps-and-charts/correlation-of-per-capita-energy?

Emma Luxton. (Published Feburary 25, 2016) *What will global GDP look like in 2030?* Retrieved November 20, 2024, from https://www.weforum.org/stories/2016/02/what-will-global-gdp-look-like-in-2030/

Energy Institute. (2024). Statistical Review of World Energy 2024 73rd Edition. Retrieved November 3, 2024, from https://www.energyinst.org/statistical-review

EPI. (2024). *Unsafe drinking water*. Retrieved November 15, 2024, from https://epi.yale.edu/measure/2024/UWD

ESPAS. (2019). *Welcome to 2030: The Mega-Trends*. Retrieved November 18, 2024, from https://ec.europa.eu/assets/epsc/pages/espas/chapter1.html?

FAO. (2024). *Water Use*, Retrieved October 24, 2024, from https://www.fao.org/aquastat/en/overview/methodology/water-use

Faruque Hossain. (2019). *Sustainable Design and Build: Building, Energy, Roads, Bridges, Water and Sewer Systems*. Butterworth-Heinemann.

Fayiah, M., Dong, S., Singh, S., & Kwaku, E. A. (2020). A review of water–energy nexus trend, methods, challenges and future prospects. *International Journal of Energy and Water Resources*, *4*(1), 91–107. DOI: 10.1007/s42108-020-00057-6

Febregas, J. (May 30, 2023). *DAF systems for oil refinery wastewater treatment*. Retrieved October 25, 2024, from https://sigmadafclarifiers.com/en/sistemas-daf-para-el-tratamiento-de-aguas-residuales-de-refinerias-de-petroleo/

Ferasso, M., Bares, L., Ogachi, D., & Blanco, M. (2021). Economic and Sustainability Inequalities and Water Consumption of European Union Countries. *Water (Basel)*, *13*(19), 2696. DOI: 10.3390/w13192696

Fotis, P., Karkalakos, S., & Asteriou, D. (2017). The relationship between energy demand and real GDP growth rate: The role of price asymmetries and spatial externalities within 34 countries across the globe. *Energy Economics*, *66*, 69–84. DOI: 10.1016/j.eneco.2017.05.027

Güney, E. (2022). Water Footprint Assessment of Mining and Processing of Gold, Unpublished Master Thesis, The Graduate School of Natural and Applied Scinces, Middle East Technical University, Ankara.

Guo, L. L. X., & Wang, L. (2022). Economic size and water use efficiency: An empirical analysis of trends across China. *Water Policy*, *24*(1), 117–131. DOI: 10.2166/wp.2021.189

Helerea, E., Calin, M. D., & Musuroi, C. (2023). Water Energy Nexus and Energy Transition—A Review. *Energies*, *16*(4), 1879. DOI: 10.3390/en16041879

Hernandez, Y., Naumann, G., Corral, S., & Barbosa, P. (2020). Water Footprint Expands with Gross Domestic Product. *Sustainability (Basel)*, *12*(20), 8741. DOI: 10.3390/su12208741

IEA. (March 3, 2020, Last updated 20 Mar 2023). *Introduction to the water-energy nexus*. Retrieved November 8, 2024, from https://www.iea.org/articles/introduction-to-the-water-energy-nexus?

IEA. (2024), *Transport*, Retrieved October 25, 2024, from https://www.iea.org/energy-system/transport

IMF. (2024), *GDP per capita, current prices*. Retrieved November 18, 2024, from https://www.imf.org/external/datamapper/NGDPDPC@WEO/OEMDC/ADVEC/WEOWORLD

INED. (July, 2024). *Projections de population (en millions) par continent de 2025 à 2100. Scénario central*. Retrieved November 19, 2024, from https://www.ined.fr/fr/tout-savoir-population/chiffres/projections-mondiales/projections-par-continent/

INED. (July, 2024). *2024: les Nations unies publient de nouvelles projections de population mondiale*. Retrieved November 13, 2024, from https://www.ined.fr/fr/tout-savoir-population/memos-demo/focus/2024-les-nations-unies-publient-de-nouvelles-projections-de-population-mondiale/

IRENA. (2023), *Status and pace of progress*. Retrieved October 25, 2024, from https://www.irena.org/Innovation-landscape-for-smart-electrification/Power-to-heat-and-cooling/Status

IRENA and Bluerisk. (2023), Water for hydrogen production, International Renewable Energy Agency, Bluerisk, Abu Dhabi, United Arab Emirates. https://www.irena.org/-/media/Files/IRENA/Agency/Publication/2023/Dec/IRENA_Bluerisk_Water_for_hydrogen_production_2023.pdf

Jin, Y., Behrens, P., Tukker, A., & Scherer, L. (2019). Water use of electricity technologies: A global meta-analysis. *Renewable & Sustainable Energy Reviews*, *115*, 109391. DOI: 10.1016/j.rser.2019.109391

Kheirinejad, S., Bozorg-Haddad, O., Singh, V. P., & Loáiciga, H. A. (2022). The effect of reducing per capita water and energy uses on renewable water resources in the water, food and energy nexus. *Scientific Reports*, *12*(1), 7582. DOI: 10.1038/s41598-022-11595-w PMID: 35534602

Kong, Y., He, W., Yuan, J., Zhang, Z., Gao, X., Zhao, Y., & Degefu, D. M. (2021). Decoupling economic growth from water consumption in the Yangtze River Economic Belt, China. *Ecological Indicators*, *123*, 107344. DOI: 10.1016/j.ecolind.2021.107344

Küçük, G. (February 06, 2025), *DeepSeek's 'efficient' AI model sparks energy market shift*, Retrieved February 08, 2025, from https://www.aa.com.tr/en/energy/general/deepseeks-efficient-ai-model-sparks-energy-market-shift/47337

Kumari, D., Shashwat, S., Verma, P.K. & Giri, A.K. (2024). Examining the nexus between carbon dioxide emissions, economic growth, fossil fuel energy use, urbanization and renewable energy towards achieving environmental sustainability in India, International *Journal of Energy Sector Management*, Vol. ahead-of-print No. ahead-of-print. https://doi.org/DOI: 10.1108/IJESM-05-2024-0007

Li, R., & Leung, G. C. K. (2021). The relationship between energy prices, economic growth and renewable energy consumption: Evidence from Europe. *Energy Reports*, *7*, 1712–1719. DOI: 10.1016/j.egyr.2021.03.030

Ma, H., Shen, M., Tong, Y., & Wang, X. (2023). Radioactive Wastewater Treatment Technologies: A Review. *Molecules (Basel, Switzerland)*, *28*(4), 1935. DOI: 10.3390/molecules28041935 PMID: 36838922

Mahmood, T., & Ahmad, E. (2018). The relationship of energy intensity with economic growth:Evidence for European economies. *Energy Strategy Reviews*, *20*, 90–98. DOI: 10.1016/j.esr.2018.02.002

Mahmoudi, A., Bostani, M., Rashidi, S., & Valipour, M. S. (2023). Challenges and opportunities of desalination with renewable energy resources in Middle East countries. *Renewable & Sustainable Energy Reviews*, *184*, 113543. DOI: 10.1016/j.rser.2023.113543

Mörk Water. (May 13, 2024). *Water use by the Australian renewable energy sector*. Retrieved October 26, 2024, from https://moerkwater.com.au/updates/water-use-in-renewable-energy-power-generation/

Pelkmans, L. (2020). Global Bioenergy, IEA Bioenergy. Retrieved October 24, 2024, from https://www.ieabioenergy.com/bioenergy-a-sustainable-solution/

Ren, S., Li, P., Yang, J., & Islam, M. A. (2023) Making AI Less "Thirsty": Uncovering and Addressing the Secret Water Footprint of AI Models, University of California, Riverside, https://doi.org//arXiv.2304.03271DOI: 10.48550

Ritchie, H., Rosado, P., & Roser, M. (2020a). *Breakdown of carbon dioxide, methane and nitrous oxide emissions by sector*. Retrieved November 20 2024, from https://ourworldindata.org/emissions-by-sector

Ritchie, H., Rosado, P., & Roser, M. (2020b). *Energy Production and Consumption*. Retrieved November 17, 2024 from, https://ourworldindata.org/energy-production -consumption

Russ, J., Zaveri, E., Desbureaux, S., Damania, R., & Rodella, A. S. (2022). The impact of water quality on GDP growth: Evidence from around the world. *Water Security, 17*, 100130. DOI: 10.1016/j.wasec.2022.100130

Soliman, A. (2024). *'Unacceptable': a staggering 4.4 billion people lack safe drinking water, study finds*. Retrieved November 5, 2024, from https://www.nature.com/articles/d41586-024-02621-0 doi: https://doi.org/DOI: 10.1038/d41586-024-02621-0

Statista (2024), *Distribution of electricity generation worldwide in 2023, by energy source*. Retrieved October 25 2024, from https://www.statista.com/statistics/269811/world-electricity-production-by-energy-source/

UN General Assembly Economic and Social Council. Progress towards the Sustainable Development Goals, Retrieved February 07, 2025, from https://unstats.un.org/sdgs/files/report/2024/SG-SDG-Progress-Report-2024-advanced-unedited-version.pdf

Water, U. N. (2021). *Progress on Level of Water Stress*. Retrieved November 15, 2024, from https://www.sdg6data.org/en/indicator/6.4.2

Water, U. N. (2024). Water Report, Water for prosperity and peace 2024, Paris.

Wieczorek-Kosmala, M. (2020). Weather Risk Management in Energy Sector: The Polish Case. *Energies, 13*(4), 945. DOI: 10.3390/en13040945

Xiong, J., & Xu, D. (2021). Relationship between energy consumption, economic growth and environmental pollution in China. *Environmental Research, 194*, 110718. DOI: 10.1016/j.envres.2021.110718 PMID: 33421428

Compilation of References

Abdin, Z. (2024). Shaping the stationary energy storage landscape with reversible fuel cells. *Journal of Energy Storage*, *86*, 111354. DOI: 10.1016/j.est.2024.111354

Abdolmaleki, L., & Berardi, U. (2024). Hybrid solar energy systems with hydrogen and electrical energy storage for a single house and a midrise apartment in North America. *International Journal of Hydrogen Energy*, *52*, 1381–1394. DOI: 10.1016/j.ijhydene.2023.11.222

Abebe, W. T., & Endalie, D. (2023). Artificial intelligence models for prediction of monthly rainfall without climatic data for meteorological stations in Ethiopia. *Journal of Big Data*, *10*(1), 2. Advance online publication. DOI: 10.1186/s40537-022-00683-3

Ablazov, N., Qodirov, A., Ibragimova, Z., & Akhmedov, K. (2024). Robo-Advisors and Investment Management. *Analyzing the Role of AI in Personal Finance.*, *1*, 1–5. DOI: 10.1109/ICKECS61492.2024.10617229

Abozaid, A. (2022). Financial Applications of Artificial Intelligence. *Sharī ah Issues and the Maqā idī Dimensions.*, *13*(2), 31–66. DOI: 10.55188/ijifarabic.v13i2.333

Abubakar, A. A., Al-Mamary, Y. H., Singh, H. P., Singh, A., Alam, F., & Agrawal, V. (2024). Exploring factors influencing sustainable human capital development: Insights from Saudi Arabia. *Heliyon*, *10*(16), e35676. DOI: 10.1016/j.heliyon.2024.e35676 PMID: 39229536

Acharya, B., Akil, A. M., Ridhanda, P., Na, Y. J., Hong, C. Y., & Park, K. W. (2013, June). Public Private Partnership in Hydropower Development: Prospects for Nepal's Development. *Conference: Energy Transition and Policy Challenges, 36th IAEE International Conference.* https://www.researchgate.net/publication/353209903_Public_Private_Partnership_in_Hydropower_Development_Prospects_for_Nepal's_Development

Ackoff, R. L. (2001). OR: After the post mortem. *System Dynamics Review, 17*(4), 341–346. DOI: 10.1002/sdr.222

Adiele, J. G., Schut, A. G. T., van den Beuken, R. P. M., Ezui, K. S., Pypers, P., Ano, A. O., Egesi, C. N., & Giller, K. E. (2020). Towards closing cassava yield gap in West Africa: Agronomic efficiency and storage root yield responses to NPK fertilizers. *Field Crops Research, 253*, 107820. DOI: 10.1016/j.fcr.2020.107820

Adinolfi, P. (2021). A journey around decision-making: Searching for the "big picture" across disciplines. *European Management Journal, 39*(1), 9–21.

Adom, P. K. (2024). The socioeconomic impact of climate change in developing countries over the next decades: A literature survey. *Heliyon, 10*(15), e35134. DOI: 10.1016/j.heliyon.2024.e35134 PMID: 39170312

Afghan, N., & Wiqar, T. (2007). *Succession in family businesses of Pakistan: Kinship culture and Islamic inheritance law*. Centre for Management and Economic Research, Lahore University of Management Sciences.

Afif, M. Z., Sulaiman, H. J., & Amelia, T. (2024). Research agenda of shariah-compliant investment using bibliometric analysis r studio. *Jurnal Ilmiah Manajemen, Ekonomi. Dan Akuntansi, 8*(3), 1186–1200. DOI: 10.31955/mea.v8i3.4471

Aguinis, H., & Glavas, A. (2019). On corporate social responsibility, sensemaking, and the search for meaningfulness through work. *Journal of Management, 45*(3), 1057–1086. DOI: 10.1177/0149206317691575

Ahmad, A. R. (2002). A Research Framework for Intelligent Decision Support in the Web Page Layout Design and Presentation Planning, *M.S. Thesis*. Management Sciences, Faculty of Engineering, University of Waterloo.

Ahmad, A. R. (2005). An Intelligent Expert System for Decision Analysis and Support in Multi-Attribute Layout Design, *Ph.D. Thesis*. Systems Design Engineering, Faculty of Engineering, University of Waterloo.

Ahmad, A. R., Basir, O. A., Hassanein, K., & Imam, M. H. (2004b, October). Decision preferences, constraints, and evaluation objectives in layout design: a review of modeling techniques. In Proc. of the 5th Int'l Conf. on Op. and Quan. Manag (ICOQM-V).

Ahmad, Z., al-Amatullah, I., & Mahboob, M. N. (2024). Utilising AI-based Chatbot for Maximising Value and Profitability of Islamic Financial Institutions. https://doi.org/DOI: 10.33001/M0104202421/125

Ahmad, A. R., Basir, O. A., & Hassanein, K. (2004, July). Intelligent Decision Support System for Layout Design. In *Proceedings of the 9th Asia-Pacific Decision Sciences Institute Conference (APDSI'04).*

Ahmad, A. R., Basir, O. A., & Hassanein, K. (2008). "An Intelligent Expert Systems' Approach to Layout Decision Analysis and Support under Uncertainty", Intelligent Decision Making: An AI-Based Approach. *Studies in Computational Intelligence, 97*, 321–364.

Ahmad, A. R., Tasadduq, I. A., Imam, M. H., & Shaban, K. B. (2015). Automated Discovery and Utilization of Tacit Knowledge in Facility Layout Planning and Optimization. *Journal of Software and Systems Development, 2015*, 27–39. DOI: 10.5171/2015369029

Ahmed, T., Khan, M. S., Thitivesa, D., Siraphatthada, Y., & Phumdara, T. (2020). Impact of employees engagement and knowledge sharing on organizational performance: Study of HR challenges in COVID-19 pandemic. *Human Systems Management, 39*(4), 589–601. DOI: 10.3233/HSM-201052

Ahmed, U., Umrani, W. A., Yousaf, A., Siddiqui, M. A., & Pahi, M. H. (2021). Developing faithful stewardship for environment through green HRM. *International Journal of Contemporary Hospitality Management, 33*(10), 3115–3133. DOI: 10.1108/IJCHM-09-2020-1066

Ahoya, D. K. D., Sawadogo-Compaore, E. M. F. W., Yabi, J. A., Zandjanakou-Tachin, M., Houngue, J. A., Houedjissin, S. S., Pita, J. S., & Ahanhanzo, C. (2024). Factors Influencing Adoption of the PlantVillage Nuru Application for Cassava Mosaic Disease Diagnosis Among Farmers in Benin. *Agriculture, 14*(11), 2001. DOI: 10.3390/agriculture14112001

Aiguobarueghian, I., Adanma, U. M., Ogunbiyi, E. O., & Solomon, N. O. (2024). An overview of initiatives and best practices in resource management and sustainability. *World Journal of Advanced Research and Reviews, 22*(2), 1734–1745. DOI: 10.30574/wjarr.2024.22.2.1519

Ait Kadi, M., & Ziyad, A. (2018). *Integrated Water Resources Management in Morocco. Water Resources Development and Management.* Springer Nature., DOI: 10.1007/978-981-10-7913-9_6

Akter, M. S. (2024). AI for Sustainability: Leveraging Technology to Address Global Environmental. *Journal of Artificial Intelligence General Science (JAIGS) ISSN:3006-4023, 3*(1), 40–48. DOI: 10.60087/jaigs.v3i1.64

Alam, M. A., & Niu, X. (2021). Impact of GHRM on organization's … Impact of GHRM on Organization's Environmental Performance: Mediating Role of Green Employee Empowerment. *The Journal of Business, 06*(01), 6–6. DOI: 10.18533/job.v6i2.202

Alford, K. R., Stedman, N. L., Bunch, J., Baker, S., & Roberts, T. G. (2025). Real-World Experiences in Higher Education: Contributing to the Developing a Systems Thinking Paradigm. *Journal of Experiential Education, 48*(1), 169–188. DOI: 10.1177/10538259241259626

Al-Hawari, M. A., Quratulain, S., & Melhem, S. B. (2021). How and when frontline employees' environmental values influence their green creativity? Examining the role of perceived work meaningfulness and green HRM practices. *Journal of Cleaner Production, 310*(August 2020), 127598-127598. https://doi.org/DOI: 10.1016/j.jclepro.2021.127598

Ali Ababneh, O. M., Awwad, A. S., & Abu-Haija, A. (2021). The association between green human resources practices and employee engagement with environmental initiatives in hotels: The moderation effect of perceived transformational leadership. *Journal of Human Resources in Hospitality & Tourism, 20*(3), 390–416. DOI: 10.1080/15332845.2021.1923918

Alsyouf, I., Kumar, U., Al-Ashi, L., & Al-Hammadi, M. (2018). Improving baggage flow in the baggage handling system at a UAE-based airline using lean Six Sigma tools. *Quality Engineering, 30*(3), 432–452. DOI: 10.1080/08982112.2018.1437180

Amarnath, G., Alahacoon, N., Smakhtin, V., & Aggarwal, P. (2017). *Mapping multiple climate-related hazards in South Asia*. International Water Management Institute (IWMI).

Amit, R., & Villalonga, B. (2014). Financial performance of family firms. The Sage handbook of family business, 157-178. DOI: 10.4135/9781446247556.n9

Amjad, F., Abbas, W., Zia-Ur-Rehman, M., Baig, S. A., Hashim, M., Khan, A., & Rehman, H. (2021). Effect of green human resource management practices on organizational sustainability: The mediating role of environmental and employee performance. *Environmental Science and Pollution Research International, 28*(22), 28191–28206. DOI: 10.1007/s11356-020-11307-9 PMID: 33527245

Anwar, N., Nik Mahmood, N. H., Yusliza, M. Y., Ramayah, T., Noor Faezah, J., & Khalid, W. (2020). Green Human Resource Management for organisational citizenship behaviour towards the environment and environmental performance on a university campus. *Journal of Cleaner Production, 256*, 120401–120401. DOI: 10.1016/j.jclepro.2020.120401

Arias Calvo, A., Feijoo Costa, G., Moreira Vilar, M. T., Nika, C. E., Vasilaki, V. K., & Katsou, E. (2024). Assessing the future prospects of emerging technologies for shipping and aviation biofuels. *Critical Review*.

Arias, C. (2024). Understanding the drivers and barriers towards sustainable consumption: An approach by linking a pedagogical strategy of active learning with adopting sustainable behaviors by students. *Journal of Sustainability Perspectives, 4*, 270–293. DOI: 10.14710/jsp.2024.25041

Armstrong, M., & Taylor, S. (2020). *Armstrong's Handbook of Human Resource Management Practice*. Kogan Page.

Arruda, E. H., Melatto, R. A. P. B., Neto, W. L. B. da S., & Conti, D. de M. (2021). Circular Economy: A Brief Literature Review (2015-2020). *Sustainable Operations and Computers, 2*, 79–86. DOI: 10.1016/j.susoc.2021.05.001

Arthur, E. E., Gyamfi, S., Gerstlberger, W., Stejskal, J., & Prokop, V. (2023). Towards circular economy: Unveiling heterogeneous effects of government policy stringency, environmentally related innovation, and human capital within OECD countries. *Sustainability (Basel), 15*(6), 4959. DOI: 10.3390/su15064959

Aryal, S., Ghimire, S., Tiwari, S., Baaniya, Y., & Pandey, V. P. (2024). Evolution and future prospects of hydropower sector in Nepal: A review. *Heliyon, 10*(10), e31139–e31139. Elsevier. DOI: 10.1016/j.heliyon.2024.e31139

Ashoka, P., Devi, B. R., Sharma, N., Behera, M., Gautam, A., Jha, A., & Sinha, G. (2024). Artificial Intelligence in Water Management for Sustainable Farming: A Review. *Journal of Scientific Research and Reports, 30*(6), 511–525. DOI: 10.9734/jsrr/2024/v30i62068

Astrachan, J. H., & Kolenko, T. A. (1994). A neglected factor explaining family business success: Human resource practices. *Family Business Review, 7*(3), 251–262. DOI: 10.1111/j.1741-6248.1994.00251.x

Awofeso, O. (2019). Managing formal organizations in the 21st century: A critique of Fredrick Taylor's scientific management theory. *Journal of Public Management Research, 5*(2), 1–11. DOI: 10.5296/jpmr.v5i2.15970

Ayhan, F., Kartal, M. T., Kılıç Depren, S., & Depren, Ö. (2023). Asymmetric effect of economic policy uncertainty, political stability, energy consumption, and economic growth on CO2 emissions: Evidence from G-7 countries. *Environmental Science and Pollution Research International, 30*(16), 47422–47437. DOI: 10.1007/s11356-023-25665-7 PMID: 36737567

Ayyub, B. M. (2001). *Elicitation of Expert Opinions for Uncertainty and Risks.* CRC Press. DOI: 10.1201/9781420040906

Azaria, A., Ekblaw, A., Vieira, T., & Lippman, A. (2016, August). Medrec: Using blockchain for medical data access and permission management. In 2016 2nd international conference on open and big data (OBD) (pp. 25-30). IEEE.

Azzumi, A. R., & Aziz, J. A. (2023). *Considering Dawam Raharjo as the Foundation of Sharia Digital Business Ethics.* Money., DOI: 10.31004/money.v1i2.15991

Bade, S. O., & Tomomewo, O. S. (2024). A review of governance strategies, policy measures, and regulatory framework for hydrogen energy in the United States. *International Journal of Hydrogen Energy*, *78*, 1363–1381. DOI: 10.1016/j.ijhydene.2024.06.338

Bade, S. O., Tomomewo, O. S., Meenakshisundaram, A., Ferron, P., & Oni, B. A. (2024). Economic, social, and regulatory challenges of green hydrogen production and utilization in the US: A review. *International Journal of Hydrogen Energy*, *49*, 314–335. DOI: 10.1016/j.ijhydene.2023.08.157

Bagheri, N., Ben Abdelaziz, F., & Rao, A. (2017). Ethical Stochastic Objectives Programming Approach for Portfolio Selection. *Ethical Stochastic Objectives Programming Approach for Portfolio Selection.*, *1*(1), 495–505. DOI: 10.30585/icabml-cp.v1i1.40

Bansal, P., & Roth, K. (2000). Why companies go green: A model of ecological responsiveness. *Academy of Management Journal*, *43*(4), 717–736. DOI: 10.2307/1556363

Barber, C. P., Cochrane, M. A., Souza, C. M.Jr, & Laurance, W. F. (2014). Roads, deforestation, and the mitigating effect of protected areas in the Amazon. *Biological Conservation*, *177*, 203–209. DOI: 10.1016/j.biocon.2014.07.004

Basil, D. Z., & Erlandson, J. (2008). Corporate social responsibility website representations: A longitudinal study of internal and external self-presentations. *Journal of Marketing Communications*, *14*(2), 125–137. DOI: 10.1080/13527260701858497

Becker-Olsen, K. L., Cudmore, B. A., & Hill, R. P. (2006). The impact of perceived corporate social responsibility on consumer behavior. *Journal of Business Research*, *59*(1), 46–53. DOI: 10.1016/j.jbusres.2005.01.001

Ben & Jerry's. (2019). *Social justice initiatives and labor rights.* Retrieved from https://www.benjerry.com/

Ben & Jerry's. (2023). Social justice initiatives and labor rights. Retrieved from https://www.benjerry.com/

Ben Saad, M., Ben Saad, R., Kertous, M., & Samir, B. M. (2024). Asymmetric impacts of renewable energy and economic development in North Africa countries: Evidence from the NARDL model. *Environmental Economics and Policy Studies*. Advance online publication. DOI: 10.1007/s10018-024-00425-0

Béres, R., Nijs, W., Boldrini, A., & van den Broek, M. (2024). Will hydrogen and synthetic fuels energize our future? Their role in Europe's climate-neutral energy system and power system dynamics. *Applied Energy*, *375*, 124053. DOI: 10.1016/j.apenergy.2024.124053

Berrone, P., Cruz, C., & Gomez-Mejia, L. R. (2012). Socioemotional wealth in family firms: Theoretical dimensions, assessment approaches, and agenda for future research. *Family Business Review*, *25*(3), 258–279. DOI: 10.1177/0894486511435355

Bersin, J. (2021). *AI in HR: The Future of Work*. Deloitte University Press.

Bessen, J. E. (2019). AI and Jobs: The role of demand. *NBER Working Paper Series*, 24235.

Bharadwaj, B. (2016). *Plastic Bag Ban in Nepal: Enforcement and Effectiveness*. South Asian Network for Development and Environmental Economics (SANDEE).

Bhatt, M. W., & Sharma, S. (2022). An Object Recognition-Based Neuroscience Engineering: A Study for Future Implementations. *Istanbul University -. Journal of Electrical and Electronics Engineering (Oradea)*, *23*(2). Advance online publication. DOI: 10.5152/electrica.2022.22125

Bhatt, P., & Joshi, K. R. (2024). Hydropower Development in Nepal: Status, Opportunities and Challenges. *Journal of UTEC Engineering Management*, *2*(01), 125–135. DOI: 10.36344/utecem.2024.v02i01.011

Bioenergy (2024). Products. Retrieved October 24, 2024, from https://bioenergy.com.tr/products/

Bisschoff, C., & Fullerton, S. (2025). Comparing green customer citizenship attitudes and behaviours in South Africa and South Korea. *Acta Commercii*, *25*(2). Advance online publication. DOI: 10.4102/ac.v25i2.1336

Blok, K., Afanador, A., van der Hoorn, I., Berg, T., Edelenbosch, O. Y., & van Vuuren, D. P. (2020). Assessment of Sectoral Greenhouse Gas Emission Reduction Potentials for 2030. *Energies*, *13*(4), 943. DOI: 10.3390/en13040943

Boretti, A., & Rosa, L. (2019). Reassessing the projections of the World Water Development Report. *NPJ Clean Water*, *2*(1), 15. DOI: 10.1038/s41545-019-0039-9

Bostrom, N. (2014). *Superintelligence: Paths, Dangers, Strategies*. Oxford University Press.

Bouyssou, D., & Vincke, P. (1998). Preference Modeling. *Annals of Operations Research, 80*, 147–180. DOI: 10.1023/A:1018988520000

Bowen, R., Dowell, D., & Morris, W. (2023). Hospitality SMEs and the circular economy: Strategies and practice post-COVID. *British Food Journal, 126*(1), 80–97. DOI: 10.1108/BFJ-10-2022-0932

Braverman, H. (1998). *Labor and monopoly capital: The degradation of work in the twentieth century*. NYU Press.

Bredariol, T. O. (March 22, 2023). *Clean energy can help to ease the water crisis, 2021 and 2030*. Retrieved October 24, 2024, from https://www.iea.org/commentaries/clean-energy-can-help-to-ease-the-water-crisis

Brundtland, G. H., & Mansour, K. (2010). World Commission on Environment and Development (WCED). 1987. *Our common future*, 2010.

Brynjolfsson, E., & McAfee, A. (2014). *The Second Machine Age: Work, Progress, and Prosperity in a Time of Brilliant Technologies*. W. W. Norton & Company.

Burns, P. (2012). *Corporate entrepreneurship: innovation and strategy in large organizations*. Macmillan International Higher Education.

Bwire, V. (2024, January 22). *Solar energy: Changing rural lives in Kenya*. Unesco.org; UNESCO. https://courier.unesco.org/en/articles/solar-energy-changing-rural-lives-kenya

Carfora, A., Pansini, R. V., & Scandurra, G. (2019). The causal relationship between energy consumption, energy prices and economic growth in Asian developing countries: A replication. *Energy Strategy Reviews, 23*, 81–85. DOI: 10.1016/j.esr.2018.12.004

Carroll, A. B. (1991). The pyramid of corporate social responsibility: Toward the moral management of organizational stakeholders. *Business Horizons, 34*(4), 39–48. DOI: 10.1016/0007-6813(91)90005-G

Carroll, A. B. (1999). Corporate social responsibility: Evolution of a definitional construct. *Business & Society, 38*(3), 268–295. DOI: 10.1177/000765039903800303

Carrubbo, L., Cosimato, S., & Gagliardi, A. R. (2025). Towards dynamic decision-making in government as service organization: insights from systems thinking. Transforming Government: People. *Process and Policy, 19*(1), 108–129.

Casady, C. B., Cepparulo, A., & Giuriato, L. (2024). Public-private partnerships for low-carbon, climate-resilient infrastructure: Insights from the literature. *Journal of Cleaner Production, 470*, 143338–143338. DOI: 10.1016/j.jclepro.2024.143338

Cascio, W. F., & Aguinis, H. (2020). *Research in Personnel and Human Resources Management: AI and the Future of Work*. Emerald Publishing Limited.

Cascio, W. F., & Montealegre, R. (2016). How Technology Is Changing Work and Organizations. *Annual Review of Organizational Psychology and Organizational Behavior, 3*(1), 349–375. DOI: 10.1146/annurev-orgpsych-041015-062352

Cembranel, P., Gewehr, L., Dal Moro, L., Fuchs, P. G., Birch, R. S., & Andrade Guerra, J. B. S. O. D. A. (2024). The pivotal role of higher education institutions in cultivating a sustainable development goals-centric culture. *International Journal of Sustainability in Higher Education, 25*(7), 1385–1411. DOI: 10.1108/IJSHE-01-2024-0057

Cerri, J., Testa, F., & Rizzi, F. (2018). The more I care, the less I will listen to you: How information, environmental concern and ethical production influence consumers' attitudes and the purchasing of sustainable products. *Journal of Cleaner Production, 175*, 343–353. DOI: 10.1016/j.jclepro.2017.12.054

Chain, S. Sustainability Report. (2023). Ethical supply chain management and responsible sourcing. Retrieved from https://www.supplychainsustainability.org/

Chakraborty, D., & Mukherjee, S. (2013). Fiscal subsidies and environmental sustainability: What does the cross-country empirical estimates suggest? *Foreign Trade Review, 48*(3), 383–397. DOI: 10.1177/0015732513496619

Chekima, B., Chekima, S., Syed Khalid Wafa, S. A., Igau, O., & Sondoh, S. L. (2015). undefined. *International Journal of Sustainable Development and World Ecology, 23*(2), 210–220. https://doi.org/ DOI: 10.1080/13504509.2015.1114043

Chekima, B., Syed Khalid Wafa, S. A., Igau, O. A., Chekima, S., & Sondoh, S. L. (2016). undefined. *Journal of Cleaner Production, 112*, 3436–3450. DOI: 10.1016/j.jclepro.2015.09.102

Chen, C. (2020, December 23). A Glimpse of the First Eight Months of the COVID-19 Literature on Microsoft Academic Graph: Themes, Citation Contexts, and Uncertainties. *Frontiers in Research Metrics and Analytics, 5*, 607286. DOI: 10.3389/frma.2020.607286 PMID: 33870064

Chen, C., Zhang, D., Zhu, L., & Zhang, F. (2024). undefined. *Resources, Conservation and Recycling, 207*, 107682. DOI: 10.1016/j.resconrec.2024.107682

Chisom, O. N., Biu, P. W., Umoh, A. A., Obaedo, B. O., Adegbite, A. O., & Abatan, A. (2024). Reviewing the Role of AI in Environmental Monitoring and conservation: A data-driven Revolution for Our Planet. *World Journal of Advanced Research and Reviews, 21*(1), 161–171. DOI: 10.30574/wjarr.2024.21.1.2720

Choudhury, P., Foroughi, C., & Larson, B. (2020). Work-from-anywhere: The productivity effects of geographic flexibility. *Strategic Management Journal, 41*(9), 1748–1770.

Chua, J. H., Chrisman, J. J., & Sharma, P. (1999). Defining the family business by behaviour. *Entrepreneurship Theory and Practice, 23*(4), 19–39. DOI: 10.1177/104225879902300402

Chu, C., Zhang, W., Yan, X., Yan, Y., Pan, J., Zohreh, S., & Moradian, J. M. (2024). Silver-incorporated NiCo metal-organic frameworks with controlled morphology for enhanced cycling in flexible supercapacitors application. *Journal of Materials Chemistry. C, Materials for Optical and Electronic Devices.*

Churchman, C. W. (1970). Operations Research as a Profession. *Management Science, 17*(2), B-37–B53. DOI: 10.1287/mnsc.17.2.B37

Churchman, C. W., Ackoff, R. L., & Arnoff, E. L. (1957). *Introduction to Operations Research.* John Wiley& Sons.

Cleveland, C., & Louf, N. (February 24, 2024). *What methods of electricity generation use the most water?* Retrieved October 25, 2024, from https://visualizingenergy .org/what-methods-of-electricity-generation-use-the-most-water/?

Contreras-Medina, D. I., Nieto, E. S. D., Medina-Cuéllar, S. E., & Aguilar-Rivera, N. (2025). Foundational concepts: An introduction to ecocentrism, knowledge management and sustainability. In *Ecocentrism for Knowledge Management and Sustainability* (pp. 1–8). Routledge.

Cooke, F. L., Schuler, R., & Varma, A. (2020). Human resource management research and practice in Asia: Past, present and future. *Human Resource Management Review, 30*(4), 100778–100778. DOI: 10.1016/j.hrmr.2020.100778

Cozzi, L.. (2023). *World Energy Outlook 2023.* IEA.

Cozzi, L.. (2024). *World Energy Outlook 2024.* IEA.

D'Agostino, A., Di Marco, G., Canini, A., & Gismondi, A. (2024). Gallic acid as a phytostimulant enhancing yield and quality of Mentha spicata L. under deficit-and well-watered conditions. *Environmental and Experimental Botany, 219,* 105656. DOI: 10.1016/j.envexpbot.2024.105656 da Silva, R. M., Schettino, S., de Lima, P. M. R., & Adriano, F. C. (2014). SMART GRID: Evaluation and Trend in Brazil. *Independent Journal of Management & Production, 5*(3), 752–776. Advance online publication. DOI: 10.14807/ijmp.v5i3.122

Daellenbach, H., & McNickle, D. (2012). *Management Science: Decision Making Through Systems Thinking* (2nd ed.). Palgrave Mcmillan. DOI: 10.1007/978-1-137-07512-3

Dai, J., Wu, S., Han, G., Weinberg, J., Xie, X., Wu, X., Song, X., Jia, B., Xue, W., & Yang, Q. (2018). Water-energy nexus: A review of methods and tools for macro-assessment. *Applied Energy, 210*, 393–408. DOI: 10.1016/j.apenergy.2017.08.243

Daspit, J. J., Madison, K., Barnett, T., & Long, R. G. (2018). The emergence of bifurcation bias from unbalanced families: Examining HR practices in the family firm using circumplex theory. *Human Resource Management Review, 28*(1), 18–32. DOI: 10.1016/j.hrmr.2017.05.003

Dauletbay, A. (2024). Transportation of Hydrogen: Hydrogen Usage. In *Hydrogen Technologies-Advances, Insights, and Applications*. IntechOpen.

Daví-Arderius, D., Sanin, M.-E., & Trujillo-Baute, E. (2017). CO2 content of electricity losses. *Energy Policy, 104*, 439–445. DOI: 10.1016/j.enpol.2017.01.011

Day, M.-Y., Cheng, T.-K., & Li, J.-G. (2018). AI robo-advisor with big data analytics for financial services. *Advances in Social Networks Analysis and Mining, 1027–1031*, 1027–1031. Advance online publication. DOI: 10.1109/ASONAM.2018.8508854

De Kok, J. M., Uhlaner, L. M., & Thurik, A. R. (2006). Professional HRM practices in family owned-managed enterprises. *Journal of Small Business Management, 44*(3), 441–460. DOI: 10.1111/j.1540-627X.2006.00181.x

Debsarma, S. K. (2009). Simulations of Storm Surges in the Bay of Bengal. *Marine Geodesy, 32*(2), 178–198. DOI: 10.1080/01490410902869458

Delbeke, J. (2024). *Delivering a Climate Neutral Europe*. Taylor & Francis. DOI: 10.4324/9781003493730

Deloitte. (2021a). *Global Human Capital Trends: The social enterprise at work*. Deloitte Insights.

Dessler, G. (2019). *Human Resource Management*. Pearson.

DeVries, P. D. (2008, January). The state of RFID for effective baggage tracking in the airline industry. *International Journal of Mobile Communications, 6*(2), 151–164. DOI: 10.1504/IJMC.2008.016574

Dey, D. (2010). The political economy of India's national solar energy mission. SSRN *Electronic Journal*. https://doi.org/DOI: 10.2139/ssrn.1633911

Dhar, V. (2020). Digital Transformation and Human Resource Management. *Journal of World Business*, *55*(1), 1–13.

Diamantopoulos, A., Schlegelmilch, B. B., Sinkovics, R. R., & Bohlen, G. M. (2003). Can socio-demographics still play a role in profiling green consumers? A review of the evidence and an empirical investigation. *Journal of Business Research*, *56*(6), 465–480. DOI: 10.1016/S0148-2963(01)00241-7

Diao, M. (2019). Towards sustainable urban transport in Singapore: Policy instruments and mobility trends. *Transport Policy*, *81*, 320–330. DOI: 10.1016/j.tranpol.2018.05.005

Dike, S. E., Davis, Z., Abrahams, A., Anjomshoae, A., & Ractham, P. (2024). Evaluation of passengers' expectations and satisfaction in the airline industry: An empirical performance analysis of online reviews. *Benchmarking*, *31*(2), 611–639. DOI: 10.1108/BIJ-09-2021-0563

Docker. (2024). Available: https://www.docker.com/company. (Accessed: 3-Oct-2024).

Dorado, S., Chen, Y., Prado, A. M., & Simon, V. (2021). Attuned HRM Systems for Social Enterprises. *Journal of Business Ethics*, •••, 1–20.

Dorri, M., Steger, M., Kanhere, S. S., & Jurdak, R. (2017, December). Steger, S. S. Kanhere and R. Jurdak, "BlockChain: A Distributed Solution to Automotive Security and Privacy,". *IEEE Communications Magazine*, *55*(12), 119–125. DOI: 10.1109/MCOM.2017.1700879

Doyle, J. (1997). The Cognitive Psychology of Systems Thinking. *System Dynamics Review*, *13*(3), 253–265. DOI: 10.1002/(SICI)1099-1727(199723)13:3<253::AID-SDR129>3.0.CO;2-H

Doz, Y. (2017). The Strategic Decisions That Caused Nokia's Failure Retrieved from https://knowledge.insead.edu/strategy/the-strategic-decisions-that-caused-nokias-failure 7766#JFeXsK5OyE6qReQU.99 last accessed on September 7, 2018.

Drews, S. (2025). Towards systems-oriented behavioral ecological economics. *Ecological Economics*, *227*, 108402. DOI: 10.1016/j.ecolecon.2024.108402

Duarte, P., Silva, S. C., Roza, A. S., & Dias, J. C. (2024). Enhancing consumer purchase intentions for sustainable packaging products: An in-depth analysis of key determinants and strategic insights. *Sustainable Futures : An Applied Journal of Technology, Environment and Society*, *7*, 100193. DOI: 10.1016/j.sftr.2024.100193

Dujak, D., & Sajter, D. (2019). Blockchain Applications in Supply Chain. In Kawa, A., & Maryniak, A. (Eds.), *SMART Supply Network. EcoProduction (Environmental Issues in Logistics and Manufacturing)*. Springer. DOI: 10.1007/978-3-319-91668-2_2

Du, L., Yang, Y., Zhou, L., & Liu, M. (2024). Greenhouse Gas Reduction Potential and Economics of Green Hydrogen via Water Electrolysis: A Systematic Review of Value-Chain-Wide Decarbonization. *Sustainability (Basel)*, *16*(11), 4602. DOI: 10.3390/su16114602

Dutta, B., & Hwang, H. (2021). Consumers purchase intentions of green electric vehicles: The influence of consumers technological and environmental considerations. *Sustainability (Basel)*, *13*(21), 12025. DOI: 10.3390/su132112025

EEA. (Published Feburary 28, 2015; Modified September 11, 2024) *Correlation of energy consumption and GDP per person*. Retrieved November 17, 2024, from https://www.eea.europa.eu/en/analysis/maps-and-charts/correlation-of-per-capita-energy?

Ehnert, I. (2009). *Sustainability and human resource management: Developing sustainable business organizations*. Springer., DOI: 10.1007/978-3-7908-2188-8

Ehnert, I. (2014). *Sustainable human resource management: A conceptual and exploratory analysis from a paradox perspective*. Springer., DOI: 10.1007/978-3-319-07623-0

Ehnert, I., Parsa, S., Roper, I., Wagner, M., & Muller-Camen, M. (2023). Sustainable human resource management: Directions for new research. *Human Resource Management Journal*, *33*(1), 1–18. DOI: 10.1002/hrmj.20402

Elahi, Y., & Abd Aziz, M. I. (2011). *New model for shariah-compliant portfolio optimization under fuzzy environment*. Springer., DOI: 10.1007/978-3-642-25462-8_18

Electric, S. (2023). Sustainability scorecard and HRM practices. Retrieved from https://www.se.com/

Elkington, J. (1997). *Cannibals with forks: The triple bottom line of 21st-century business*. Capstone Publishing.

Emerald Insight. The artificial intelligence-enabled customer experience in tourism: a systematic literature review.

Emma Luxton. (Published Feburary 25, 2016) *What will global GDP look like in 2030?* Retrieved November 20, 2024, from https://www.weforum.org/stories/2016/02/what-will-global-gdp-look-like-in-2030/

Energy Institute. (2024). Statistical Review of World Energy 2024 73rd Edition. Retrieved November 3, 2024, from https://www.energyinst.org/statistical-review

Entezari, A., Aslani, A., Zahedi, R., & Noorollahi, Y. (2023). *Artificial intelligence and machine learning in energy systems: A bibliographic perspective. Energy Strategy Reviews, 45(101017), 101017.* Elsevier., DOI: 10.1016/j.esr.2022.101017

EPI. (2024). *Unsafe drinking water*. Retrieved November 15, 2024, from https://epi.yale.edu/measure/2024/UWD

ESPAS. (2019). *Welcome to 2030: The Mega-Trends*. Retrieved November 18, 2024, from https://ec.europa.eu/assets/epsc/pages/espas/chapter1.html?

Evans, R. G., & Sadler, E. J. (2008). Methods and technologies to improve efficiency of water use. *Water Resources Research*, *44*(7), 2007WR006200. Advance online publication. DOI: 10.1029/2007WR006200

Faccia, A., Le Roux, C. L., & Pandey, V. (2023). Innovation and e-Commerce models, the technology catalysts for sustainable development: The emirate of Dubai case study. *Sustainability (Basel)*, *15*(4), 3419. DOI: 10.3390/su15043419

Faheem, M. A., Anwer, S., Rayhan, Z., Ullah, M. A., Paudel, R., Ahmed, M. F., & Khan, H. (2024). AI-Driven Innovation In HRM And Its Impact On Business Management: An In-Depth Study Of Technology Advancement And Strategic Implementation. *Nanotechnology Perceptions*, *20*, 1174–1204.

Fair Trade International. (2023). Ethical sourcing and supply chain sustainability. Retrieved from https://www.fairtrade.net/

Fam, A., & Fam, S. (2024). Review of the US 2050 long term strategy to reach net zero carbon emissions. *Energy Reports*, *12*, 845–860. DOI: 10.1016/j.egyr.2024.06.031

Fan, Z., Ghaddar, B., Wang, X., Xing, L., Zhang, Y., & Zhou, Z. (2024). Artificial intelligence for operations research: Revolutionizing the operations research process. arXiv preprint arXiv:2401.03244.

FAO. (2024). *Water Use*, Retrieved October 24, 2024, from https://www.fao.org/aquastat/en/overview/methodology/water-use

Farooq, M., Ramzan, M., & Yen, Y. (2025a). *AI in Consumer Behavior Management*. IGI Global., DOI: 10.4018/979-8-3693-4322-7.ch011

Farooq, M., Ramzan, M., & Yen, Y. (2025b). *Exploring the Impact of Artificial Intelligence on Brand Management: A Multi-Dimensional Analysis*. IGI Global., DOI: 10.4018/979-8-3693-4322-7.ch006

Faruque Hossain. (2019). *Sustainable Design and Build: Building, Energy, Roads, Bridges, Water and Sewer Systems*. Butterworth-Heinemann.

Faturohman, T., & Nugraha, T. (2022). Islamic stock portfolio optimization using deep reinforcement learning. *Journal of Islamic Monetary Economics and Finance*, *8*(2). Advance online publication. DOI: 10.21098/jimf.v8i2.1430

Fauquet, C., & Fargette, D. (1990). African Cassava Mosaic Virus: Etiology, Epidemiology, and Control. *Plant Disease*, *74*(6), 404. DOI: 10.1094/PD-74-0404

Fayiah, M., Dong, S., Singh, S., & Kwaku, E. A. (2020). A review of water–energy nexus trend, methods, challenges and future prospects. *International Journal of Energy and Water Resources*, *4*(1), 91–107. DOI: 10.1007/s42108-020-00057-6

Febregas, J. (May 30, 2023). *DAF systems for oil refinery wastewater treatment*. Retrieved October 25, 2024, from https://sigmadafclarifiers.com/en/sistemas-daf-para-el-tratamiento-de-aguas-residuales-de-refinerias-de-petroleo/

Ferasso, M., Bares, L., Ogachi, D., & Blanco, M. (2021). Economic and Sustainability Inequalities and Water Consumption of European Union Countries. *Water (Basel)*, *13*(19), 2696. DOI: 10.3390/w13192696

Feroz, A. K., & Stines, A. (2024). Utilization of AI, big data, and internet of things (IoT) in sustainable innovation and their role in the emergence of novel business models. *Issues in Information Systems*, *25*(1).

Ficko, M., Berus, L., Palčič, I., & Klančnik, S. (2021). Design of Facility Layout for Industry 4.0. In *Research Anthology on Cross-Industry Challenges of Industry 4.0* (pp. 219–244). IGI Global. DOI: 10.4018/978-1-7998-8548-1.ch012

Firnando, H. G., & Wahyudi, M. A. (2024). The Role of Artificial Intelligence in Shaping the Islamic Worldview of the Digital Economy. *Journal of Islamic Economics and Philanthropy*, *6*(3), 231–249. DOI: 10.21111/jiep.v6i3.11386

Foroglou, G., & Tsilidou, A. L. (2015, May). Further applications of the blockchain. In 12th student conference on managerial science and technology (pp. 1-8).

Forrester, J. W. (1971). The counterintuitive behavior of social systems. *Technology Review*, *73*(3), 52–68.

Forum, W. E. (2018). The Future of Jobs Report 2018. *World Economic Forum*.

Fotis, P., Karkalakos, S., & Asteriou, D. (2017). The relationship between energy demand and real GDP growth rate: The role of price asymmetries and spatial externalities within 34 countries across the globe. *Energy Economics*, *66*, 69–84. DOI: 10.1016/j.eneco.2017.05.027

Foulds, L., & Thachenkary, C. (2001). Empower to the people–how decision support systems and IT can aid the OR analyst. OR/MS Today, June.

Fransisca, V., & Ningsih, W. (2023). The Advancement of Technology and its Impact on Social Life in Indonesia. *Devotion Journal of Community Service*, *4*(3), 860–864. DOI: 10.36418/devotion.v4i3.445

Fredershausen, S., Meyer-Larsen, N., & Klumpp, M. (2024, February). Comprehensive Sustainability Evaluation Concept for Offshore Green Hydrogen from Wind Farms. In *International Conference on Dynamics in Logistics* (pp. 309-321). Cham: Springer Nature Switzerland. DOI: 10.1007/978-3-031-56826-8_24

Freeman, R. E. (2010). *Strategic Management: A Stakeholder Approach*. Cambridge University Press. DOI: 10.1017/CBO9781139192675

Freeman, R. E., Harrison, J. S., & Zyglidopoulos, S. (2023). *Stakeholder theory: Concepts and strategies*. Cambridge University Press., DOI: 10.1017/9781009092254

Frieden, F., & Leker, J. (2024). Future costs of hydrogen: A quantitative review. *Sustainable Energy & Fuels*, *8*(9), 1806–1822. DOI: 10.1039/D4SE00137K

Gao, C., Cui, X., & Matsumura, J. (2024). Multidimensional Exploration of Wood Extractives: A Review of Compositional Analysis, Decay Resistance, Light Stability, and Staining Applications. *Forests*, *15*(10), 1782. DOI: 10.3390/f15101782

Gao, J., Siddik, A. B., Khawar Abbas, S., Hamayun, M., Masukujjaman, M., & Alam, S. S. (2023). undefined. *Sustainability*, *15*(2), 1594. DOI: 10.3390/su15021594

Garey, M. R., & Johnson, D. S. (1979). *Computers and Intractability*. W.H. Freeman Press.

Garrett, K. (2023). *The Adaptive Capacity to implement Climate Migration Policy in response to Sea-Level Rise in the Maldives*. https://digitalcollections.sit.edu/capstones/3275?utm_source=digitalcollections.sit.edu%2Fcapstones%2F3275&utm_medium=PDF&utm_campaign=PDFCoverPages

Gartner. (2022). Top Trends in Human Resource Management and Artificial Intelligence. *Gartner Research*.

Garvin, D. A. (1993). Building a learning organization. *Harvard Business Review*, *71*(4), 78–91. PMID: 10127041

Gass, S. I., & Assad, A. A. (2005). *An annotated timeline of operations research: An informal history* (Vol. 75). Springer Science & Business Media.

Geisbauer, A., & Berchthold, C. (2024). Will Power and Waste-to-X enable the worldwide energy transition to renewables?. *Monatshefte für Chemie-Chemical Monthly*, 1-14.

Geissdoerfer, M., Savaget, P., Bocken, N. M. P., & Hultink, E. J. (2017). The circular economy–A new sustainability paradigm? *Journal of Cleaner Production, 143*, 757–768. DOI: 10.1016/j.jclepro.2016.12.048

George, J. M., & Jones, G. R. (2024). *ISE Contemporary Management* (13th ed.). McGraw-Hill.

Ghorbian, M., & Ghorbian, S. (2025). Empowering Clinical Decision-Making with Generative AI in Intelligent Decision Support Systems. In *Generative Artificial Intelligence (AI) Approaches for Industrial Applications* (pp. 241–259). Springer Nature Switzerland. DOI: 10.1007/978-3-031-76710-4_12

Ghosh, B. (2016). Impact of packaging on consumers' buying Behaviour:A case study of mother dairy, Kolkata. *Parikalpana: KIIT Journal of Management, 12*(1), 63. DOI: 10.23862/kiit-parikalpana/2016/v12/i1/133042

Gleim, M. R., Smith, J. S., Andrews, D., & Cronin, J. J.Jr. (2013). Against the green: A multi-method examination of the barriers to green consumption. *Journal of Retailing, 89*(1), 44–61. DOI: 10.1016/j.jretai.2012.10.001

Global Compact, U. N. (2023). Corporate social responsibility and HRM alignment. Retrieved from https://www.unglobalcompact.org/

Goel, M. (2024). 3 'Ps' of Hydrogen Economy in India: Production Pathways, Policies, and Perspectives. In *Climate Action and Hydrogen Economy: Technologies Shaping the Energy Transition* (pp. 11–37). Springer Nature Singapore. DOI: 10.1007/978-981-99-6237-2_2

Goldstein, N. J., Cialdini, R. B., & Griskevicius, V. (2008). A room with a viewpoint: Using social norms to motivate environmental conservation in hotels. *The Journal of Consumer Research, 35*(3), 472–482. DOI: 10.1086/586910

Gomez-Mejia, L. R., Haynes, K. T., Nunez-Nickel, M., Jacobson, K. J., & Moyano-Fuentes, J. (2007). Socioemotional wealth and business risks in family-controlled firms: Evidence from Spanish olive oil mills. *Administrative Science Quarterly, 52*(1), 106–137. DOI: 10.2189/asqu.52.1.106

Gong, Q., Fan, D., & Bartram, T. (2025). Integrating artificial intelligence and human resource management: A review and future research agenda. *International Journal of Human Resource Management, 36*(1), 103–141. DOI: 10.1080/09585192.2024.2440065

Goodfellow, I., Bengio, Y., & Courville, A. (2016). *Deep Learning*. MIT Press.

Göthel, J., Burkmann, K. J., & Volkova, O. (2024). Review and Evaluation of Ceramic-Stabilized Iron Oxides for Use as Energy Storage Based on Iron-Steam Process. *steel research international*, 2400265.

Greenberg, J. (2010). *Behavior in Organizations* (10th ed.). Pearson Education.

Griggs, G., & Reguero, B. G. (2021). Coastal Adaptation to Climate Change and Sea-Level Rise. *Water (Basel), 13*(16), 2151. DOI: 10.3390/w13162151

Guerci, M., & Carollo, L. (2016). A paradox view on green human resource management: Insights from the Italian context. *International Journal of Human Resource Management, 27*(2), 212–238. DOI: 10.1080/09585192.2015.1033641

Gunaratne, M. S., Radin Firdaus, R. B., & Rathnasooriya, S. I. (2021). Climate change and food security in Sri Lanka: Towards food sovereignty. *Humanities & Social Sciences Communications, 8*(1), 229. Advance online publication. DOI: 10.1057/s41599-021-00917-4

Gunathilake, C., Soliman, I., Panthi, D., Tandler, P., Fatani, O., Ghulamullah, N. A., Marasinghe, D., Farhath, M., Madhujith, T., Conrad, K., Du, Y., & Jaroniec, M. (2024). A comprehensive review on hydrogen production, storage, and applications. *Chemical Society Reviews, 53*(22), 10900–10969. DOI: 10.1039/D3CS00731F PMID: 39421882

Güney, E. (2022). Water Footprint Assessment of Mining and Processing of Gold, Unpublished Master Thesis, The Graduate School of Natural and Applied Scinces, Middle East Technical University, Ankara.

Guo, L. L. X., & Wang, L. (2022). Economic size and water use efficiency: An empirical analysis of trends across China. *Water Policy, 24*(1), 117–131. DOI: 10.2166/wp.2021.189

Gupta, S., Modgil, S., Bhattacharyya, S., & Bose, I. (2022). Artificial intelligence for decision support systems in the field of operations research: Review and future scope of research. *Annals of Operations Research, 308*(1), 215–274. DOI: 10.1007/s10479-020-03856-6

Guven, D. (2024). Offshore wind-driven green hydrogen: Bridging environmental sustainability and economic viability. *International Journal of Hydrogen Energy*, *72*, 661–676. DOI: 10.1016/j.ijhydene.2024.05.335

Hamadou, I., Yumna, A., Hamadou, H., & Jallow, M. S. (2024). Unleashing the power of artificial intelligence in Islamic banking: A case study of Bank Syariah Indonesia (BSI). *Modern Finance*, *2*(1), 131–144. DOI: 10.61351/mf.v2i1.116

Hammer, M. (2004). Deep Change: How Operational Innovation Can Transform Your Company. *Harvard Business Review*, *82*(4), 84–93. DOI: 10.1109/EMR.2004.25106 PMID: 15077369

Handler, W. C. (1989). Methodological issues and considerations in studying family businesses. *Family Business Review*, *2*(3), 257–276. DOI: 10.1111/j.1741-6248.1989.00257.x

Handoko, B., Franty, Y. K., & Indrayatna, F. (2025). Quantifying Uncertainty of Insurance Claims Based on Expert Judgments. *Mathematics*, *13*(2), 245. DOI: 10.3390/math13020245

Han, H., Hsu, L., & Lee, J.-S. (2009). Empirical investigation of the roles of attitudes toward green behaviors, overall image, gender, and age in hotel customers' eco-friendly decision-making process. *International Journal of Hospitality Management*, *28*(4), 519–528. DOI: 10.1016/j.ijhm.2009.02.004

Haq, I., & Ahmad, A. R. (2025) "Shared Resources for Blockchain-based Baggage Tracking System for Sustainable Airport Operations: System Design, Implementation, and Testing", https://drive.google.com/file/d/1OpKgNGvLDa2wbpyHdido fSQrTOULBZ-k/view

Harris, O. O., Bialous, S. A., Muench, U., Chapman, S., & Dawson-Rose, C. (2022). Climate change, public health, health policy, and nurses training. *American Journal of Public Health*, *112*(S3), S321–S327. DOI: 10.2105/AJPH.2022.306826 PMID: 35679559

Hassan, Q., Algburi, S., Jaszczur, M., Al-Jiboory, A. K., Al Musawi, T. J., Ali, B. M., Viktor, P., Fodor, M., Ahsan, M., Salman, H. M., & Sameen, A. Z. (2024a). Hydrogen role in energy transition: A comparative review. *Process Safety and Environmental Protection*, *184*, 1069–1093. DOI: 10.1016/j.psep.2024.02.030

Hassan, Q., Algburi, S., Sameen, A. Z., Jaszczur, M., & Salman, H. M. (2024b). Hydrogen as an energy carrier: Properties, storage methods, challenges, and future implications. *Environment Systems & Decisions*, *44*(2), 327–350. DOI: 10.1007/s10669-023-09932-z

Hassan, Q., Algburi, S., Sameen, A. Z., Salman, H. M., & Al-Jiboory, A. K. (2024c). A review of green hydrogen production by renewable resources. *Energy Harvesting and Systems, 11*(1), 20220127. DOI: 10.1515/ehs-2022-0127

Hassan, Q., Algburi, S., Sameen, A. Z., Salman, H. M., & Jaszczur, M. (2024d). Green hydrogen: A pathway to a sustainable energy future. *International Journal of Hydrogen Energy, 50*, 310–333. DOI: 10.1016/j.ijhydene.2023.08.321

Hassan, S., Dhali, M., Zaman, F., & Tanveer, M. (2021). Big data and predictive analytics in healthcare in Bangladesh: Regulatory challenges. *Heliyon, 7*(6), e07179. DOI: 10.1016/j.heliyon.2021.e07179 PMID: 34141936

Hatta, M. F. (2022). Financial robo-advisor savings2u for shariah-compliant savings and investment application. *Journal of Information Systems and Technology Management, 7*(29), 70–85. DOI: 10.35631/JISTM.729006

Haugen, J. B., Link, J. S., Cribari, K., Bundy, A., Dickey-Collas, M., Leslie, H. M., Hall, J., Fulton, E. A., Levenson, J. J., Parsons, D. M., Hassellöv, I.-M., Olsen, E., DePiper, G. S., Gentry, R. R., Clark, D. E., Brainard, R. E., Mateos-Molina, D., Borja, A., Gelcich, S., & Agnalt, A. L. (2024). Marine ecosystem-based management: Challenges remain, yet solutions exist, and progress is occurring. *Npj Ocean Sustainability, 3*(1), 7. DOI: 10.1038/s44183-024-00041-1

Heidaria, S. M., & Eshtiaghib, M. (2024). Women's Barriers to Utilizing Renewable Energy and Achieving the Sustainable Development Goals (SDGs) 2. *energy, 24*, 25.

Helerea, E., Calin, M. D., & Musuroi, C. (2023). Water Energy Nexus and Energy Transition—A Review. *Energies, 16*(4), 1879. DOI: 10.3390/en16041879

Helo, P., & Hao, Y. (2022). Artificial intelligence in operations management and supply chain management: An exploratory case study. *Production Planning and Control, 33*(16), 1573–1590. DOI: 10.1080/09537287.2021.1882690

Hemmet, A. (2023). *Harmonizing Artificial Intelligence with Islamic Values - A Thoughtful Analysis of Religious*. Social, and Economic Impacts of Technological Advancements., DOI: 10.54536/ajsts.v2i2.2239

Hendarti, Y., Winarno, B., & Aprilianto, M. P. (2024). *Use of Blockchain Technology and AI in Sharia Financial Risk Management*. https://doi.org/DOI: 10.62885/ekuisci.v1i3.165

Hendrix, M. (2012). Water in Ethiopia: Drought, Disease and Death. *Global Majority E-Journal, 3*(2), 110–120. https://www.american.edu/cas/economics/ejournal/upload/hendrix_accessible.pdf

Heriyati, P., Yadav, N., & Tamara, D. (2024). Accomplishing sustainable development goals through international management system standards and multinational supply chains. *Business Strategy and the Environment, 33*(6), 5214–5231. DOI: 10.1002/bse.3752

Hernández-Linares, R., López-Fernández, M. C., Memili, E., Mullins, F., & Patel, P. C. (2021). High-performance work practices, socioemotional wealth preservation, and family firm labor productivity. *BRQ Business Research Quarterly*, 23409444211002521.

Hernandez, Y., Naumann, G., Corral, S., & Barbosa, P. (2020). Water Footprint Expands with Gross Domestic Product. *Sustainability (Basel), 12*(20), 8741. DOI: 10.3390/su12208741

Hirai, Y. (2017). Defining the Ethereum Virtual Machine for Interactive Theorem Provers. In Brenner, M. (Eds.), Lecture Notes in Computer Science: Vol. 10323. *Financial Cryptography and Data Security. FC 2017*. Springer. DOI: 10.1007/978-3-319-70278-0_33

Hofstede, G., & Bond, M. H. (1984). Hofstede's culture dimensions: An independent validation using Rokeach's value survey. *Journal of Cross-Cultural Psychology, 15*(4), 417–433. DOI: 10.1177/0022002184015004003

Hölbl, M., Kompara, M., Kamišalić, A., & Nemec Zlatolas, L. (2018). A Systematic Review of the Use of Blockchain in Healthcare. *Symmetry, 10*(10), 470. DOI: 10.3390/sym10100470

Hooda, J., Singh, A., Kazaure, M. A., & Singh, R. (2023). *Shariah-Compliant Robo-Advisors*. IGI Global., DOI: 10.4018/979-8-3693-1038-0.ch015

Horne, R. E. (2009). Limits to labels: The role of eco-labels in the assessment of product sustainability and routes to sustainable consumption. *International Journal of Consumer Studies, 33*(2), 175–182. DOI: 10.1111/j.1470-6431.2009.00752.x

Hosterman, H., & Smith, J. (2025). *Economic Costs and Benefits of Climate Change Impacts and Adaptation to the The Maldives Tourism Industry*. Tourism.gov.mv;Ministry of Tourism. Male', Republic of Maldives. https://resources.tourism.gov.mv/wp-content/uploads/2024/09/Economic-Costs-and-benefits-of-Climate-Change-Impacts-and-Adaptation-to-the-Maldives-Tourism-Industry.pdf

Huang, S., Yue, C., Uvdal, K., & Hu, Z. (2024). Recent advances in irradiation-mediated synthesis and tailoring of inorganic nanomaterials for photo-/electrocatalysis. *Nanoscale Advances*. PMID: 39610792

Hyperledger Fabric. (2024). Available: https://hyperledger-fabric.readthedocs.io/en/latest/whatis.html. (Accessed: 7-Oct-2024).

Ibegbulam, C., & Ibegbulam, A.Ibegbulam C.MAigbovbiosa, O.JOlowonubi, J.AF-atounde, S.A. (2023). Role Of Artificial Intelligence In Electrification Of Africa. *Engineering Science & Technology Journal*, *4*(6), 456–472. DOI: 10.51594/estj.v4i6.667

IBM. (2023). Sustainable HR analytics and business performance. Retrieved from https://www.ibm.com/

IEA. (2024), *Transport*, Retrieved October 25, 2024, from https://www.iea.org/energy-system/transport

IEA. (March 3, 2020, Last updated 20 Mar 2023). *Introduction to the water-energy nexus*. Retrieved November 8, 2024, from https://www.iea.org/articles/introduction-to-the-water-energy-nexus?

Ignizio, J. P., & Cavalier, T. M. (2019). *Linear Programming*. Pearson.

IKEA. (2019). *Sustainable sourcing and energy efficiency initiatives*. Retrieved from https://www.ikea.com/

Imam, M. H., Tasadduq, I. A., & Ahmad, A. R. (2015). A Hybrid Algorithm for Optimising Facility Layout. *South African Journal of Industrial Engineering*, *26*(1).

IMF. (2024), *GDP per capita, current prices*. Retrieved November 18, 2024, from https://www.imf.org/external/datamapper/NGDPDPC@WEO/OEMDC/ADVEC/WEOWORLD

INED. (July, 2024). *2024: les Nations unies publient de nouvelles projections de population mondiale*. Retrieved November 13, 2024, from https://www.ined.fr/fr/tout-savoir-population/memos-demo/focus/2024-les-nations-unies-publient-de-nouvelles-projections-de-population-mondiale/

INED. (July, 2024). *Projections de population (en millions) par continent de 2025 à 2100. Scénario central*. Retrieved November 19, 2024, from https://www.ined.fr/fr/tout-savoir-population/chiffres/projections-mondiales/projections-par-continent/

INFORMS. (2025). Institute for Operations Research and the Management Sciences, https://www.informs.org/, accessed on Feb. 25, 2025.

International Labour Organization (ILO). (2023). Global labor rights and fair working conditions. Retrieved from https://www.ilo.org/

Irawan, B., Maimun, M., Faizal, L., & Zaki, M. (2023). Tinjauan Hukum Ekonomi Syariah Tentang Penggunaan Robo Advisor Dalam Islamic Wealth Management (Studi Pustaka Robo Advisor di Indonesia Tahun 2022). *Al-Manhaj : Jurnal Hukum Dan Pranata Sosial Islam.* https://doi.org/DOI: 10.37680/almanhaj.v5i2.3214

IRENA and Bluerisk. (2023), Water for hydrogen production, International Renewable Energy Agency, Bluerisk, Abu Dhabi, United Arab Emirates. https://www.irena.org/-/media/Files/IRENA/Agency/Publication/2023/Dec/IRENA_Bluerisk_Water_for_hydrogen_production_2023.pdf

IRENA. (2023), *Status and pace of progress.* Retrieved October 25, 2024, from https://www.irena.org/Innovation-landscape-for-smart-electrification/Power-to-heat-and-cooling/Status

Irfan, M., & Kismawadi, E. R. (2024). Sharia Stock Transactions Utilizing Technology and Artificial Intelligence. *Advances in Business Strategy and Competitive Advantage Book Series*, 79–98. https://doi.org/DOI: 10.4018/979-8-3693-3980-0.ch004

Ishak, M. N., & Mohamed, A. (2023). *Harmonization of Islamic Economics With Artificial Intelligence: Towards an Ethical and Innovative Economic Paradigm.* https://doi.org/DOI: 10.24256/kharaj.v5i4.4387

Jabbour, C. J. C. (2008). Environmental training and environmental management maturity in companies. *The case of Brazil.Journal of Cleaner Production*, *16*(3), 216–226.

Jabbour, C. J. C., & Santos, F. C. A. (2008). The central role of human resource management in the search for sustainable organizations. *International Journal of Human Resource Management*, *19*(12), 2133–2154. DOI: 10.1080/09585190802479389

Jackson, M. C. (2016). *Systems thinking: Creative holism for managers.* John Wiley & Sons, Inc.

Jackson, M. C., & Keys, P. (1985). Introduction. *The Journal of the Operational Research Society*, *36*(9), 753–755. DOI: 10.1057/jors.1985.140

Jackson, S. E., & Renwick, D. W., & Jabbour, C. J. C. (2011). Sustainable human resource management: A comprehensive review and research agenda. *Business &amp. Society*, *50*(2), 299–323.

Jackson, S. E., Renwick, D. W., Jabbour, C. J., & Muller-Camen, M. (2011). State-of-the-art and future directions for green human resource management: Introduction to the special issue. *German Journal of Human Resource Management*, *25*(2), 99–116. DOI: 10.1177/239700221102500203

Jain, A. (2012). Shariah-Compliant Portfolio Management: A More Suitable And-Balanced Risk-Return Matrix? *Social Science Research Network*. https://doi.org/ DOI: 10.2139/SSRN.2285927

Jaradat, M., Almashaileh, S., Bendea, C., Juaidi, A., Bendea, G., & Bungau, T. (2024). Green Hydrogen in Focus: A Review of Production Technologies, Policy Impact, and Market Developments. *Energies*, *17*(16), 3992. DOI: 10.3390/en17163992

Jatobá, A. M. B., Pimentel, M. D. G., & Alvarenga, M. A. (2019). Applications of artificial intelligence in human resource management: A systematic review. *Brazilian Journal of Operations & Production Management*, *16*(3), 509–524.

Jawadi, F., Pondie, T., & Cheffou, A. (2024). New challenges for green finance and sustainable industrialization in developing countries: A panel data analysis. https:// doi.org/DOI: 10.2139/ssrn.4916541

Jiang, W., Qian, K., Fan, C., Ding, W., & Li, Z. (2024). Applications of generative AI-based financial robot advisors as investment consultants. *Applied and Computational Engineering*, *67*(1), 28–33. DOI: 10.54254/2755-2721/67/2024MA0057

Jin, Y., Behrens, P., Tukker, A., & Scherer, L. (2019). Water use of electricity technologies: A global meta-analysis. *Renewable & Sustainable Energy Reviews*, *115*, 109391. DOI: 10.1016/j.rser.2019.109391

Jokhio, M. N., & Jaffer, M. A. (2024). Generative AI in Shariah Advisory in Islamic Finance: An Experimental Study. *Business Review*, *19*(2), 74–92. DOI: 10.54784/1990-6587.1665

Jolly, S., Ahmad, N., & Scott, M. (Eds.). (2024). *Climate-Related Human Mobility in Asia and the Pacific* (1st ed.). Springer Nature. DOI: 10.1007/978-981-97-3234-0

Jordan, P. G. (2014). Federal and state energy policies. *Solar Energy Markets*, 65-109. https://doi.org/DOI: 10.1016/B978-0-12-397174-6.00006-4

Joshi, Y., & Rahman, Z. (2015). Factors affecting green purchase behaviour and future research directions. *International Strategic Management Review*, *3*(1-2), 128–143. DOI: 10.1016/j.ism.2015.04.001

Joshi, Y., & Rahman, Z. (2019). undefined. *Ecological Economics*, *159*, 235–243. DOI: 10.1016/j.ecolecon.2019.01.025

Kahneman, D., Slovi, P., & Tversky, A. (1982). *Judgment Under Uncertainty: Heuristics and Biases*. Cambridge University Press. DOI: 10.1017/CBO9780511809477

Kahn, W. A. (1990). Psychological conditions of personal engagement and disengagement at work. *Academy of Management Journal*, *33*(4), 692–724. DOI: 10.2307/256287

Kandwal, N. P., Nautiyal, A., & Juyal, S. A. (2024, December). Promoting Health and Well-being through Integrative Wellness Practices: Pathways to Sustainable Development Goal 3. In *5th International Conference on the Role of Innovation, Entrepreneurship and Management for Sustainable Development (ICRIEMSD-2024)* (pp. 362-375). Atlantis Press. DOI: 10.2991/978-94-6463-612-3_24

Kaplan, A., & Haenlein, M. (2019). Siri, Siri, in my hand: Who's the fairest in the land? On the interpretations, illustrations, and implications of artificial intelligence. *Business Horizons*, *62*(1), 15–25. DOI: 10.1016/j.bushor.2018.08.004

Karim, R., Pk, M. A. B., Dey, P., Akbar, M. A., & Osman, M. S. (2024). A study about the prediction of population growth and demographic transition in Bangladesh. *Journal of Umm Al-Qura University for Applied Sciences*. DOI: 10.1007/s43994-024-00150-0

Karray, F., & De Silva, C. (2004). *"Soft Computing and Intelligent Systems Design: Theory, Tools and Applications", Readings*. Addison-Wesley.

Kasim, R. S. R. (2023). *A smart shariah-based decision-making in islamic fintech*. https://doi.org/DOI: 10.33102/iiecons.v10i1.77

Kassa, H., Abiyu, A., Hagazi, N., Mokria, M., Kassawmar, T., & Gitz, V. (2022). Forest landscape restoration in Ethiopia: Progress and challenges. *Frontiers in Forests and Global Change*, *5*, 796106. Advance online publication. DOI: 10.3389/ffgc.2022.796106

Kato, Z. (2015). SEA: Practical Application of Science. Academic Search Ultimate. Vol. 3 Issue 2, p229-234. 6p.

Kaushal, N., Kaurav, R. P. S., Sivathanu, B., & Kaushik, N. (2023). Artificial intelligence and HRM: Identifying future research Agenda using systematic literature review and bibliometric analysis. *Manag Rev Q*, *73*(2), 455–493. DOI: 10.1007/s11301-021-00249-2

Kellermanns, F. W., & Eddleston, K. A. (2006). Corporate entrepreneurship in family firms: A family perspective. *Entrepreneurship Theory and Practice*, *30*(6), 809–830. DOI: 10.1111/j.1540-6520.2006.00153.x

Kelling, S., Gerbracht, J., Fink, D., Lagoze, C., Wong, W.-K., Yu, J., Damoulas, T., & Gomes, C. (2013). A Human/Computer Learning Network to Improve Biodiversity Conservation and Research. *AI Magazine*, *34*(1), 10–10. DOI: 10.1609/aimag.v34i1.2431

Khakwani, S., & Case, P. (2012). Worldly leadership in Pakistan Seth organizations: an empirical challenge to the concept of global leadership. In *Worldly Leadership* (pp. 192–205). Palgrave Macmillan. DOI: 10.1057/9780230361720_11

Khan, S., & Rabbani, M. R. (2021). *Artificial Intelligence and NLP -Based Chatbot for Islamic Banking and Finance. 11*(3), 65–77. https://doi.org/DOI: 10.4018/IJIRR.2021070105

Khan, U., Bilal, M., Adil, H. M., Darlington, N., Khan, A., Khan, N., & Ihsanullah, I. (2024). Hydrogen from sewage sludge: Production methods, influencing factors, challenges, and prospects. *The Science of the Total Environment*, *919*, 170696. DOI: 10.1016/j.scitotenv.2024.170696 PMID: 38340850

Kharrim, M. (2024). Quantitative framework for shariah-compliant multi-period fuzzy portfolio optimization: A sustainable approach to ethical investing. *Journal of Islamic Accounting and Business Research.* Advance online publication. DOI: 10.1108/JIABR-02-2024-0052

Kheirinejad, S., Bozorg-Haddad, O., Singh, V. P., & Loáiciga, H. A. (2022). The effect of reducing per capita water and energy uses on renewable water resources in the water, food and energy nexus. *Scientific Reports*, *12*(1), 7582. DOI: 10.1038/s41598-022-11595-w PMID: 35534602

Kieseberg, P., Leithner, M., Mulazzani, M., Munroe, L., Schrittwieser, S., Sinha, M., & Weippl, E. *Proceedings of the 8th International Conference on Advances in Mobile Computing and Multimedia*, 430-435, 2010

Kingsbury, N., & Scanzoni, J. (2009). Structural-Functionalism. In Boss, P., Doherty, W. J., LaRossa, R., Schumm, W. R., & Steinmetz, S. K. (Eds.), *Sourcebook of Family Theories and Methods*. Springer.

Kismawadi, E. R. (2024). *Sustainable Islamic Financial Inclusion*. IGI Global., DOI: 10.4018/979-8-3693-1565-1.ch013

Kismawadi, E. R., Irfan, M., & Shah, S. M. A. R. (2024). *Revolutionizing Islamic Finance*. IGI Global., DOI: 10.4018/979-8-3693-0082-4.ch011

Kleespies, M. W., & Dierkes, P. W. (2022). The importance of the Sustainable Development Goals to students of environmental and sustainability studies—A global survey in 41 countries. *Humanities & Social Sciences Communications*, *9*(1), 218. DOI: 10.1057/s41599-022-01242-0

Kobbacy, K. A., Vadera, S., & Rasmy, M. H. (2007). AI and OR in management of operations: History and trends. *The Journal of the Operational Research Society*, *58*(1), 10–28.

Konar, A. (2018). *"Artificial Intelligence and Soft Computing: Behavioral and Cognitive Modeling of the Human Brain", Readings*. CRC Press.

Koneczna, R., & Cader, J. (2024). Towards effective monitoring of hydrogen economy development: A European perspective. *International Journal of Hydrogen Energy*, *59*, 430–446. DOI: 10.1016/j.ijhydene.2024.02.036

Kong, Y., He, W., Yuan, J., Zhang, Z., Gao, X., Zhao, Y., & Degefu, D. M. (2021). Decoupling economic growth from water consumption in the Yangtze River Economic Belt, China. *Ecological Indicators*, *123*, 107344. DOI: 10.1016/j.ecolind.2021.107344

Kosba, A. Miller, E. Shi, Z. Wen and C. Papamanthou, "Hawk: The Blockchain Model of Cryptography and Privacy-Preserving Smart Contracts," *2016 IEEE Symposium on Security and Privacy (SP)*, San Jose, CA, 2016, pp. 839-858. DOI: 10.1109/SP.2016.55

Koswatte, I., Iddawala, J., Kulasekara, R., Ranaweera, P., Dasanayaka, C. H., & Abeykoon, C. (2024). Can Sri Lanka be a Net-Zero Nation by 2050? – Current Renewable Energy Profile, Opportunities, Challenges, and Recommendations. *Cleaner Energy Systems*, *8*, 100126. Advance online publication. DOI: 10.1016/j.cles.2024.100126

Kotter, J. P. (2023). *Leading change in a sustainable world*. Harvard Business Review Press.

Kramar, R. (2023). Beyond strategic human resource management: Is sustainable human resource management the next approach? *International Journal of Human Resource Management*, *24*(13), 2553–2578. DOI: 10.1080/09585192.2013.816863

Kshetri, N. (2017). Can Blockchain Strengthen the Internet of Things? *IT Professional*, *19*(4), 68–72. DOI: 10.1109/MITP.2017.3051335

Küçük, G. (February 06, 2025), *DeepSeek's 'efficient' AI model sparks energy market shift*, Retrieved February 08, 2025, from https://www.aa.com.tr/en/energy/general/deepseeks-efficient-ai-model-sparks-energy-market-shift/47337

Kuir-Ayius, D. (2024). Impact of climate change and accessing services in Papua New Guinea. In Jolly, S., Ahmad, N., & Scott, M. (Eds.), *Climate-related human mobility in Asia and the Pacific*. Sustainable Development Goals Series. Springer., DOI: 10.1007/978-981-97-3234-0_11

Kumar Apat, S., & Mishra, J. (2023). An Artificial Intelligence-based Crop Recommendation System using Machine Learning. *Journal of Scientific and Industrial Research, 82*(05), 558–567. DOI: 10.56042/jsir.v82i05.1092

Kumar, S., & Chandra, N. (2010). Sustainability concerns in Indian agriculture: An analysis of subsidies and investment. SSRN *Electronic Journal*. https://doi.org/ DOI: 10.2139/ssrn.1709898

Kumari, D., Shashwat, S., Verma, P.K. & Giri, A.K. (2024). Examining the nexus between carbon dioxide emissions, economic growth, fossil fuel energy use, urbanization and renewable energy towards achieving environmental sustainability in India, International *Journal of Energy Sector Management*, Vol. ahead-of-print No. ahead-of-print. https://doi.org/DOI: 10.1108/IJESM-05-2024-0007

Kumar, S., Nanan-Surujbally, A., Sharma, D. P., & Pathak, D. (2024). Hydrogen safety/standards (national and international document standards on hydrogen energy and fuel cell). In *Towards Hydrogen Infrastructure* (pp. 315–346). Elsevier. DOI: 10.1016/B978-0-323-95553-9.00011-X

Labaki, R. (2005). The Family Relationship Factor: Its Theoretical Contribution to Family Business Performance. In FBN-IFERA World Academic Research Forum, EHSAL Brussels.

Labaki, R., Michael-Tsabari, N., & Zachary, R. K. (2013). *31 Emotional dimensions within the family business: towards a conceptualization. Handbook of Research on Family Business* (2nd ed.). Elgar Publishing.

Lakritz, T. (2019, September 10). *These 11 sinking cities are in danger of disappearing*. World Economic Forum. https://www.weforum.org/stories/2019/09/11-sinking-cities-that-could-soon-be-underwater/

Lam, S. H., & Toan, T. D. (2006). Land Transport Policy and Public Transport in Singapore. *Transportation, 33*(2), 171–188. DOI: 10.1007/s11116-005-3049-z

Lane, D. (1994). With Little Efforts from our Friends: How System Dynamics and "Soft" O.R. can learn from each other. *System Dynamics Review, 10*(2-3), 101–134. DOI: 10.1002/sdr.4260100203

Lawler, E. E., & Worley, C. G. (2006). *Built to Change: How to Achieve Sustained Organizational Effectiveness*. John Wiley & Sons.

Lawson, D. (2005). *Engineering Disasters – Lessons to be learned, Readings*. ASME Press.

Lee, J. (2006). Impact of family relationships on attitudes of the second generation in the family business. *Family Business Review*, *19*(3), 175–191. DOI: 10.1111/j.1741-6248.2006.00069.x

Lehtonen, M. (2024). Unveiling obstacles and assessing the EU policy and funding instruments for low-carbon hydrogen investments.

Léopold, E. (2018). Blockchain in Aviation White Paper. Retrieved April 26, 2024, from https://www.iata.org/en/publications/blockchain/

Liben, F., Abera, W., Tariku Chernet, M., Ebrahim, M., Tilaye, A., Erkossa, T., Tibebe Degefie, D., Mponela, P., Kihara, J., & Tamene, L. (2024). Site-specific fertilizer recommendations using data-driven machine learning enhanced wheat productivity and resource use efficiency. *Field Crops Research*, *313*, 109413–109413. DOI: 10.1016/j.fcr.2024.109413

Lim, D. T., Goh, K. W., & Lee, L. H. (2023). A new class of Shariah-compliant portfolio optimization model: Diversification analysis. *AIMS Mathematics*, *8*(9), 20933–20965. DOI: 10.3934/math.20231066

Lin, X. Y., Ho, T. W., Fang, C. C., Yen, Z. S., Yang, B. J., & Lai, F. (2015, August). A mobile indoor positioning system based on iBeacon technology. In 2015 37th Annual International Conference of the IEEE Engineering in Medicine and Biology Society (EMBC) (pp. 4970-4973). IEEE.

Lindsey, R. (2023, August 22). *Climate change: Global sea level* (R. Lumpkin, G. Johnson, P. Thompson, & W. Sweet, Eds.). Climate.gov; National Oceanic and Atmospheric Administration. https://www.climate.gov/news-features/understanding-climate/climate-change-global-sea-level

Li, R., & Leung, G. C. K. (2021). The relationship between energy prices, economic growth and renewable energy consumption: Evidence from Europe. *Energy Reports*, *7*, 1712–1719. DOI: 10.1016/j.egyr.2021.03.030

Liu, L., Xu, J., Qian, C., Zhong, Y., & Qin, L. (2024). Robo-Advisors: Revolutionizing Wealth Management through the Integration of Big Data and Artificial Intelligence in Algorithmic Trading Strategies. *Journal of Knowledge Learning and Science Technology*, *3*(3), 33–45. DOI: 10.60087/jklst.vol3.n3.p33-45

Lubatkin, M. H., Schulze, W. S., Ling, Y., & Dino, R. N. (2005). The effects of parental altruism on the governance of family-managed firms. Journal of Organizational Behavior: The International Journal of Industrial. *Journal of Organizational Behavior*, *26*(3), 313–330. DOI: 10.1002/job.307

Luo, X., & Bhattacharya, C. (2006). Corporate social responsibility, customer satisfaction, and market value. *Journal of Marketing*, *70*(4), 1–18. DOI: 10.1509/jmkg.70.4.001

Madison, K., Daspit, J. J., Turner, K., & Kellermanns, F. W. (2018). Family firm human resource practices: Investigating the effects of professionalization and bifurcation bias on performance. *Journal of Business Research*, *84*, 327–336. DOI: 10.1016/j.jbusres.2017.06.021

Ma, H., Shen, M., Tong, Y., & Wang, X. (2023). Radioactive Wastewater Treatment Technologies: A Review. *Molecules (Basel, Switzerland)*, *28*(4), 1935. DOI: 10.3390/molecules28041935 PMID: 36838922

Mahmood, T., & Ahmad, E. (2018). The relationship of energy intensity with economic growth:Evidence for European economies. *Energy Strategy Reviews*, *20*, 90–98. DOI: 10.1016/j.esr.2018.02.002

Mahmoudi, A., Bostani, M., Rashidi, S., & Valipour, M. S. (2023). Challenges and opportunities of desalination with renewable energy resources in Middle East countries. *Renewable & Sustainable Energy Reviews*, *184*, 113543. DOI: 10.1016/j.rser.2023.113543

Mahmoud, M. A., Tsetse, E. K., Tulasi, E. E., & Muddey, D. K. (2022). Green packaging, environmental awareness, willingness to pay and consumers' purchase decisions. *Sustainability (Basel)*, *14*(23), 16091. DOI: 10.3390/su142316091

Malhotra, P. (2024). Harmonizing Financial Returns With Ethical Standards. *Advances in Healthcare Information Systems and Administration Book Series*, 163–172. https://doi.org/DOI: 10.4018/979-8-3693-7452-8.ch010

Maqbool, S., Zafeer, H. M. I., Zeng, P., Maqbool, S., Draissi, Z., & Javed, S. (2024). Inventive leadership styles and their impact for achieving sustainable development goals in education at secondary schools: A case study from Multan, Pakistan. *Humanities & Social Sciences Communications*, *11*(1), 1–11. DOI: 10.1057/s41599-024-03086-2

Marakas, G. M. (2015). *Decision Support Systems*. Pearson India.

March, J. G., & Simon, H. A. (1958). *Organizations*. Wiley.

Mariappanadar, S. (2020). *Sustainable human resource management: Strategies, practices, and outcomes*. Springer., DOI: 10.1007/978-3-030-27972-1

Martinez, M., & Aldrich, H. (2014). Sociological theories applied to family businesses. The Sage handbook of family business, 83-105. DOI: 10.4135/9781446247556.n5

Martinho, G., Pires, A., Portela, G., & Fonseca, M. (2015). Factors affecting consumers' choices concerning sustainable packaging during product purchase and recycling. *Resources, Conservation and Recycling*, *103*, 58–68. DOI: 10.1016/j.resconrec.2015.07.012

Martinho, V. J. P. D., & Rodrigues, R. N. (2024). Bioenergy relations with agriculture, forestry and other land uses: Highlighting the specific contributions of artificial intelligence and co-citation networks. *Heliyon*, *10*(4), e26267. DOI: 10.1016/j.heliyon.2024.e26267 PMID: 38379976

Marzouk, O. A. (2024). Expectations for the role of hydrogen and its derivatives in different sectors through analysis of the four energy scenarios: IEA-STEPS, IEA-NZE, IRENA-PES, and IRENA-1.5 C. *Energies*, *17*(3), 646. DOI: 10.3390/en17030646

Masri, H. (2018). A Shariah-compliant portfolio selection model. *The Journal of the Operational Research Society*, *69*(10), 1568–1575. DOI: 10.1057/s41274-017-0223-6

Matthews, R., & Vivoda, V. (2023). "Water Wars": Strategic implications of the grand Ethiopian Renaissance Dam. *Conflict Security and Development*, *23*(4), 1–34. DOI: 10.1080/14678802.2023.2257137

McIntyre, R. J. (1966). Multiplication noise in uniform avalanche diodes. *IEEE Transactions on Electron Devices*, *ED-13*(1), 164–168. DOI: 10.1109/T-ED.1966.15651

Meadows, D. H., Meadows, D. L., & Randers, J. (1992). Beyond the limits. [Human populations and resource depletion]. World Watch;(United States), 6(1).

Merton, R. K. (1957). *Social theory and social structure*. Free Press.

Microsoft. (2020). *Carbon reduction and sustainability goals*. Retrieved from https://www.microsoft.com/

Microsoft. (2023). Sustainability and HR technology solutions. Retrieved from https://www.microsoft.com/

Milliman, J., & Clair, J. (2017). Best environmental HRM practices in the US. In *Greening people* (pp. 49–73). Routledge. DOI: 10.4324/9781351283045-2

Mirza, N., Tudor, C., Horobeţ, A., & Belaşcu, L. (2025). *Optimizing global risk-conscious portfolios: the strategic role of Sharia-compliant and ESG investments*. Sustainability Accounting, Management and Policy Journal., DOI: 10.1108/SAMPJ-08-2024-0879

Mohamad, S. N. A., Dali, N. R. S. M., & Basah, M. Y. (2024). *Navigating the Moral Compass of Halal FinTech in Islamic Finance: Harvesting Benefits While Curbing Risks* (pp. 157–171). https://doi.org/DOI: 10.1007/978-981-97-1375-2_10

Mohamed, H. (2021). *I-FinTech and Its Value Proposition for Islamic Asset and Wealth Management*. Palgrave Macmillan., DOI: 10.1007/978-3-030-45827-0_14

Mohamed, H., & Sari, N. (2020). *Artificial Intelligence and Blockchain-Driven Islamic Capital Markets*. IGI Global., DOI: 10.4018/978-1-7998-0218-1.ch044

Molden, D., Oweis, T., Steduto, P., Bindraban, P., Hanjra, M. A., & Kijne, J. (2010). Improving agricultural water productivity: Between optimism and caution. *Agricultural Water Management*, *97*(4), 528–535. DOI: 10.1016/j.agwat.2009.03.023

Mörk Water. (May 13, 2024). *Water use by the Australian renewable energy sector*. Retrieved October 26, 2024, from https://moerkwater.com.au/updates/water-use-in-renewable-energy-power-generation/

Mrisho, L. M., Mbilinyi, N. A., Ndalahwa, M., Ramcharan, A. M., Kehs, A. K., McCloskey, P. C., Murithi, H., Hughes, D. P., & Legg, J. P. (2020). Accuracy of a Smartphone-Based Object Detection Model, PlantVillage Nuru, in Identifying the Foliar Symptoms of the Viral Diseases of Cassava-CMD and CBSD. *Frontiers in Plant Science*, *11*, 590889. DOI: 10.3389/fpls.2020.590889 PMID: 33391304

Muhammad Nadhirin, N. F. S., Ishak, N., & Elias, S. M. (2020).. . *Performance of Shariah-Compliant Equity Portfolio Using Model-Based Return and Risk Estimation.*, *7*(2), 104–119. DOI: 10.31580/jei.v7i2.1439

Muksalmina, M., Idroes, G. M., & Maulana, A. (2024). Artificial Intelligence in Islamic Finance: Forecasting Stock Indices with Neural Prophet. *Indatu Journal of Management and Accounting*, 2(2), 68–80. DOI: 10.60084/ijma.v2i2.232

Munir, M. T., Naqvi, M., Li, B., Raza, R., Khan, A., Taqvi, S. A. A., & Nizami, A. S. (2024). From Waste to Watts: Emerging role of waste lignin-derived materials for energy storage. *Journal of Energy Storage*, *82*, 110477. DOI: 10.1016/j.est.2024.110477

Münster, M., Bramstoft, R., Kountouris, I., Langer, L., Keles, D., Schlautmann, R., ... & Rasmusson, H. (2024). Perspectives on green hydrogen in Europe—during an energy crisis and towards future climate neutrality. *Oxford Open Energy, 3*, oiae001.

Nabi, Ijaz. (2012) "Pakistan's Quest for a New Growth Vent: Lessons from History." The Lahore Journal of Economics 17: SE (September 2012): pp. 243–269

Nabi, S. (2017). *Four approaches to unleashing Pakistan's growth.* International Growth Centre Blog.

Nagarajan, M., Saha, R., Kumar, R., & Sathasivam, D. (2022). Impact of peer influence and environmental knowledge on green consumption. *International Journal of Social Ecology and Sustainable Development, 13*(6), 1–16. DOI: 10.4018/IJSESD.292039

Nath, O., Kumar, S., Sharma, V., Lal Meena, M., & Jain, R. (2023). Design of strategic framework for green sustainable practices in e-Commerce companies. *Materials Today: Proceedings.* Advance online publication. DOI: 10.1016/j.matpr.2023.03.032

Nations, U. Goal 8: Decent Work and Economic Growth. .

Nawi, A., Khamis, N. Y., Yaakob, M. F. M., Samuri, M. A. A., & Zakaria, G. A. N. (2023). Exploring Opportunities and Risks of Artificial Intelligence Research for Islamic Ethical Guidelines. *Afkar, 25*(2), 1–34. DOI: 10.22452/afkar.vol25no2.1

Needhi, J. (2024). Enhancing Financial Intelligence: AI Robo-Advisors for Strategic Investment Decisions.

Needhi, J., Prasath, G. R., Riffath, K. M., & Manokar, S. (2024). *Enhancing Financial Intelligence: AI Robo-Advisors for Strategic Investment Decisions.* https://doi.org/DOI: 10.31219/osf.io/3xena

Negnevitsky, M. (2019). *Artificial Intelligence: A Guide to Intelligent Systems* (3rd ed.). Pearson India.

Nichifor, B., Zai , L., & Timiras, L. (2025). Drivers, barriers, and innovations in sustainable food consumption: A systematic literature review. https://doi.org/DOI: 10.20944/preprints202502.0273.v1

Node.js. (2024). About Page. Available: https://nodejs.org/en/about/. (Accessed: 1-Oct- 20240.

Noe, R. A., Hollenbeck, J. R., Gerhart, B., & Wright, P. M. (2021). *Fundamentals of Human Resource Management.* McGraw-Hill Education.

Norman, D. A. (1993). *Thing That Make Us Smart: Defending Human Attributes in the Age of Machines.* Persius Books Group.

Nosé, L., Korunka, C., Frank, H., & Danes, S. M. (2017). Decreasing the effects of relationship conflict on family businesses: The moderating role of family climate. *Journal of Family Issues, 38*(1), 25–51. DOI: 10.1177/0192513X15573869

Nwokolo, S. C., Singh, R., Khan, S., Kumar, A., & Luthra, S. (2023). Technological pathways to net-zero goals in Africa. In *Africa's Path to Net-Zero: Exploring Scenarios for a Sustainable Energy Transition* (pp. 93–210). Springer., DOI: 10.1007/978-3-031-44514-9_5

Nyborg, K., Anderies, J. M., Dannenberg, A., Lindahl, T., Schill, C., Schlüter, M., Adger, W. N., Arrow, K. J., Barrett, S., Carpenter, S., Chapin, F. S.III, Crépin, A., Daily, G., Ehrlich, P., Folke, C., Jager, W., Kautsky, N., Levin, S. A., Madsen, O. J., & De Zeeuw, A. (2016). Social norms as solutions. *Science*, *354*(6308), 42–43. DOI: 10.1126/science.aaf8317 PMID: 27846488

Nyland, C., & Bruce, K. (2025). Taylorism, trade unions and participatory management. *Journal of Management History*. Advance online publication. DOI: 10.1108/JMH-09-2024-0136

Nyokabi, Jesse & Kamau, Kabera & Oyugi, Noah & Mbugua, Githae & Hiuhu, Antony. (2021). Machine Learning Model for Energy Analysis of a Wellhead Geothermal Plant: A Case of Kenya.

Odume, B. W. (2024). Artificial Intelligence in Agriculture: Application in Developing Countries. *Journal of Agricultural Science (Toronto)*, *16*(12), 60–67. DOI: 10.5539/jas.v16n12p60

Ogutu, M. O., Akor, J., Mulindwa, M. S., Heshima, O., & Nsengimana, C. (2023). Implementing circular economy and sustainability policies in Rwanda: Experiences of Rwandan manufacturers with the plastic ban policy. *Frontiers in Sustainability*, *4*, 1092107. Advance online publication. DOI: 10.3389/frsus.2023.1092107

Okanlawon, S. U., Zagoon-Sayeed, H., & Salisu, T. (2024). *Diagnosing Qur'ānic Precepts on Wealth Creation in Digital Age*. https://doi.org/DOI: 10.21154/invest.v4i1.8422

Oláh, J., Popp, J., Khan, M. A., & Kitukutha, N. (2023). Sustainable e-Commerce and environmental impact on sustainability. *Economics & Sociology (Ternopil)*, *16*(1), 85–105. DOI: 10.14254/2071-789X.2023/16-1/6

Olando, S., Mwenda, M. N., Kikwatha, R. W., & Rambo, C. M. (2024). Effect Of Public Private Partnership Financing Structure On Completion Of Geothermal Energy Development Projects In Kenya. *International Journal of Professional Business Review*, *9*(11), e05048–e05048. DOI: 10.26668/businessreview/2024.v9i11.5048

Olawade, D. B., Wada, O. Z., Ige, A. O., Egbewole, B. I., Olojo, A., & Oladapo, B. I. (2024). Artificial Intelligence in Environmental Monitoring: Advancements, Challenges, and Future Directions. *Hygiene and Environmental Health Advances, 12*, 100114–100114. DOI: 10.1016/j.heha.2024.100114

Olson, D. (2011). FACES IV and the circumplex model: A validation study. *Journal of Marital and Family Therapy, 37*(1), 64–80. DOI: 10.1111/j.1752-0606.2009.00175.x PMID: 21198689

Olson, D. H., Sprenkle, D. H., & Russell, C. S. (1979). Circumplex model of marital and family systems: I. Cohesion and adaptability dimensions, family types, and clinical applications. *Family Process, 18*(1), 3–28. DOI: 10.1111/j.1545-5300.1979.00003.x PMID: 437067

Ossenbruggen, P. J. (1994). *Fundamental Principles of Systems Analysis and Decision-Making*. John Wiley.

Othman, A. (2024, November). *AI in Climate Change Modeling and Prediction*. ResearchGate. https://www.researchgate.net/publication/385985206_AI_in_Climate_Change_Modeling_and_Prediction

Pachar, S., & Singh, R. (2021). Role of Sustainable Development Goals and Corporate Social Responsibility in India's Growth: Opportunities and Challenges. *The Empirical Economics Letters*, 20(2).

Pal, A., Kakran, S., Kumar, A., Youssef, A. B., Singh, U. P., & Sidhu, A. (2024). Powering squarely into the future: A strategic analysis of hydrogen energy in QUAD nations. *International Journal of Hydrogen Energy, 49*, 16–41. DOI: 10.1016/j.ijhydene.2023.06.169

Pal, T., Irfan, M., Bukhari, S. A. C., & Kassim, S. (2023). *Harnessing the Power of Artificial Intelligence and Machine Learning in the Islamic Capital Market*. IGI Global., DOI: 10.4018/979-8-3693-1038-0.ch006

Pandey, S., & Regmi, N. (2017). Changing connectivities and renewed priorities: Status and challenges facing Nepali Internet. *First Monday, 23*(1). Advance online publication. DOI: 10.5210/fm.v23i1.8071

Papadopoulos, T., Sivarajah, U., Spanaki, K., Despoudi, S., & Gunasekaran, A. (2022). Artificial Intelligence (AI) and data sharing in manufacturing, production and operations management research. *International Journal of Production Research, 60*(14), 4361–4364. DOI: 10.1080/00207543.2021.2010979

Parra-Domínguez, J., David, F., & Azevedo, T. (2021). Family firms and coupling among CSR disclosures and performance. *Administrative Sciences, 11*(1), 30. DOI: 10.3390/admsci11010030

Parsons, T. (1965). The normal American family. In S. Farber, P. Mustacchi, & R. Wilson (Eds.), Man and civilization: The family search for survival (pp. 3 1- 50). New York: McGraw Hill.

Parsons, T., & Shils, E. (Eds.). (1951). *Toward a general theory of action*. Harvard University Press. DOI: 10.4159/harvard.9780674863507

Pashchenko, D. (2024). Green hydrogen as a power plant fuel: What is energy efficiency from production to utilization? *Renewable Energy, 223*, 120033. DOI: 10.1016/j.renene.2024.120033

Patagonia. (2021). *Environmental responsibility and sustainable practices*. Retrieved from https://www.patagonia.com/

Patagonia. (2023). Environmental responsibility and sustainable practices. Retrieved from https://www.patagonia.com/

Pathak, M. (2010). Climate Change: Uncertainty for Hydropower Development in Nepal. *Hyrdo Nepal, 6*, 31–34. DOI: 10.3126/hn.v6i0.4190

Pathan, M. S. K., Khoso, A. A., & Ahmed, M. (2022). Digital Model Anecdotes Through Artificial Intelligence in Socioeconomic and Islamic Investments. *International Research Journal of Education and Innovation, 3*(2), 195–209. DOI: 10.53575/irjei.v3.02(22)21.195-209

Patkowski, W., Zybert, M., Ronduda, H., Albrecht, A., Moszyński, D., Fidler, A., ... & Raróg-Pilecka, W. (2024). Toward green ammonia synthesis–exploring the influence of lanthanide oxides as supports on the cobalt catalysts properties. *Journal of CO2 Utilization, 80*, 102699.

Peattie, K. (2010). Green consumption: Behavior and norms. *Annual Review of Environment and Resources, 35*(1), 195–228. DOI: 10.1146/annurev-environ-032609-094328

Pelkmans, L. (2020). Global Bioenergy, IEA Bioenergy. Retrieved October 24, 2024, from https://www.ieabioenergy.com/bioenergy-a-sustainable-solution/

Peters, R., Pasel, J., & Samsun, R. C. (2024). Technology transfer from fuel processing for fuel cells to fuel synthesis from hydrogen and carbon dioxide. *International Journal of Hydrogen Energy, 49*, 943–961. DOI: 10.1016/j.ijhydene.2023.08.022

Petropoulos, F., Laporte, G., Aktas, E., Alumur, S. A., Archetti, C., Ayhan, H., Battarra, M., Bennell, J. A., Bourjolly, J.-M., Boylan, J. E., Breton, M., Canca, D., Charlin, L., Chen, B., Cicek, C. T., Cox, L. A.Jr, Currie, C. S. M., Demeulemeester, E., Ding, L., & Zhao, X. (2024). Operational Research: Methods and applications. *The Journal of the Operational Research Society*, *75*(3), 423–617. DOI: 10.1080/01605682.2023.2253852

Pidd, M. (2001). The Futures of O.R. *The Journal of the Operational Research Society*, *52*(11), 1181–1190. DOI: 10.1057/palgrave.jors.2601268

Pierro, M. D. (2017). What Is the Blockchain? *Computing in Science & Engineering*, *19*(5), 92–95. DOI: 10.1109/MCSE.2017.3421554

Piracci, G., Boncinelli, F., & Casini, L. (2023). Investigating consumer preferences for sustainable packaging through a different behavioural approach: A random regret minimization application. *Environmental and Resource Economics*, *86*(1-2), 1–27. DOI: 10.1007/s10640-023-00785-3

Porter, M. E., & Kramer, M. R. (2011). Creating shared value. *Harvard Business Review*, *89*(1/2), 62–77.

Pothen, F., & Hundt, C. (2024). *European post-consumer steel scrap in 2050: A review of estimates and modeling assumptions* (Vol. 2024, No. 1). Jenaer Beiträge zur Wirtschaftsforschung.

Prakash, G., Choudhary, S., Kumar, A., Garza-Reyes, J. A., Khan, S. A., & Panda, T. K. (2019). Do altruistic and egoistic values influence consumers' attitudes and purchase intentions towards eco-friendly packaged products? An empirical investigation. *Journal of Retailing and Consumer Services*, *50*, 163–169. DOI: 10.1016/j.jretconser.2019.05.011

Priya, A. K., Devarajan, B., Alagumalai, A., & Song, H. (2023). Artificial intelligence enabled carbon capture: A review. *The Science of the Total Environment*, *886*, 163913. DOI: 10.1016/j.scitotenv.2023.163913 PMID: 37150463

Prodromou, A., Papademetriou, C., & Savvides, D. (2025). The Strategic Role of Green HRM in Hospitality Industry for Reaching Sustainable Development Goals. In Sustainability Through Green HRM and Performance Integration (pp. 1-38). IGI Global.

Provenance. (2025). Transparency for Your Products. Available online: https://www.provenance.org. (Accessed: 7-Feb-2025)

Puthal, D., Malik, N., Mohanty, S. P., Kougianos, E., & Yang, C. (2018, March). The Blockchain as a Decentralized Security Framework – Future Directions. *IEEE Consumer Electronics Magazine*, *7*(2), 18–21. DOI: 10.1109/MCE.2017.2776459

Qasem, N. A., & Abdulrahman, G. A. (2024). A recent comprehensive review of fuel cells: History, types, and applications. *International Journal of Energy Research*, *2024*(1), 7271748. DOI: 10.1155/2024/7271748

Qureshi, S., Mian, S. A., & Rana, A. I. (2010). Bahavioral Characteristics of Entrepreneurs in the Gujrat, Gujranawala and Slalkot Industrial Clusters of Pakistan: A Comparisn of First, Second and Third Generation Family Firms. *International Journal of Business and Social Science*, *1*(2).

Rahman, S., Sharmin, N., Rahat, A., Rahman, M., & Rahman, M. (2024). Tropical cyclone warning and forecasting system in Bangladesh: Challenges, prospects, and future direction to adopt artificial intelligence. *Computational Urban Science*, *4*(1), 4. Advance online publication. DOI: 10.1007/s43762-023-00113-x

Raina, G., & Sinha, S. (2019). Outlook on the Indian scenario of solar energy strategies: Policies and challenges. *Energy Strategy Reviews*, *24*, 331–341. DOI: 10.1016/j.esr.2019.04.005

Raisch, S., & Krakowski, S. (2021). Artificial intelligence and management: The automation-augmentation paradox. *Academy of Management Review*, *46*(1), 192–210. DOI: 10.5465/amr.2018.0072

Rajab, A., Farman, H., Islam, N., Syed, D., Elmagzoub, M. A., Shaikh, A., Akram, M., & Alrizq, M. (2023). Flood Forecasting by Using Machine Learning: A Study Leveraging Historic Climatic Records of Bangladesh. *Water (Basel)*, *15*(22), 3970. DOI: 10.3390/w15223970

Ramírez-Márquez, C., Posadas-Paredes, T., Raya-Tapia, A. Y., & Ponce-Ortega, J. M. (2024). Natural resource optimization and sustainability in society 5.0: A comprehensive review. *Resources*, *13*(2), 19. DOI: 10.3390/resources13020019

Rand, G. K. (1998). Re-Engineering the Manufacturing System: Applying the Theory of Constraints. *The Journal of the Operational Research Society*, *49*(7), 772–773.

Raquib, A., Channa, B., Zubair, T., & Qadir, J. (2022). Islamic virtue-based ethics for artificial intelligence. *Discover Artificial Intelligence*, *2*(1), 11. Advance online publication. DOI: 10.1007/s44163-022-00028-2

Ratouis, T. M. P., Snæbjörnsdóttir, S. Ó., Voigt, M. J., Sigfússon, B., Gunnarsson, G., Aradóttir, E. S., & Hjörleifsdóttir, V. (2022). Carbfix 2: A transport model of long-term CO2 and H2S injection into basaltic rocks at Hellisheidi, SW-Iceland. *International Journal of Greenhouse Gas Control, 114*, 103586. DOI: 10.1016/j.ijggc.2022.103586

Reay, T. (2019). Family routines and next-generation engagement in family firms. *Entrepreneurship Theory and Practice, 43*(2), 244–250. DOI: 10.1177/1042258718796083

Ren, S., Li, P., Yang, J., & Islam, M. A. (2023) Making AI Less "Thirsty": Uncovering and Addressing the Secret Water Footprint of AI Models, University of California, Riverside, https://doi.org//arXiv.2304.03271DOI: 10.48550

Renwick, D. W., Redman, T., & Maguire, S. (2013). Green human resource management: A review and research agenda. *International Journal of Management Reviews, 15*(1), 1–14. DOI: 10.1111/j.1468-2370.2011.00328.x

Revinova, S., Lazanyuk, I., Gabrielyan, B., Shahinyan, T., & Hakobyan, Y. (2024). Hydrogen in Energy Transition: The Problem of Economic Efficiency, Environmental Safety, and Technological Readiness of Transportation and Storage. *Resources, 13*(7), 92. DOI: 10.3390/resources13070092

Rhodes, J. S. (2000). "Usability can save you money", http://www.webword.com/moving/savecompany.html, accessed on Nov. 02 2004, December 2000.

Ritchie, H., Rosado, P., & Roser, M. (2020a). *Breakdown of carbon dioxide, methane and nitrous oxide emissions by sector*. Retrieved November 20 2024, from https://ourworldindata.org/emissions-by-sector

Ritchie, H., Rosado, P., & Roser, M. (2020b). *Energy Production and Consumption*. Retrieved November 17, 2024 from, https://ourworldindata.org/energy-production-consumption

Robertson, I. (1989). *Sociology*. Published by Worth Publishers.

Rogers, M. M. (2021). *Understanding family dynamics in a cross-cultural sample* (Doctoral dissertation, Mississippi State University).

Rohde, F., Wagner, J., Meyer, A., Reinhard, P., Voss, M., Petschow, U., & Mollen, A. (2024). Broadening the perspective for sustainable artificial intelligence: Sustainability criteria and indicators for Artificial Intelligence systems. *Current Opinion in Environmental Sustainability, 66*, 101411. DOI: 10.1016/j.cosust.2023.101411

Rosa Susila, J. K., Laksono, P., & Afit, M.ROSA SUSILA. (2022). Rancang Bangun Robo-Advisor untuk Pendanaan Rumah Syariah Berbasis Aplikasi Bergerak. *MIND (Multimedia Artificial Intelligent Networking Database). Journal, 7*(1), 98–110. DOI: 10.26760/mindjournal.v7i1.98-110

Rosenberg, D. (2019). AI and HR: New Partners in Talent Management. *Harvard Business Review, 97*(2), 42–50.

Roszko-Wójtowicz, E., Deep Sharma, G., Dańska-Borsiak, B., & Grzelak, M. M. (2024). Innovation-driven e-Commerce growth in the EU: An empirical study of the propensity for online purchases and sustainable consumption. *Sustainability (Basel), 16*(4), 1563. DOI: 10.3390/su16041563

Russell, S., & Norvig, P. (2020). *Artificial Intelligence: A Modern Approach.* Prentice Hall.

Russ, J., Zaveri, E., Desbureaux, S., Damania, R., & Rodella, A. S. (2022). The impact of water quality on GDP growth: Evidence from around the world. *Water Security, 17*, 100130. DOI: 10.1016/j.wasec.2022.100130

Sage, V., Patel, J., Hazewinkel, P., Yasin, Q. U. A., Wang, F., Yang, Y., Kozielski, K., & Li, C. E. (2024). Recent progress and techno-economic analysis of liquid organic hydrogen carriers for Australian renewable energy export–A critical review. *International Journal of Hydrogen Energy, 56*, 1419–1434. DOI: 10.1016/j.ijhydene.2023.12.269

Saha, P., Akash, F. A., Shovon, S. M., Monir, M. U., Ahmed, M. T., Khan, M. F. H., Sarkar, S. M., Islam, M. K., Hasan, M. M., Vo, D.-V. N., Aziz, A. A., Hossain, M. J., & Akter, R. (2024). Grey, blue, and green hydrogen: A comprehensive review of production methods and prospects for zero-emission energy. *International Journal of Green Energy, 21*(6), 1383–1397. DOI: 10.1080/15435075.2023.2244583

Sahu, S. K., & Gonnade, S. K. (2013). QR code and application in India. *Int J Recent Technol Eng, 2*, 26–28.

Salahi, F. (2024). 12 Application of Solar. *Hydrogen Production from Renewable Resources and Wastes*, 281.

Saleem, S. (2008). SMEDA SME Policy Paper 2007-A Critical Review. Available at *SSRN* 1092050. DOI: 10.2139/ssrn.1092050

Sales, M., de Bruin, S., Herold, M., Kyriakidis, P., & Souza, C.Jr. (2017). A spatio-temporal geostatistical hurdle model approach for short-term deforestation prediction. *Spatial Statistics, 21*, 304–318. DOI: 10.1016/j.spasta.2017.06.003

Sánchez-Marín, G., Meroño-Cerdán, Á. L., & Carrasco-Hernández, A. J. (2019). Formalized HR practices and firm performance: An empirical comparison of family and non-family firms. *International Journal of Human Resource Management, 30*(7), 1084–1110. DOI: 10.1080/09585192.2017.1289547

SAP. (2023). HR analytics for sustainable workforce management. Retrieved from https://www.sap.com/

Sarvajayakesavalu, S. (2015). Addressing challenges of developing countries in implementing five priorities for sustainable development goals. *Ecosystem Health and Sustainability, 1*(7), 1–4. DOI: 10.1890/EHS15-0028.1

Saxena, A., Ramaswamy, M., Beale, J., Marciniuk, D., & Smith, P. (2021). Striving for the United Nations (UN) Sustainable Development Goals (SDGs): What will it take? *Discover Sustainability, 2*(1), 20. DOI: 10.1007/s43621-021-00029-8

Schultz, P. W., Nolan, J. M., Cialdini, R. B., Goldstein, N. J., & Griskevicius, V. (2018). The constructive, destructive, and reconstructive power of social norms: Reprise. *Perspectives on Psychological Science, 13*(2), 249–254. DOI: 10.1177/1745691617693325 PMID: 29592653

Science of Better. (2015). Operations Research: The Science for Better, http://www.scienceofbetter.org

Sensor, C. S., Branden, P. S., Clary-Muronda, V., Hawkins, J. E., Fitzgerald, D., Shimek, A. M., Al-Itani, D., Madigan, E. A., & Rosa, W. E. (2021). Nurses achieving sustainable development goals: The United Nations and Sigma. *The American Journal of Nursing, 121*(4), 65–68. DOI: 10.1097/01.NAJ.0000742544.07615.db PMID: 33755634

Setyowati, W., & Rahayu, I. S. (2023). Sector Analysis of Islamic Capital Markets and Artificial Intelligence Functioning as Sharia Advisors. *Sector Analysis of Islamic Capital Markets and Artificial Intelligence Functioning as Sharia Advisors., 1*(2), 236–244. DOI: 10.33050/italic.v1i2.334

Seuyong, F. T., Silwal, A. R., Begazo, T., Newhouse, D., & N'Ghauran, A. (2023). The Size and Distribution of Digital Connectivity Gaps in Sub-Saharan Africa. *The World Bank*. https://documents1.worldbank.org/curated/en/099241003142325200/pdf/IDU0cb2e42f3050260484d0b8370b84eee303ecf.pdf

Shafer, G. (1976). *A Mathematical Theory of Evidence*. Princeton University Press. DOI: 10.1515/9780691214696

Shalhoob, H. (2025). The role of AI in enhancing shariah compliance: Efficiency and transparency in Islamic finance. *Journal of Infrastructure. Policy and Development*, *9*(1), 11239. DOI: 10.24294/jipd11239

Sharda, R., Delen, D., & Turban, E. (2020). *Analytics, Data Science, & Artificial Intelligence: Systems for Decision Support*. Pearson.

Sheth, J. N., Sethia, N. K., & Srinivas, S. (2010). Mindful consumption: A customer-centric approach to sustainability. *Journal of the Academy of Marketing Science*, *39*(1), 21–39. DOI: 10.1007/s11747-010-0216-3

Shoaei, M., Noorollahi, Y., Hajinezhad, A., & Moosavian, S. F. (2024). A review of the applications of artificial intelligence in renewable energy systems: An approach-based study. *Energy Conversion and Management*, *306*, 118207–118207. DOI: 10.1016/j.enconman.2024.118207

Shrestha, T. A., Matin, M., Chitale, V., & Thomas, S. (2022). Exploring the potential of deep learning for classifying camera trap data: A case study from Nepal. In *ICIMOD.org*. International Centre for Integrated Mountain Development (ICIMOD). https://lib.icimod.org/record/35997/files/HimalDoc2022_WP_ExploringDeepLearning_ForApproval.pdf

Shrestha, A. B., & Aryal, R. (2011). Climate change in Nepal and its impact on Himalayan glaciers. *Regional Environmental Change*, *11*(S1), 65–77. DOI: 10.1007/s10113-010-0174-9

Shull, C., & Sękowska, A. (n.d.). Proxiair. Retrieved April 27, 2020, from https://proxiair.com/

Silva, R. F., Fava, M. C., Saraiva, A. M., Mendiondo, E. M., Cugnasca, C. E., & Delbem, A. C. B. (2021). A Theoretical Framework for Multi-Hazard Risk Mapping on Agricultural Areas Considering Artificial Intelligence, IoT, and Climate Change Scenarios. *Engineering Proceedings*, *9*(1), 39. DOI: 10.3390/engproc2021009039

Simanjuntak, D. A., Safitri, K., Rahayu, D., Nainggolan, N. S., & Amelia, R. (2023). Islamic law perspective on investment. *Cashflow Current Advanced Research On Sharia Finance And Economic Worldwide*, *2*(3), 418–423. DOI: 10.55047/cashflow.v2i3.644

Simon, H. A. (1955). A Behavioral Model of Rational Choice. *The Quarterly Journal of Economics*, *69*(1), 99–118. DOI: 10.2307/1884852

Simon, H. A. (1957). *Administrative behavior*. Simon and Schuster.

Singh, R., Bhattarai, N., Prajapati, A., & Shakya, S. R. (2022). Impact of Variation in Climatic Parameters on Hydropower Generation: A Case of Hydropower Project in Nepal. *Heliyon, 8*(12), e12240. ScienceDirect. https://doi.org/DOI: 10.1016/j.heliyon.2022.e12240

Singh, R., Khan, S., & Centobelli, P. (2022). Investigating the Interplay between Social Performance and Organisational Factors Supporting Circular Economy Practices. *Sustainability (Basel), 14*(24), 16781. DOI: 10.3390/su142416781

SITA. (2023). Baggage IT Insights 2023. Retrieved from https://www.sita.aero/resources/reports/baggage-it-insights-2023

Soliman, A. (2024). *'Unacceptable': a staggering 4.4 billion people lack safe drinking water, study finds.* Retrieved November 5, 2024, from https://www.nature.com/articles/d41586-024-02621-0 doi: https://doi.org/DOI: 10.1038/d41586-024-02621-0

Sousa, J., Bessani, A., & Vukolic, M. (2018, June). A byzantine fault-tolerant ordering service for the hyperledger fabric blockchain platform. In 2018 48th annual IEEE/IFIP international conference on dependable systems and networks (DSN) (pp. 51-58). IEEE.

Stainton, R. S. (1984). Applicable Systems Thinking. *European Journal of Operational Research, 18*(2), 145–154. DOI: 10.1016/0377-2217(84)90180-2

Starbucks. (2020). *Employee benefits and ethical business practices.* Retrieved from https://www.starbucks.com/

Starbucks. (2023). Ethical sourcing and employee benefits. Retrieved from https://www.starbucks.com/

Statista (2024), *Distribution of electricity generation worldwide in 2023, by energy source.* Retrieved October 25 2024, from https://www.statista.com/statistics/269811/world-electricity-production-by-energy-source/

Sterman, J. D. (1994). Judgment under Uncertainty: Heuristics and Biases. *System Dynamics Review, 10*(2-3), 291–330. DOI: 10.1002/sdr.4260100214

Sterman, J. D. (2017). *Business Dynamics: An Introduction.* Mc-Graw Hill.

Sterner, T., Barbier, E. B., Bateman, I., van den Bijgaart, I., Crépin, A.-S., Edenhofer, O., Fischer, C., Habla, W., Hassler, J., Johansson-Stenman, O., Lange, A., Polasky, S., Rockström, J., Smith, H. G., Steffen, W., Wagner, G., Wilen, J. E., Alpízar, F., Azar, C., & Robinson, A. (2019). Policy design for the Anthropocene. *Nature Sustainability, 2*(1), 14–21. DOI: 10.1038/s41893-018-0194-x

Stevens, C. (2010). Linking sustainable consumption and production: The government role. *Natural Resources Forum*, *34*(1), 16–23. DOI: 10.1111/j.1477-8947.2010.01273.x

Stewart, A., & Hitt, M. A. (2012). Why can't a family business be more like a nonfamily business? Modes of professionalization in family firms. *Family Business Review*, *25*(1), 58–86. DOI: 10.1177/0894486511421665

Stone, D. L., Deadrick, D. L., Lukaszewski, K. M., & Johnson, R. (2015). The influence of technology on the future of human resource management. *Human Resource Management Review*, *25*(2), 216–231. DOI: 10.1016/j.hrmr.2015.01.002

Sudheer Varma Surimalla, S. (2018). Intelligent Traffic Management Solution(ITMS) through the implementation of Google Maps and Artificial Intelligence. *International Journal of Engineering, Science and Mathematics*. https://www.researchgate.net/publication/327020444_Intelligent_Traffic_Management_SolutionITMS_through_the_implementation_of_Google_Maps_and_Artificial_Intelligence

Suman, S. (2022). Augmented Intelligence for National Security and Development. *Unity Journal*, *3*(01), 245–252. DOI: 10.3126/unityj.v3i01.43329

Swapnil Mahadik, S., Gedam, M., & Shah, D. (2024). Environment sustainability with smart grid sensor. *Frontiers in Artificial Intelligence*, *7*, 1510410. Advance online publication. DOI: 10.3389/frai.2024.1510410 PMID: 40045982

Sypniewska, B., Baran, M., & Kłos, M. (2023). Work engagement and employee satisfaction in the practice of sustainable human resource management–based on the study of Polish employees. *The International Entrepreneurship and Management Journal*, *19*(3), 1069–1100.

Taguchi, G., Elsayed, E. A., & Hsiang, T. (1989). *Quality Engineering in Production Systems*. McGraw Hill.

Taha, H. A. (2022). Operations Research: An Introduction, 11th Kindle Ed., Pearson.

Tahir, S. H., Sabir, H. M., & Shah, S. (2016). Impact of family ownership on financing decisions: a comparative analysis of companies listed at the KSE. Pakistan Business Review, 17(4).

Tanner, C., & Wölfing Kast, S. (2003). Promoting sustainable consumption: Determinants of green purchases by Swiss consumers. *Psychology and Marketing*, *20*(10), 883–902. DOI: 10.1002/mar.10101

Tayachi, T., Brahimi, T., & Essafi, Y. (2022). Artificial Intelligence in the Islamic Finance: A bibliometric. *Analysis*.

Taylor, B. W. (2018). *Introduction to Management Science* (13th ed.). Pearson.

Tesla. (2021). *Clean energy initiatives and sustainable practices.* Retrieved from https://www.tesla.com/

Th⊘gersen, J. (2009). The motivational roots of norms for environmentally responsible behavior. *Basic and Applied Social Psychology, 31*(4), 348-362. https://doi.org/DOI: 10.1080/01973530903317144

Thomas, W. (2015). *Rational action: The sciences of policy in Britain and America, 1940-1960.* MIT Press. DOI: 10.7551/mitpress/9997.001.0001

Tiwari, S., Bashir, S., Sarker, T., & Shahzad, U. (2024). Sustainable pathways for attaining net zero emissions in selected South Asian countries: Role of green energy market and pricing. *Humanities & Social Sciences Communications, 11*(1), 93. DOI: 10.1057/s41599-023-02552-7

Todorof, M. (2018). Robo Advice in Islamic Finance. *Social Science Research Network.* https://doi.org/DOI: 10.2139/SSRN.3337712

TRACE ME Luggage Tracker Ltd. (2025). TRACE ME Smart ID. https://www.tmlt.co.uk. (Accessed: 7-Feb-2025).

Triantaphyllou, E., & Lin, C. (1996). Development and Evaluation of Five Fuzzy Multi-Attribute Decision-Making Methods. *International Journal of Approximate Reasoning, 14*(4), 281–310. DOI: 10.1016/0888-613X(95)00119-2

Triantaphyllou, E., Shu, B., Nieto Sanchez, S., & Ray, T. (1998). Multi-Criteria Decision Making: An Operations Research Approach. In Webster, J. D. (Ed.), *Encyclopedia of Electrical and Electronics Engineering*. John Wiley.

Turban, E., & Aronson, J. E. (2005). *Decision Support Systems and Intelligent Systems*. Prentice Hall Inc.

Turkanović, M., Hölbl, M., Košič, K., Heričko, M., & Kamišalić, A. (2018). EduCTX: A Blockchain-Based Higher Education Credit Platform. *IEEE Access : Practical Innovations, Open Solutions, 6*, 5112–5127. DOI: 10.1109/ACCESS.2018.2789929

Tversky, A., & Kahneman, D. (1974). Judgment under Uncertainty: Heuristics and Biases. *Science, 185*(4157), 1124–1131. DOI: 10.1126/science.185.4157.1124 PMID: 17835457

ud Din, M., & Ahmad, A. (2023). Environmental Sustainability Through Green Human Resource Management Practices: An Analyses Of Industries Of Lahore, Pakistan. *Journal of Positive School Psychology*, 415-426.

ud Din, M., Ahmad, A., Samad, S., Lee, S., & Han, H. (2025). Sowing the seeds of sustainability: A sociological exploration of environmental sustainability within SME sector. *Climate Services, 37*, 100533.

Udeagha, M. C., & Muchapondwa, E. (2022). Environmental sustainability in South Africa: Understanding the criticality of economic policy uncertainty, fiscal decentralization, and green innovation. *Sustainable Development (Bradford), 31*(3), 1638–1651. DOI: 10.1002/sd.2473

UN General Assembly Economic and Social Council. Progress towards the Sustainable Development Goals, Retrieved February 07, 2025, from https://unstats.un.org/sdgs/files/report/2024/SG-SDG-Progress-Report-2024-advanced-unedited-version.pdf

UNFCCC. (2016). *The Paris Agreement.* Unfccc.int; UNFCCC. https://unfccc.int/process-and-meetings/the-paris-agreement

Unilever. (2023). *The Sustainable Living Plan.* Retrieved from https://www.unilever.com/

United Nations. (2015). *Transforming our world: The 2030 agenda for sustainable development.*

United Nations. (n.d.). Sustainable Development Goals. United Nations. https://sdgs.un.org/goals

Van den Bergh, J. C. (2008). Environmental regulation of households: An empirical review of economic and psychological factors. *Ecological Economics, 66*(4), 559–574. DOI: 10.1016/j.ecolecon.2008.04.007

Vats, G., & Mathur, R. (2022). A net-zero emissions energy system in India by 2050: An exploration. *Journal of Cleaner Production, 352*, 131417. DOI: 10.1016/j.jclepro.2022.131417

Veen, S., Moorten, R., & Durani, W. (2018). Marriage Decision-Making: A family affair Insight into marriage decision-making processes and social norms around child marriage in Larkana and Shikarpur in Sindh and Lodhran and Muzaffargarh in Punjab, Pakistan. oxfamilibrary.openrepository.com

Venizelou, V., & Poullikkas, A. (2024). Comprehensive Overview of Recent Research and Industrial Advancements in Nuclear Hydrogen Production. *Energies, 17*(12), 2836. DOI: 10.3390/en17122836

Verver, M., & Koning, J. (2017). Toward a Kinship Perspective on Entrepreneurship. *Entrepreneurship Theory and Practice.*

Vrontis, D., Christofi, M., & Makrides, A. (2021). Artificial intelligence, robotics, advanced technologies and human resource management: A systematic review. *International Journal of Human Resource Management, 32*(21), 4599–4627.

Wamba, S. F., Akter, S., Edwards, A., Chopin, G., & Gnanzou, D. (2017). How 'big data' can make big impact: Findings from a systematic review and a longitudinal case study. *International Journal of Production Economics, 165*, 234–246. DOI: 10.1016/j.ijpe.2014.12.031

Wang, Y. (2024). *Ethical considerations of AI in financial decision. 2*(1), 1290. https://doi.org/DOI: 10.59400/cai.v2i1.1290

Want, R. (2006, January-March). An introduction to RFID technology. *IEEE Pervasive Computing, 5*(1), 25–33. DOI: 10.1109/MPRV.2006.2

Warfield, J. N., & Staley, S. M. (1996). Structural Thinking: Organizing Complexity Through Disciplined Activity. *Systems Research, 13*(1), 47–67. DOI: 10.1002/(SICI)1099-1735(199603)13:1<47::AID-SRES27>3.0.CO;2-A

Water, U. N. (2021). *Progress on Level of Water Stress.* Retrieved November 15, 2024, from https://www.sdg6data.org/en/indicator/6.4.2

Water, U. N. (2024). Water Report, Water for prosperity and peace 2024, Paris.

Weder, F., Golob, U., & Podnar, K. (2025). Sustainable consumption in context: A cross-cultural study of social representations. *Sustainability (Basel), 17*(4), 1531. DOI: 10.3390/su17041531

Wehrmeyer, W. (2017). *Greening people: Human resources and environmental management.* Routledge. DOI: 10.4324/9781351283045

White, K., Habib, R., & Hardisty, D. J. (2019). How to SHIFT consumer behaviors to be more sustainable: A literature review and guiding framework. *Journal of Marketing, 83*(3), 22–49. DOI: 10.1177/0022242919825649

White, K., & Simpson, B. (2013). When do (and don't) normative appeals influence sustainable consumer behaviors? *Journal of Marketing, 77*(2), 78–95. DOI: 10.1509/jm.11.0278

Widya, W., & Syafi'i, A. (2024). *Application of Islamic Business Ethics in the Fintech Industry: an Analysis of the Foundation of Sharia Maqashid.* .DOI: 10.58631/ajemb.v2i12.68

Wieczorek-Kosmala, M. (2020). Weather Risk Management in Energy Sector: The Polish Case. *Energies, 13*(4), 945. DOI: 10.3390/en13040945

Wiggin, T. (2025). Weaponizing the Workplace: How Algorithmic Management Shaped Amazon's Antiunion Campaign in Bessemer, Alabama. *Socius: Sociological Research for a Dynamic World, 11*, 23780231251318389. DOI: 10.1177/23780231251318389

Williams, C. C., Shahid, M. S., & Martínez, A. (2016). Determinants of the level of informality of informal micro-enterprises: Some evidence from the city of Lahore, Pakistan. *World Development, 84*, 312–325. DOI: 10.1016/j.worlddev.2015.09.003

Williams, R. M.Jr. (1961). The sociological theory of Talcott Parsons. In Black, M. (Ed.), *The social theories of Talcott Parsons. A critical examination* (pp. 64–99). Prentice-Hall.

Wolf, J., Pellumbi, K., Haridas, S., Kull, T., Kleinhaus, J. T., Wickert, L., Apfel, U.-P., & Siegmund, D. (2024). Electroplated Electrodes for Continuous and Mass-Efficient Electrochemical Hydrogenation. *Chemistry (Weinheim an der Bergstrasse, Germany), 30*(18), e202303808. DOI: 10.1002/chem.202303808 PMID: 38100290

World Commission on Environment and Development. (1987). *Our common future*. Oxford University Press. https://sustainabledevelopment.un.org/content/documents/5987our-common-future.pdf

World Economic Forum. (2023). Future of sustainable business practices. Retrieved from https://www.weforum.org/

Xiao, Z., & Zhang, Y. (2024). The influence of information sharing on supply chain performance in Internet of Things. Journal of Computational Methods in Sciences and Engineering, 14727978241300832.

Xie, Z., Jin, Q., Su, G., & Lu, W. (2024). A Review of Hydrogen Storage and Transportation: Progresses and Challenges. *Energies, 17*(16), 4070. DOI: 10.3390/en17164070

Xiong, J., & Xu, D. (2021). Relationship between energy consumption, economic growth and environmental pollution in China. *Environmental Research, 194*, 110718. DOI: 10.1016/j.envres.2021.110718 PMID: 33421428

Yachachín, O. M. L., Ocoruro, E. A. V., & La Riva, M. M. (2025). Expectation and Perception of Quality of Care from the Perspective of User Satisfaction in a Private Clinic: Wellness and Health. *Journal of Lifestyle and SDGs Review, 5*(2), e04249–e04249. DOI: 10.47172/2965-730X.SDGsReview.v5.n02.pe04249

Yangka, D., Rauland, V., & Newman, P. (2018). Carbon neutral policy in action: The case of Bhutan. *Climate Policy, 19*(6), 672–687. DOI: 10.1080/14693062.2018.1551187

Yang, R., Hu, J., Li, Z., Mu, J., Yu, T., Xia, J., Li, X., Dasgupta, A., & Xiong, H. (2024). Interpretable machine learning for weather and climate prediction: A review. *Atmospheric Environment, 338*, 120797–120797. DOI: 10.1016/j.atmosenv.2024.120797

Yenneti, K. (2016). The grid-connected solar energy in India: Structures and challenges. *Energy Strategy Reviews, 11-12*, 41–51. DOI: 10.1016/j.esr.2016.06.002

Young, W., Hwang, K., McDonald, S., & Oates, C. J. (2009). Sustainable consumption: Green consumer behaviour when purchasing products. *Sustainable Development (Bradford), 18*(1), 20–31. DOI: 10.1002/sd.394

Zahra, S. A. (2018). Entrepreneurial risk-taking in family firms: The wellspring of the regenerative capability. *Family Business Review, 31*(2), 216–226. DOI: 10.1177/0894486518776871

Zappacosta, M., & Hollema, S. (2017). *FAO/WFP crop and food security assessment mission to Sri Lanka*. Food and Agriculture Organization of the United Nations (FAO). https://openknowledge.fao.org/server/api/core/bitstreams/a7fbafb5-41db-48e2-97e7-e8f415b8df0e/content

Zellweger, T. M., & Astrachan, J. H. (2008). On the emotional value of owning a firm. *Family Business Review, 21*(4), 347–363. DOI: 10.1177/08944865080210040106

Zellweger, T. M., Kellermanns, F. W., Chrisman, J. J., & Chua, J. H. (2012). Family control and family firm valuation by family CEOs: The importance of intentions for transgenerational control. *Organization Science, 23*(3), 851–868. DOI: 10.1287/orsc.1110.0665

Zellweger, T. M., Nason, R. S., Nordqvist, M., & Brush, C. G. (2013). Why do family firms strive for nonfinancial goals? An organizational identity perspective. *Entrepreneurship Theory and Practice, 37*(2), 229–248. DOI: 10.1111/j.1540-6520.2011.00466.x

Zerssa, G., Feyssa, D., Kim, D.-G., & Eichler-Löbermann, B. (2021). Challenges of Smallholder Farming in Ethiopia and Opportunities by Adopting Climate-Smart Agriculture. *Agriculture, 11*(3), 192. DOI: 10.3390/agriculture11030192

Zhang, S., Ding, Q., & Ding, J. (2023). Return strategy of e-Commerce platform based on green and sustainable development. *Sustainability (Basel), 15*(14), 11188. DOI: 10.3390/su151411188

Zhang, T., Ouyang, Y., & He, Y. (2008). Traceable Air Baggage Handling System Based on RFID Tags in the Airport. *Journal of Theoretical and Applied Electronic Commerce Research, 3*(1), 106–115. DOI: 10.3390/jtaer3010011

Zhang, X., & Dong, F. (2020). Why do consumers make green purchase decisions? Insights from a systematic review. *International Journal of Environmental Research and Public Health, 17*(18), 6607. DOI: 10.3390/ijerph17186607 PMID: 32932797

Zha, X. F. (2003). Soft Computing Framework for Intelligent Human–Machine System Design, Simulation and Optimization. *Soft Computing, 7*(3), 184–198. DOI: 10.1007/s00500-002-0196-4

Zheng, Z., Xie, S., Dai, H., Chen, X., & Wang, H. "An Overview of Blockchain Technology: Architecture, Consensus, and Future Trends," 2017 IEEE International Congress on Big Data (BigData Congress), Honolulu, HI, 2017, pp. 557-564. DOI: 10.1109/BigDataCongress.2017.85

Zhou, L., & Alam, G. M. (2024). Commercial higher education strategies for recruiting international students in China: A catalyst or obstacle for sustainable education and learning? *Discover Sustainability, 5*(1), 33. DOI: 10.1007/s43621-024-00216-3

Zhou, Z., Syamsunur, D., Muhamad, N. S., Ghazalli, A. Z., Azahar, M. A., & Jusoh, M. N. H. B. (2025). A Tentative Analysis of the Interactions Between the Building Sector and SDGs. *Journal of Lifestyle and SDGs Review, 5*(2), e03212–e03212. DOI: 10.47172/2965-730X.SDGsReview.v5.n02.pe03212

Zuhara, S., Mackey, H. R., Al-Ansari, T., & McKay, G. (2024). A review of prospects and current scenarios of biomass co-pyrolysis for water treatment. *Biomass Conversion and Biorefinery, 14*(5), 6053–6082. DOI: 10.1007/s13399-022-03011-0 PMID: 35855911

Zybert, M., Ronduda, H., Patkowski, W., Ostrowski, A., Sobczak, K., & Raróg-Pilecka, W. (2024). Improving the catalytic performance of Co/BaCeO 3 catalyst for ammonia synthesis by Y-modification of the perovskite-type support. *RSC Advances, 14*(49), 36281–36294. DOI: 10.1039/D4RA06251E PMID: 39534043

About the Contributors

Abdul Rehman Khan is a Professor of Supply Chain Management. In 2020, he completed his Postdoctoral fellowship at Tsinghua University, ranked 16th worldwide. He has published nearly 150+ research articles in top-ranked international journals. Dr. Syed has also published 16 academic books by influential international publishers. He has received £150,000 research funds from different funding bodies in last 5 years. Moreover, Dr. Khan has 15+ years of experience in the industry and academia. He is also serving as an Editor-in-Chief and Associate Editor for different well-known International Journals. Dr. Khan is a regular contributor in International workshops and conference around the globe.

Muhammad Tanveer is a distinguished academic and research scholar, holding a Ph.D. in Management from the renowned Lincoln University College (LUC), Malaysia. Currently, he serves as an Assistant Professor in the College of Business, Department of Business Administration at Imam Mohammad Ibn Saud Islamic University (IMSIU) in Riyadh, Saudi Arabia. His diverse research interests span critical domains such as Management, Circular Economy, Supply Chain Management, AI, Sustainability, and Environmental Studies. Dr. Muhammad has made significant contributions to academia, publishing over 70 impactful research articles in prestigious journals with high impact factors, alongside numerous book chapters. His work has been recognized globally, with around 2,000 citations, highlighting his influence in the academic community. In addition to his scholarly publications, Dr. Muhammad has presented his research at several international conferences and delivered keynote speeches. His achievements have earned him accolades, including the Active Researcher Certificate from Prince Sultan University and the Global Publishing Award from IMSIU.

***

Abdul-Rahim Ahmad is a Professor of Intelligent Systems and Innovation Management at the College of Computer Science and Information Technology, King Faisal University, Saudi Arabia, as well as an adjunct Professor of Intelligent Systems and Artificial Intelligence at the department of Systems Design Engineering, University of Waterloo, Canada. He has interdisciplinary research and teaching interests in such faculties as Intelligent Tools for Decision-Making (Machine Intelligence, Knowledge Systems, Decision Support Systems, Expert Systems) and Decision Modeling (Multi-Echelon Operations, Supply Chain Management, and Mathematical Programming). He is an award-winning academician and researcher who has published or presented 70+ research papers in internationally reputed and high impact journals and conferences.

Yousuf Ahmad is affiliated with the Department of Computer Science, University of British Columbia, His interests include intelligent systems design, algorithms design, and gamification of business decisions.

Asrar Haque, Ph.D. is an Assistant Professor in the Computer Science Department at the College of Computer Science and Information Technology, King Faisal University, Saudi Arabia. He received his Ph.D. in Computer Science from Kent State University, USA,. Dr. Haque's research interests include Massively Parallel Computer Architecture, Swarm Intelligence, Wireless and Sensor Networks, Blockchain, Distributed Algorithms, and Combinatorial Optimization. He has published in these areas, contributing to international journals and conferences. He teaches advanced undergraduate and graduate courses and actively participates in curriculum development and accreditation processes.

Deeksha Gupta Ganguly is an Assistant Professor, School of Business & Management. She is a Post-graduate in Commerce & Management (Human Resource Management & Marketing). She has also cleared UGC NET in Management. Prof. Ganguly has a rich academic & industry experience of 15 plus years and worked for various Educational institutions & universities in Delhi NCR and Chennai. She is very much involved in academic research and has extensively published research papers in various Scopus-indexed academic journals.

Oindrilla Ghosh is a Research Scholar hailing from the School of Law at Christ (Deemed to be University) Bangalore, India, focuses her research on health of tribal communities. Her work includes publications on Neuroscience on Organizational Behavior at IGI Global and also Environmental Disasters of which 8 publications that are indexed in SCOPUS and Web of Sciences conference proceedings. She completed her early education at Loreto Day School, Bowbazar and went on to achieve distinction in her undergraduate (B.A/LL.B) and postgraduate (LL.M) studies at Calcutta University. Having served as an Assistant Professor of Law at Kingston Law College,Barasat, Kolkata, Ghosh has actively participated in numerous National Conferences, International Conferences, UN Global Summits, webinars, and seminars organized by prestigious academic institutions. Her expertise lies in health perspectives, environmental studies, human behavior, and related areas. Additionally, she has conducted various workshops, lectures in educational institutions, and is engaged with NGO's to help the needy

Mohammad Irfan is presently working as an Associate Professor at School of Business and Management, Christ University, Bengaluru, India. Dr. Irfan has done his Ph.D. from the Central University of Haryana. He is MBA (Finance), M.Com (Account and Law), and MA (Economics). He has qualified UGC-JRF/SRF/NET in Management and Commerce. Dr. Irfan certified NSEs (NCFM) and BSEs certification. He has experience of more than sixteen years in the area of SAPM, Artificial Intelligence, Machine Learning, Blockchain, Cryptocurrency, Financial Engineering, Fintech, Green Finance, and Alternative Finance. He has to his credit more than 60+ Scopus Indexed articles, includes The Journal of Economic Cooperation and Development (Q2), International Journal of Business Excellence (IJBEX), International Journal of Economics and Management (IJEM), Montenegrin Journal of Economics (Q2), Cogent Business & Management (Taylor & Francis) (Q2), Indian Journal of Finance (IJF) and Journal of Islamic Monetary Economics and Finance (JIMF). His citations reached 400+ along with 13 H-index. Dr. Irfan has published 9 books in Springer, IGI Global Publication (Scopus indexed).

Early Ridho Kismawadi, S.E.I, MA is a lecturer at the Department of Islamic Banking, Faculty of Islamic Economics and Business IAIN Langsa, Aceh, Indonesia, he has been a lecturer since 2013, he has completed a doctoral program in 2018 majoring in Sharia economics at the State Islamic University of North Sumatra. He was appointed head of the Islamic economics Law study program (2023) Islamic banking study program (2020) and Islamic financial management study program (2019) at Langsa State Islamic Institute (IAIN Langsa), Aceh, Indonesia. His research interests include financial economics, applied econometrics, Islamic economics, banking, and finance. He has published articles in national and international journals. In addition, he is also a reviewer of several reputable international journals such as Finance Research Letters, Financial Innovation, Cogent Business &; Management, Journal of Islamic Accounting and Business Research. He has also presented his papers at various local and international seminars. Dr. Early Ridho Kismawadi, S.E.I, MA, can be contacted at kismawadi@iainlangsa.ac.id.

Binod Kumar is an undergraduate student known for his academic excellence and active involvement in extracurricular activities, with 4 publications of his research being indexed in SCOPUS Database. Hailing from Ranchi, Jharkhand in India, Binod is currently pursuing a Bachelor of Law degree at the prestigious Radha Govind University. His passion for technology and innovation along with climate changes is evident through his high academic performance and numerous projects. Binod has a keen interest in artificial intelligence, machine learning, neuropsychology, climate changes and disaster management, fields in which he has undertaken several research projects. He is particularly fascinated by the potential of AI to solve real-world problems and is dedicated to learning and applying new techniques in these areas. Binod's technical skills are complemented by his strong analytical abilities and attention to detail. In addition to his academic and technical pursuits, Binod is passionate about community service. He volunteers at local NGOs, teaching underprivileged children basic computer skills. With a strong foundation in law and a commitment to making a positive impact, Binod Kumar is poised to become a future leader in the legal industry.

Index

A

B

C

D

E

W

www.ingramcontent.com/pod-product-compliance
Lightning Source LLC
LaVergne TN
LVHW080609300926
849162LV00005B/1097